THE FUNCTIONAL INTEGRATION OF CELLS IN ANIMAL TISSUES

THE FIFTH SYMPOSIUM OF
THE BRITISH SOCIETY FOR CELL BIOLOGY

THE FUNCTIONAL
INTEGRATION OF
CELLS IN ANIMAL TISSUES

EDITED BY

JOHN D. PITTS AND
MALCOLM E. FINBOW

Cancer Research Campaign Cell Interactions Group,
Institutes of Genetics, Virology and Biochemistry,
University of Glasgow

CAMBRIDGE UNIVERSITY PRESS

CAMBRIDGE

LONDON   NEW YORK   NEW ROCHELLE

MELBOURNE   SYDNEY

Published by the Press Syndicate of the University of Cambridge
The Pitt Building, Trumpington Street, Cambridge CB2 1RP
32 East 57th Street, New York, NY 10022, USA
296 Beaconsfield Parade, Middle Park, Melbourne 3206, Australia

First published 1982

Printed in Great Britain at the University Press, Cambridge

Library of Congress catalogue card number: 81-10213

British Library Cataloguing in Publication Data
The functional integration of cells in animal
tissues – (The British Society for Cell
Biology symposia; 5th)
1. Histology – Congresses
I. Pitts, John D   II. Finbow, Malcolm E.
III. Series
591.8′2   QM550
ISBN 0 521 24199 5

CONTENTS

v

PREFACE

The evolution of multicellular, differentiated organisms depended on the development of various mechanisms of cell–cell interaction. These interactions allow the control of the proliferation, the differentiation and the activities of the resultant interdependent component cells. The particular activities of the component cells are produced not only by their specific patterns of gene expression, but also by their relationships and interactions with neighbouring cells in the same tissue and distant cells in other tissues.

The cellular interactions which give rise to 'The functional integration of cells in animal tissues' fall broadly into two classes. One involves extracellular signal molecules (e.g. hormones, neurotransmitters, matrix molecules, surface components) while the other involves the formation of intercellular junctions which provide channels of direct communication between the interiors of neighbouring cells. Cells in an organism can therefore communicate via pathways which cross the cell membranes or by pathways which are entirely within a plasma membrane bounded compartment. This compartment can be the whole organism (in the case of early embryos of at least some species), a whole tissue or, perhaps, a specifically restricted part of a tissue.

The functions of the classical 'long-range' hormones and 'short-range' neurotransmitters are, in many cases, well understood but there is a growing realization that this trans-membrane pathway of communication also plays important, though as yet ill-defined roles in local cell interactions within tissues. Junctional communication has an obvious role in excitable tissues such as heart where it allows the spread of action potentials, but little is known about its functions in other tissues. Furthermore, nothing is known about the interplay between these different forms of interactions.

This book describes the different types of cell–cell interaction and some of the different approaches which are being used to understand their functions. It does not provide answers to many of the questions which are raised but gives a broadly based summary of the progress made so far towards understanding how cells are organized, in a functional sense, in animal tissues.

John D. Pitts
Malcolm E. Finbow
April 1981

A review of junctional mediated intercellular communication

MALCOLM E. FINBOW

CRC Cell Interactions Group, Institutes of Genetics, Virology and Biochemistry, University of Glasgow, Glasgow G12 8QQ, Scotland

INTRODUCTION

It is now some 20 years since the discovery of pathways of direct communication between cells and of the structure thought to provide these pathways, the gap junction (or nexus). It thus seems an appropriate time to take stock of what has since proved to be a fundamental property of cells in animal tissues. This review will deal with the salient features of junctional communication including recent information on structure, composition, formation and turnover as well as the 'functional assays' of communication (i.e. electrical, dye and metabolic coupling) and their correlation with the observed presence of gap junctions. Other chapters in this book deal more specifically with the roles of junctional communication in tissues.

For the purpose of this review it would be helpful to briefly present the current ideas of gap junction structure and function. Gap junctions appear to be universal features of metazoan animals and are thought to be sites of hydrophilic channels which join the cytoplasms of neighbouring cells. There is considerable evidence that these channels are permeable to perhaps all water-soluble cytoplasmic components up to MW 1000; this would include the vast majority of metabolites and inorganic ions but exclude macromolecules (e.g. proteins and nucleic acids). As a result of their permeability to ions, gap junctions are used in certain excitable tissues to transmit action potentials between cells, that is they act as electrical synapses (Bennett & Goodenough, 1978). This is clearly not their function in non-excitable tissues. Here gap junctions are thought to integrate various cellular activities such as metabolism (Sheridan, Finbow & Pitts, 1979), growth (Loewenstein, 1979) and responses to external stimuli (Lawrence, Beers & Gilula,

1978) as well as function as a possible transport system for nutrients (Bennett, 1973; Sheridan, 1976) and low molecular weight control molecules (Lawrence *et al.*, 1978). However, most of these roles in non-excitable tissues are at the moment little more than speculation with little firm evidence to support them (Finbow & Yancey, 1981).

STRUCTURE AND COMPOSITION OF THE GAP JUNCTION

Of the various types of intercellular junctions (e.g. tight junctions, various forms of adherens junctions and septate junctions), the gap junction is perhaps structurally the least complex. By freeze-fracture replication the gap junction can be recognised as an aggregate of generally uniform-sized intramembrane particles (IMPs) around 8 to 10 nm in diameter forming a plaque of intercellular contact between the plasma membranes of adjacent pairs of cells (Fig. 1; McNutt & Weinstein, 1970; Goodenough & Revel, 1970; see Larsen, 1977 for a review). In the freeze-fracture process, the membranes are split in half to reveal the hydrophobic regions of the inner leaflet (P-face) and outer leaflet (E-face) (Branton *et al.*, 1975). There is now accumulating evidence that the IMPs represent proteins buried within the membrane (see Finbow & Yancey, 1981 for references).

Tangential views of lanthanum-infiltrated thin-sectioned junctions show the particles to also span the small 2 to 4 nm space or 'gap' separating the outer leaflets of the paired plasma membranes (Revel & Karnovsky, 1967). Similar views are also obtained from negatively stained fractions of isolated gap junctions (Fig. 1). Thin-section studies also show that there is no significant peri-junctional cytoplasmic differentiation as has been observed for other classes of intercellular junctions, notably those of the adherens type (Hull & Staehelin, 1979).

Of the two electron microscopy techniques for examining tissues, thin sectioning and freeze-fracture replication, the latter provides the better method of detecting the presence of gap junctions because of the extensive *en face* views of the plasma membranes that are exposed. However, conventional thin section often allows for unambiguous identification of the cell types joined by gap junctions which is not always the case for freeze-fracture replication.

The basic structural unit of the gap junction would thus appear to be the proteinaceous particles that are embedded in juxtaposed plasma membranes and which span the small 2 to 4 nm gap. Is each junctional

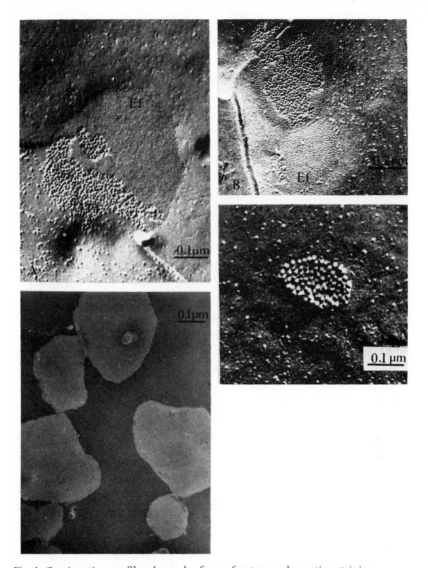

Fig. 1. Gap junction profiles shown by freeze-fracture and negative staining.

(A–C) Freeze-fracture profiles of gap junctions from foetal rat liver (A) and *Aplysia* bag cell neurons (B & C). In both vertebrates and molluscs the connexon particles fracture to the P-face (Pf) to leave pits on the E-face (Ef).

(D) Negatively stained fraction of gap junctions isolated from rat liver by conventional methods involving first the purification of plasma membranes followed by extensive trypsin and collagenase treatment before extraction with the detergent sarkosyl to remove non-junctional membranes. Note the hexagonal packing of the connexons, particle made from two halves, one from either membrane, or is it derived from one cell? Perfusion of a variety of tissues, notably liver, with hypertonic sucrose solutions has shown that gap junctions can simply be split in half (Goodenough & Gilula, 1974). This supports the belief of the dual nature of each junctional particle; i.e. being constructed from two halves which therefore abut end-to-end across the gap. Each half has been termed a 'connexon'.

The detailed study of connexon structure has been made possible by the isolation of gap junctions from plasma membrane fractions due to their insolubility to detergents. Two complementary studies using isolated rodent liver gap junctions have come up with basically the same picture. Caspar, Goodenough, Makowski & Phillips (1977; Makowski, Caspar, Phillips & Goodenough, 1977) employed X-ray diffraction techniques on pellets of gap junctions whereas the study of Unwin & Zampighi (1980) used electron diffraction analysis on negatively stained gap junctions in the electron microscope. Their results suggest that the connexon is a hollow cylinder of protein which traverses the plasma membrane joining end-to-end with a connexon from the adjoining membrane and so together forming a channel of around 1.5 to 2 nm in diameter which links the cytoplasms. Although neither of these two studies show conclusively the existence of a continuous central channel, the fact that two quite different approaches have obtained basically the same structure offers strong support for the channel hypothesis. Thus, in essence, gap junctions form networks of proteinaceous channels which join the interiors of cells.

It is not possible here to give a full account of the distribution of gap junctions (but see Table 1) as now many hundreds of instances related to their occurrence have been reported in the literature. From these many varied studies however, it seems reasonable to suppose that gap junctions are universal features of metazoan animals (i.e. from coelenterates onwards). That is not to say that every cell is joined to its neighbouring cells by these structures. Indeed it is generally thought that the majority of excitable cell types (e.g. most neurons and possibly all skeletal muscle cells of the adult) do not form gap junctions at all, presumably because the electrical communication provided by gap junctions would impair their function (Bennett, 1973). Furthermore, not all non-excitable cell types possess gap junctions. For example the epithelial cells from the distal kidney tubule and loop of Henlé are not joined by gap junctions whereas those from the glomerulus and proximal tubule are (Kühn, Reale & Wermbter, 1975; Kühn & Reale, 1975). Likewise, the endothelial cells lining the arteries and arterioles are joined by gap junctions but in at least certain instances, those lining the capillaries and venules are not (Simionescu, Simionescu & Palade, 1975; chapter by Sheridan & Larson, this volume). With regard to circulating blood cells, they appear not to form gap junctions (Bennett, 1973) and little is known about junctions between blood forming cells and the immune system in general. Thin sections have shown the presence of gap junctions between reticular cells, macrophages and a variety of early differentiated blood cell types in the bone marrow (Shaklai & Tauassoli, 1979) but a freeze-fracture study failed to detect gap junctions (Campbell, 1980). In the main though, it would seem that the vast majority of non-excitable cell types possess gap junctions.

One would like to know how many putative connexon channels are formed between the average cell pair. There appears though, to be considerable variation in the extent of gap junction formation between different cell types making such an average not too meaningful. Moreover, very few morphometric studies have been carried out which allow such a calculation. Nevertheless, estimations on rat liver hepatocytes show about 2% of the total surface area to be occupied by gap junctions (Yee & Revel, 1978; Yancey, Easter & Revel, 1979) which is the equivalent of approximately 200 000 channels per cell. As each hepatocyte is joined to on average six other hepatocytes (Revel, Yancey, Meyer & Nicholson, 1980) this would mean that there are around 30 000–40 000 channels per hepatocyte cell pair. A recent estimate on the circular smooth muscle cells from the guinea pig ileum suggests they each possess 70 000–80 000 such channels (Gabella & Blundell, 1979). In contrast though, the muscle cells of the longitudinal musculature appear not to form gap junctions. As has been pointed out by Sheridan (1973), cell volume is also important in the degree of communication. That is, a pair of large cells would need eight times as many channels to communciate to the same extent as a pair of smaller cells of half the diameter.

The accepted view from earlier freeze-fracture and lanthanum tracer studies was that the connexon particles were packed into a hexagonal array forming a honeycomb-like structure (see Fig. 1). Recent morphological studies using the rapid freeze technique at liquid helium temperatures instead of conventional fixation and infiltration with cryoprotective agents, suggest the 'crystalline' arrangement of connexon particles only to occur in damaged tissues or under unhealthy conditions (Raviola, Goodenough & Raviola, 1980). It appears that the connexons in normal situations are randomly packed with little or no crystallinity. Moreover, there is considerable variation in connexon packing density from cell type to cell type and sometimes within the same cell type. For example, in mouse liver hepatocytes the connexon particles are tightly packed (centre-to-centre spacing being 8 to 10 nm) but between the cells of the rabbit ciliary epithelium they are very loosely packed indeed being no denser than non-junctional IMPs and the junctions only recognisable on their larger uniform particle size (Raviola *et al.*, 1980).

Not only is there pleomorphism in the packing density but also in the arrangement of connexons. In most situations they are organised into plaque-like features but occasionally the connexons align side by side forming linear arrays which sometimes can produce rather elaborate circular structures (Kensler, Brink & Dewey, 1977). Often, in larger gap junctions, the connexons are arranged into rows separated by particle-free aisles (Albertini, Fawcett & Olds, 1975; Larsen, 1977). It might be thought that this arrangement represents a situation where the connexons have been widely dispersed but then aggregated into rows during fixation prior to freeze-fracture. However, such arrangements have also been observed in rapidly frozen tissues (Shibata, Nakata & Page, 1980).

As yet it is not known what all these variations in gap junctional structure mean. It is possible that the pleomorphisms are merely a consequence of the constraints placed upon them by cell architecture and the position they occupy between cells rather than any inherent differences of the various junctions.

The simple structure of the gap junction is reflected in its simple protein composition with there being perhaps only one major species of protein. However, there has been a great deal of controversy on the size of the junctional protein as seen on SDS polyacrylamide gel electrophoresis (SDS-PAGE) but over the past two years a consensus of opinion has arisen. There are a number of problems with attempting to isolate pure fractions of gap junctions. Firstly, there is no definitive criterion of purity except for subjective analysis of junctionally enriched fractions on the electron microscope. Secondly, they probably only account for around 0.01% of the total protein in rodent liver (Finbow, unpublished results), the most commonly used source. Thirdly, proteases such as collagenase and trypsin were often used in earlier investigations to remove contaminants. Whilst the gap junctions appeared morphologically unaltered (Fig. 1), there was degradation of the protein.

After extensive treatments of intact gap junctions with proteases, the major protein component has an apparent molecular weight of 12 000–16 000 often resolving into a doublet (Goodenough, 1976; Hertzberg & Gilula, 1979; Henderson, Eibl & Weber, 1979; Finbow, Yancey, Johnson & Revel, 1980). Indeed this is the only band seen in fractions that are particularly pure as judged from electron micros-copy. If protease treatment is not carried out but other methods are used to remove contaminants (e.g. extraction with urea), the major protein component has a molecular weight of 27 000. Both these mole-cular weight species are absent from the appropriately isolated junc-tion fractions from regenerating weanling rat liver at the time it is known that gap junctions are greatly reduced in area between hepato-cytes (Finbow *et al.*, 1980), the major junction forming cell type in liver. The point that is somewhat difficult to grasp is how the gap junctions remain virtually unaltered morphologically and yet there is such an apparent shift in the molecular weight of the protein com-ponent on protease treatment.

An unusual property of at least the 27 000 species is its ability to form dimers in the presence of SDS to give molecular weight species on SDS-PAGE of 48 000–50 000 (Henderson *et al.*, 1979; Hertzberg, 1980). We have recently devised a new method for isolating gap junctions by solubilising crude liver homogenates or monolayers of cultured cells with Triton in salt conditions that keep nuclei and cytoskeletal com-ponents intact (Finbow & Pitts, unpublished results). Gap junctions isolated in this manner show a major band at 16 000 if solubilised at ambient temperatures in SDS but if boiled, a band of 27 000 appears with the concomitant reduction of the 16 000 component (Fig. 2, tracks 2 & 3). Thus, it would seem that the 'native' 27 000 species may itself be a dimer. This perhaps could explain the unalteration of junction structure upon protease treatment (i.e. the protein is only degraded from a 16 000 component to a 12 000 component). In this 16 000 form, the gap junction appears to be much more resistant to trypsin. Some-what surprisingly, gap junctions isolated by conventional means through a plasma membrane step and purified by trypsin and col-lagenase treatment exhibit a complex profile on SDS-PAGE when solubilised at ambient temperatures suggestive of monomers, dimers, trimer and tetramers (and possibly higher; Fig. 2, track 4). It is only when these fractions are solubilised by heating in the presence of

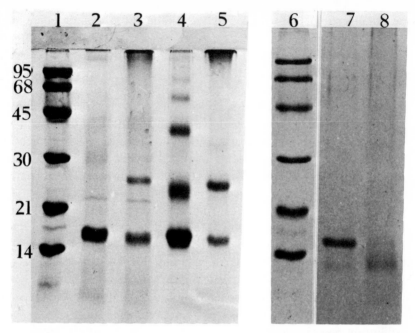

Fig. 2. SDS-PAGE of gap junctional fractions from mouse liver and cultured BRL cells.
Tracks 1 & 6 – standard protein markers (molecular weight given in thousands).

Tracks 2, 3 & 7 – gap junction fractions isolated from BRL cells (2 & 3) and mouse liver (7) using direct extraction of monolayer cultures or liver homogenates with triton followed by a trypsin treatment and a second detergent extraction with sarkosyl (see text).

Tracks, 4, 5 & 8 – gap junction fractions isolated from mouse liver using conventional methods involving plasma membrane purification as described in the legend to Fig. 1D.

Tracks 2 & 3 and tracks 4 & 5 were from the same preparation of gap junctions of BRL cells and mouse liver respectively and solubilised in SDS sample buffer either at 100 °C for 5 minutes (3 & 5) or at ambient temperatures (2 & 4) in the absence of reducing agents. The samples for tracks 7 & 8 were solubilised by briefly boiling in the presence of reducing agents. As can be seen the SDS-PAGE profile obtained depends not only on the manner of preparation of the gap junction fraction but also on solubilisation conditions.

reducing agents that the characteristic doublet of 12 000–16 000 is produced (Fig. 2, track 8) although extensive boiling again results in aggregation and a considerable amount of material remaining unresolved at the top of the running gel (Fig. 2, tracks 3 & 5).

It thus seems that the basic structural protein component of the gap junction has an apparent molecular weight of 16 000 and that its unusual aggregation properties have compounded the difficulties of attempting to identify this component. Unlike other plasma membrane proteins, the junctional protein appears not to be glycosylated and studies on the lipid composition of isolated junctions suggest they contain no unusual lipids (Hertzberg & Gilula, 1979; Henderson *et al.*, 1979).

As yet it is not known if there are different pleomorphic forms of the gap junction protein. A protein of molecular weight 26 000 from lens was thought to be of junctional origin but partial proteolytic mapping and immunological approaches suggest it is not related to the liver junctional protein (Hertzberg, 1980).

FUNCTIONAL TESTS FOR CELL COMMUNICATION AND CORRELATION WITH GAP JUNCTIONS

Two very different lines of research have shown that many varied types of neighbouring cells can communicate through the exchange of low molecular weight substances between their cytoplasms. The first and perhaps most widely used method uses electrophysiological techniques. Neighbouring cells are impaled with salt-filled (generally KCl) glass micropipettes and a pulse of current passed through one of them. If the cells are in communication with one another then there will be a potential change of similar time course in both cells, i.e. the cells are 'electrically coupled'. Electrical coupling is most likely due to the permeability of the junctional membranes to cytoplasmically located inorganic ions which, as mentioned above, allows gap junctions to be used as an alternative to the chemical synapse in excitable tissues for impulse transmission.

Permeability studies using electrophysiological techniques have been greatly extended by microinjection of fluorescently labelled molecules into the cytoplasms of cells (Fig. 3A). Such studies have been useful in determining the size limit on transfer in various systems. Thus for insect salivary gland cells the cut off point is between molecular weight 1500 and 2000 (Simpson, Rose & Loewenstein, 1977; Loewenstein, 1979) and for cultured rat liver cells it is around 900 (Flagg-Newton, Simpson & Loewenstein, 1979).

The other major lines of investigation use a variety of biochemical approaches. The main two techniques are 'metabolic co-operation' (Subak-Sharpe, Bürk & Pitts, 1969) and 'uridine nucleotide transfer' (Pitts & Simms, 1977; Fig. 3B). These approaches have been extensively

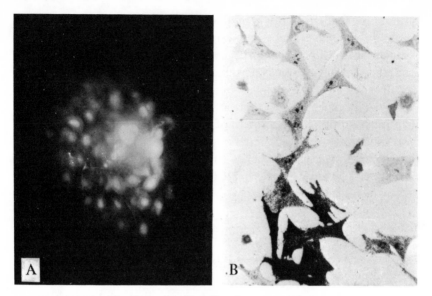

Fig. 3. Assays of junctional communication.
(A) Dye coupling showing the intercellular transfer of Lucifer Yellow between the cluster of bag cell neurons on the abdominal ganglion of *Aplysia* (cf. Fig 1B & C). Picture courtesy of Dr L. K. Kaczmarek.
(B) Uridine nucleotide transfer between 'donor' BHK cells and 'recipient' BHK cells in culture (cf. Pitts and Simms, 1977).

used in tissue culture systems and rely on the transfer of radioactive metabolic intermediates (e.g. ^{3}H-uridine nucleotides) from one cell to another cell in contact. Using a variety of adaptations, the permeability studies from electrophysiological investigations have been confirmed with a wide variety of metabolites up to molecular weight 900 to 1000 shown to pass between contacting cells but not cellular macromolecules (Pitts & Simms, 1977; Finbow & Pitts, 1981).

The advantages of electrophysiological methods over the biochemical approaches is that they can be used to investigate cell communication in whole tissues. As yet the biochemical approaches have been confined to tissue culture, except for the recent experiments of Moor, Smith & Dawson (1980) on the oocyte-cumulus complex, although they offer the great advantage of being quickly able to screen large populations of cell types not possible with electrophysiology and do not require sophisticated equipment. Also, from model metabolic co-operation systems in tissue culture, it has been possible to substantiate the predictions from the permeability studies that junctional communication would allow for metabolic (Sheridan *et al.*, 1979) and ionic integration (Ledbetter & Lubin, 1979).

Intuitively, one needs to invoke some specialised structure containing channels which joins adjacent cytoplasms to account for this wide permeability and the obvious candidate on a structural basis is the gap junction. Indeed the connexon channel diameter (1.5 to 2 nm) would be expected to produce a cut-off point of molecular weight around 1000 (Flagg-Newton *et al.*, 1979). However, there are other types of intercellular junctions and before it is accepted that the gap junction, and not these other intercellular junctions, is the morphological basis of permeability several questions need to be answered. Firstly, where examined do all three functional assays – electrical coupling, dye coupling and metabolic coupling – occur together? That is, is there more than one pathway of intercellular permeability? Secondly, is the gap junction always present whenever pairs or groups of cells have been shown to communicate with each other by any one of these functional assays? And thirdly, do changes in the ability to communicate correlate with changes in the extent of gap junctions?

In Table 1 an attempt has been made to draw together all the relevant studies to answer these questions. Many instances of intercellular communication have not been included because there is no morphological data available in those systems.

In all instances where dye coupling and metabolic coupling occur so does electrical coupling. However, there are some situations where electrical coupling occurs in the apparent absence of dye coupling. This appears to be more of a sensitivity difference in the two techniques rather than separate pathways of ionic and dye transfer (Bennett, Spira & Spray, 1978; Sheridan, 1976; also see the chapter by Lo for an example of lack of dye transfer in the presence of electrical coupling).

With regard to the second question, there is very good correlation of functional communication with the presence of gap junctions. Indeed, in many of the seventy or so examples in Table 1, gap junctions are the only cell surface specialisation found. However, there are two instances where they do not appear to go hand-in-hand, the epidermal and endodermal cells of *Hydra* and cultured mouse L cells.

For *Hydra* it has been well established in a variety of species that these cells are joined by gap junctions and so it came as a surprise to find lack of electrical and dye coupling. For other coelenterates though, notably the *Hydramedusae* (jellyfish), electrical coupling has been

Table 1. Correlation of Gap junctions with intercellular communication.

In most instances the presence of gap junctions has been established both by thin-section and freeze-fracture. EC, electrical coupling; DC, dye coupling; MC, metabolic coupling. Electrical coupling was established in most instances by impalement of neighbouring cells with microelectrodes and in the other instances by the electrical properties of the tissue (e.g. cardiac and smooth muscle). The diagonal lines (/) signify that no data are available. The superscript numbers refer to references listed at the end of the table.

Cell type	Gap junctions	Functional assays			Comments
		EC	DC	MC	
I INVERTEBRATES					
(A) COELENTERATES					
Hydra epithelial cells	+[1-3]	−[4]	−[4]	/	Various species shown to have gap junctions
Excitable endo-dermal cells of *Hippopodius*	+[5]	+[5]	/	/	
Excitable epithelial cells of *Euphysa*	+[6]	+[6]	/	/	
Excitable endo-dermal cells of *Polyorchis*	+[7]	+[7]	/	/	
(B) ANNELIDS					
Leech glial cells	+[8]	+[9]	/	/	
Septum of lateral giant earthworm	+[10]	+[11]	+[11]	/	Permeable to a wide range of fluorescein anions. Possible fixed

					Notes
axon					charge affecting permeability[12]
Leech Retzius cells	/	+[13]	/	+[14]	Not known if gap junctions are present
(C) MOLLUSCS					
Anterior byssus retractor muscle of *Mytilis*	+[15]	+[16]	/		
Aplysia bag cell neurons	+[17]	+[17]	+[17]	/	
(D) ARTHROPODS					
Septa of crayfish lateral giant axons	+[18,19]	+[18,19]	+[18]	+[20]	Treatments which cause increase in septal resistance also cause reduction in septal contact[19, 21] and change in juctional morphology[22]
Synapse between lateral giant axons and motor neurons of crayfish	+[23]	+[24]	/		First example of electrical coupling and also exhibits 'rectification'
Chironomus salivary gland epithelial cells	+[25]	+[26]	+[27]	/	Junctions shown to be permeable to fluorescently labelled oligopeptides up to MW 1500
Epidermal cells of larval beetle	+[28]	+[28]	+[28]		A dye of MW 1000 found not to be transferred
Cultured TN cells (*Lepidoptera*)	+[29]	+[29]	/		

Table 1 (cont.)

Cell type	Gap junctions	Functional assays			Comments
		EC	DC	MC	
(D) ARTHROPODS					
Cultured AC cells (*Homoptera*)	+[29]	+[29]	/	/	
II VERTEBRATES					
(A) NON-EXCITABLE TISSUES					
Mouse liver hepatocytes	+[30]	+[31]		/	
Rat liver hepatocytes	+[32]	+[33–36]	+[35, 36]	/	Preliminary evidence suggests reduced dye and electrical coupling in regenerating liver when gap junctions are greatly reduced in area[35, 36]
Mouse brown fat cells	+[37]	+[38]		/	
Rat thyroid follicular cells	+[39]	+[40]		/	
Human thyroid follicular cells	+[41]	+[40]		/	
Rat salivary gland acinar cells	+[42]	+[43]	+[43]	/	

Mouse salivary gland acinar cells	+[44]	+[44]	+[44]	/	
Mouse pancreatic acinar cells	+[45]	+[46]	+[47]	/	
Foetal rabbit adrenal cortical cells	+[48]	+[48]	/	/	
Rat cumulus–oocyte complex	+[49]	+[49]	+[49]	/	Evidence for metabolic support of oocvte through junctions from sheep oocyte–cumulus complex[50]
Bufo marinus urinary bladder epithelial cells	+[51]	+[52]	/	/	
Rana pippiens skin epidermal cells	+[42]	+[53]	/	/	
Rabbit lens fibre cells	+[54]	/	/	+[54]	
(B) EXCITABLE TISSUES					
Club endings of goldfish Mauthner cells	+[55]	+[56, 57]	/	/	
Supramedullary neurons of puffer and sargassum fish	+[58]	+[58]	/	/	

Table 1 (cont.)

Cell type	Gap junctions	Functional assays			Comments
		EC	DC	MC	
(B) EXCITABLE TISSUES					
Spinal and medullary electromotor nuclei of mormyrid fish	+[59]	+[59]	/	/	
Giant electromotor neurons of *Malapterurus electricus*	+[60]	+[60]	/	/	
Medullary electromotor nuclei of gymnotid fish	+[61]	+[61]	/	/	
Motorneurons of toadfish swim bladder muscle	+[62]	+[62]	/	/	
Reticulospinal neurons of lamprey	+[63,64]	+[63]	/	/	
Chick ciliary ganglion neurons	+[65]	+[66]	/	/	
Hair and supporting cells of alligator lizard	+[67]	+[68]	/	/	

17

				Comments
Retinal photo-receptor cells of *Bufo marinus*	+[69]	+[70]	/	
Cardiac muscle cells of *Rana pipiens*	+[71]	+[72]	/	
Cardiac muscle cells of rat	+[73]	+[74]	/	
Cardiac muscle cells of rabbit	+[75]	+[76]	/	
Smooth muscle cells from guinea pig taenia coli	+[77]	+[78]	/	Considerable variation in extent of gap junctions between different types of smooth muscles
(C) EMBRYONIC TISSUES				
Re-aggregated *Fundulus* blastomeres	+[79]	+[80]	+[80]	
Deep cells of *Fundulus* gastrula	+[81]	+[82]	/	
Late cleavage and early blastula of *Ambyostoma mexicanum*	+[83]	+[83]	/	
Xenopus 2nd cleavage to primitive blastocoel	+[84]	+[85]	+[86]	Dye coupling experiments carried out on later stages

Table 1 (*cont.*)

Cell type	Gap junctions	Functional assays			Comments
		EC	DC	MC	
(C) EMBRYONIC TISSUES					
8-cell stage to post-implantation stage of mouse embryo	+[87,88]	+[87,88]	+[87,88]		
Chick neural plate	+[89]	+[90]	/	/	
Chick neural tube	+[91]	+[90]	/	/	
Chick limb bud ectodermal cells	+[92]	+[90]	/	/	
Myotubes of *Axolotl* anterior myotomes	+[93]	+[93]	/	/	
(D) TISSUE CULTURE					
Novikoff hepatoma cells	+[94]	+[94]	+[94]	+[95]	Formation of gap junctions correlates with onset of electrical and metabolic coupling [95,96]
BHK-21/C13 (hamster fibroblasts)	+[37]	+[97]	+[97]	+[98]	Shown to exchange a variety of metabolites up to MW 1000 but not nucleic acids and proteins [99,100]
Don chinese hamster cells	+[101]	+[101]	/	+[101]	

Cell type					Comments
3T3 cells and SV3T3 cells	+[102,103]	+[97,103]	/	+[99]	Shown to transfer a variety of fluorescent probes up to MW 1000
RL cells (rat liver epithelial cells)	+[104]	+[105]	+[105]	/	
H4IIE cells (rat liver tumor)	+[106]	+[107]	/	+[108]	Low levels of electrical and metabolic coupling correlate with small gap junctions
BRL cells (rat liver epithelial cells)	+[109]	/	/	+[110]	
IMR 90 (human embryo fibroblasts)	+[111]	/	/	+[111]	
B9CR/MIR-4 (rat fibroblastoma)	+[103]	+[103]	/	/	
RE cells	-[103]	-[103]	/	/	
Hela cells	-[103]	-[103]	/	+[99,112] -[112]	Low level of metabolic coupling and different clonal strains exhibit different coupling abilities
L cells (and various derivatives)	-[101,113] +/-[115]	-[101,114] +/-[116]	-[114]	-[98,99,101] +/-[112,116,117]	See text
Human skin fibroblasts	+[113]	+[114]	+[114]	+[118]	Segregant hybrids of human skin fibroblasts and L cells which exhibit reduced levels of coupling also show reduced incidence of gap junctions[113,114]

Table 1 (*cont.*)

	Gap junctions	Functional assays			Comments
Cell type		EC	DC	MC	
(D) TISSUE CULTURE					
Human primary skin keratinocytes	+[119]	+[120]	/	+[121]	
Chick primary myocardial cells	+[122]	+[123]	/	/	
Mouse primary myocardial cells	+[124]	+[124]	/	+[124]	
Mouse primary granulosa cells	+[124]	+[124]	/	+[124]	
Chick primary myoblasts	+[125]	+[125]	/	+[125]	
Mouse primary mammary epithelial cells from pregnant mice or spontaneous tumors	+[126]	+[127]	/	/	
Rat primary pancreatic islet B-cells	+[128]	/	+[129]	+[129, 130]	Mouse pancreatic islet B-cells are electrically coupled *in situ*[131]
PC13 embryonal carcinoma cells	+[132]	/	/	+[133]	R5/3 subline of PC13 shows reduced metabolic coupling and

| PCC4Azal embryonal carcinoma cells | $+$ [134,135] | $+$ [136] | $+$ [136] | $+$ [136,137] | incidence of gap junctions whereas H2T12 revertant exhibits normal parental level of both | Fab fragments of anti-embryonal carcinoma cell serum reduces level of metabolic coupling (as measured by an indirect assay) and incidence of gap junctions[135, 137] |

1, Hand & Gobel (1972); 2, Filshie & Flower (1977); 3, Wood (1977); 4, De Laat, Tertoolen & Grimmelikhuijzen (1980); 5, Mackie (1976); 6, Josephson & Schwab (1979); 7, King & Spencer (1979); 8, Coggeshall (1974); 9, Kuffler & Potter (1964); 10, Kensler, Brink & Dewey (1979); 11, Brink & Dewey (1978); 12, Brink & Dewey (1980); 13, Hagiwara & Morita (1962); 14, Rieske, Schubert & Kreutzberg (1975); 15, Brink, Kensler & Dewey (1979); 16, Twarog, Dewey & Hidaka (1973); 17, Kaczmarek, Finbow, Revel & Strumwasser (1979); 18, Payton, Bennett & Pappas (1969); 19, Pappas, Asada & Bennett (1971); 20, Hermann, Rieske, Kreutzberg & Lux (1975); 21, Asada & Bennett (1971); 22, Perrachia & Dulhunty (1976); 23, Hanna, Keeter & Pappas (1978); 24, Furshpan & Potter (1959); 25, Rose (1971); 26, Rose & Loewenstein (1971); 27, Simpson, Rose & Loewenstein (1977); 28, Caveney & Podgorski (1975); 29, Epstein & Gilula (1977); 30, Goodenough & Revel (1970); 31, Penn (1966); 32, Yancey, Easter & Revel (1979); 33, Loewenstein & Penn (1967); 34, Graf & Peterson (1978); 35, Revel, Yancey, Meyer & Finbow (1980); 36, Revel, Yancey, Meyer & Nicholson (1980); 37, Revel, Yee & Hudspeth (1971); 38, Sheridan (1971); 39, Luciano, Thiele & Reale (1979); 40, Jamakosmanovic & Loewenstein (1968); 41, Thiele & Reale (1976); 42, Dewey & Barr (1964); 43, Hammer & Sheridan (1978); 44, Kater & Galvin (1978); 45, Meda (unpublished results); 46, Iwatsuki & Peterson (1978); 47, Iwatsuki & Peterson (1979); 48, Joseph, Slack & Gould (1973); 49, Gilula, Epstein & Beers (1978); 50, Moor, Smith & Dawson (1980); 51, Wade (1978); 52, Loewenstein et al. (1965); 53, Fisher, Erlij & Helman (1980); 54, Goodenough, Dick & Lyons (1980); 55, Robertson (1963); 56, Furshpan (1964); 57, Furukawa, Fukami & Asada (1963); 58, Bennett, Nakajima & Pappas (1967a); 59, Bennett, Pappas, Aljure & Nakajima (1967); 60, Bennett, Nakajima & Pappas (1967b); 61, Bennett, Pappas, Giménez & Nakajima (1967); 62, Pappas & Bennett (1966); 63, Rovainen (1974); 64, Pfenninger & Rovainen (1974); 65, Cantino & Mugnaini (1975); 66, Martin & Pilar (1964); 67, Nadol, Mulroy, Goodenough & Weiss (1976); 68, Weiss, Mulroy & Altmann (1974); 69, Gold & Dowling (1979); 70, Gold (1979); 71, Kensler, Brink & Dewey (1977);

Notes to Table 1 (*cont.*)

72, Barr, Dewey & Berger (1965); 73, Matter (1973); 74, Dreiffus, Girardier & Forsmann (1966); 75, Baldwin (1979); 76, De Mello (1975); 77, Fry, Devine & Burnstock (1977); 78, Barr, Berger & Dewey (1968); 79, Ne'eman, Spira & Bennett (1980); 80, Bennett, Spira & Spray (1978); 81, Hogan & Trinkaus (1977); 82, Bennett & Trinkaus (1970); 83, Hanna *et al.* (1980); 84, Sanders & DiCaprio (1976); 85, DiCaprio, French & Sanders (1976); 86, Sheridan (1971); 87, Lo & Gilula (1979a); 88, Lo & Gilula (1979b); 89, Revel & Brown (1976); 90, Sheridan (1968); 91, Schoenwolf & Kelley (1980); 92, Fallon & Kelley (1977); 93, Keeter, Pappas & Model (1975); 94, Johnson & Sheridan (1971); 95, Pederson, Johnson & Sheridan (1980); 96, Sheridan, Hammer-Wilson, Preuss & Johnson (1978); 97, Furshpan & Potter (1968); 98, Pitts (1971); 99, Pitts & Simms (1977); 100, Finbow & Pitts (1981); 101, Gilula, Reeves & Steinbach (1972); 102, Montesano & Philippeaux (1981); 103, Hülser & Dempsey (1973); 104, Flagg-Newton & Loewenstein (1979); 105, Flagg-Newton, Simpson & Loewenstein (1979); 106, Porvaznik, Johnson & Sheridan (1976); 107 Sheridan (1973); 108, Finbow (1979); 109, Finbow & Pitts (unpublished results); 110, Pitts & Bürk (1976); 111, Kelley *et al.* (1979); 112, Cox, Krauss, Balis & Dancis (1974); 113, Larsen, Azarnia & Loewenstein (1977); 114, Azarnia & Loewenstein (1977); 115, Nelson & Peacock (1972); 116, Kohen & Kohen (1977); 117, Gaunt & Subak-Sharpe (1979); 118, Cox, Krauss, Balis & Dancis (1970); 119, Flaxman (1972); 120, Flaxman & Cavato (1973); 121, Pitts (un-published results); 122, Griepp & Bernfield (1978); 123, Griepp, Peacock, Bernfield & Revel (1978); 124, Lawrence, Beers & Gilula (1978); 125, Kalderon, Epstein & Gilula (1977); 126, Pickett, Pitelka, Hamamoto & Misfeldt (1975); 127, Shen, Hamamoto & Pitelka (1976); 128, Meda, Perrelet & Orci (1979); 129, Meda, Perrelet & Orci, chapter 6, this book; 130, Kohen *et al.* (1979); 131, Eddlestone & Rojas (1980); 132, Hooper & Parry (1980); 133, Slack, Morgan & Hooper (1978); 134, Lo & Gilula (1980a); 135, Dunia *et al.* (1979); 136, Lo & Gilula (1980b); 137, Nicolas, Kemler & Jacob (1981).

established to occur between cells that form gap junctions. Although the studies of Mackie (1976) and Josephson & Schwab (1979) on the excitable epithelia of *Hippopodius* and *Euphysa* respectively only tentatively identified gap junctions by thin section, the study of King & Spencer (1979) clearly demonstrated these structures by lanthanum-infiltration techniques in the excitable epithelium of *Polyorchis*. Moreover a function can be ascribed to the gap junctions in these excitable epithelia, i.e. conduction of action potential between cells. Thus, the situation of the *Hydra* may be a special case. Either the gap junctions exist in an impermeable state and are only open at particular times, or the impalement of the cells with microelectrodes caused sufficient damage to 'uncouple' the cells. There is now mounting evidence that cells can break communication with each other rapidly and reversibly due to intracellular acidification (Spray, Harris & Bennett, 1981) and rises in the free concentration of cytoplasmic calcium (Rose & Loewenstein, 1976).

Mouse L cells have now been in continuous culture for well over thirty years and so it is perhaps not too surprising to find variability in their communication abilities in different laboratories. Indeed, different clonal strains of Hela cells exhibit different abilities to participate in metabolic co-operation (Cox, Krauss, Balis & Dancis, 1974) and partially metabolic co-operation deficient strains of various cultured cells have been isolated (Wright, Goldfarb & Subak-Sharpe, 1976; Slack, Morgan & Hooper, 1978). Of the two studies where morphological analysis was carried out in tandem with the functional assays on L cells, both showed the absence of intercellular communication and the lack of gap junctions (Gilula, Reeves & Steinbach, 1972; Azarnia & Loewenstein, 1977; Larsen, Azarnia & Loewenstein, 1977).

Moving now to the third question, in every case of where gap junctional frequency has been increased or decreased there is attendant increase or decrease in the level of intercellular communication. On the type of correlative evidence presented in Table 1 and of course structurally, it must be concluded that the gap junction is the morphological basis of intercellular permeability.

However, a permeability function cannot be excluded from other classes of intercellular junctions, notably the tight junction (Sheridan, 1978). As is shown by Sheridan & Larson (see their chapter) there appears to be good dye coupling between the endothelial cells of capillaries and venules from the rat omentum where gap junctions are known to be absent as defined structures but tight junctions are present (Simionescu *et al.*, 1975). Their particular point is discussed in more detail in their chapter.

FORMATION AND TURNOVER OF GAP JUNCTIONS

Because gap junctions are sites of cell–cell interaction as well as being integral plasma membrane components, it would be of considerable interest to understand the process of their formation, the manner of their degradation and their rate of turnover. In quite a few different re-aggregation systems *in vitro* functional communication and gap junctions have been found to form rapidly between pairs or groups of cells over a time course of minutes to an hour or so (Johnson, Hammer, Sheridan & Revel, 1974; Pederson, Johnson & Sheridan, 1980; Sheridan, Hammer-Wilson, Preuss & Johnson, 1978; Loewenstein, Kanno & Socolar, 1978; Ne'eman, Spira & Bennett, 1980).

The many morphological studies to date in a wide variety of different systems suggest that gap junction formation proceeds by a process of recruitment of connexon particles into growing gap junctions rather than complete structures being directly inserted into opposing plasma membranes (Johnson *et al.*, 1974; Decker & Friend, 1974; Benedetii, Dunia & Bloemendahl, 1974; Decker, 1976; Elias & Friend, 1976; Dahl & Berger, 1978; Yee & Revel, 1978; Yancey *et al.*, 1979; Ginzberg & Gilula, 1979; Schneider-Picard, Carpentier & Orci, 1980; Ne'eman *et al.*, 1980; Montesano, 1980; Lane & Swales, 1980). This manner of formation provides further support for the dual composition of each basic structural unit of the gap junction (i.e. the connexon) as each cell needs to be competent. Such formation predicts that 'precursor connexons' are present in the plasma membrane which must therefore be in a closed – or at least partially closed – conformation. It would also seem likely that such free precursors would show little affinity for each other in the same plasma membrane if aggregation were not to occur until after being 'paired' with opposing connexons. Thus, there must be a considerable degree of conformational change of the junctional proteins taking place on the pairing of the connexons.

Each connexon itself appears to be made of six polypeptides as revealed by the structural studies on isolated gap junctions (see above). The apparent aggregation properties of the junctional protein in SDS suggest they have a high affinity for each other and so it would seem likely that the connexons would self-assemble *en route* from the endoplasmic reticulum (the site of synthesis of plasma membrane proteins).

The observations from the freeze-fracture studies are also consistent with junction formation *per se* in the main being a self-assembly process. Although of course, freeze-fracture can only reveal morphological events taking place within the membrane, and biochemical analysis of the precursor connexons will be needed to elucidate the mechanism of their interaction.

The first discernible features observed by freeze-fracture consist of relatively smooth area of membrane, often slightly raised, containing a loose aggregate of uniform sized IMPs which are slightly larger than the connexon particles of the formed junction. These areas have been termed 'formation plaques' (Johnson *et al.*, 1974; Decker, 1976). At later times these areas can be seen to contain centrally located small gap junctions consisting of a few connexon particles although in some systems these are the first observable stages (Elias & Friend, 1976; Yancey *et al.*, 1979; Ne'eman *et al.*, 1980). As formation proceeds the junctions gradually increase in size by accretion of new connexons.

The investigation on re-aggregated Novikoff hepatoma cells has shown that at early times the formation plaques may be separated by an extracellular space exceeding 10 nm compared to the 2 to 4 nm of the mature gap junction. It would be of great advantage if opposing connexons could interact and subsequently align over these distances thus overcoming thermodynamic and steric problems of plasma membranes coming very close together.

One might ask whether there is a need for paired connexons to be aggregated at all into a definable structure and instead exist freely in the plasma membranes. It might be difficult to recognise such paired connexons in freeze-fracture replicas and this could account, for example, for the apparent lack of gap junctions in the dye coupling experiments of Sheridan and Larson (see their chapter). However, there may be compelling reasons why paired connexons aggregate. Isolated paired connexons would not be expected to withstand forces applied to them through normal cellular activity as well as an aggregated structure and thus might not provide a sound basis for intercellular communication. Also interaction of two connexons from opposing membranes might be a sufficiently rare event that it would be favourable to allow such paired connexons to act as a focus for junction formation through aggregation. In this context it is worth noting that gap junctions preferentially form in close association with existing tight junctions (Elias & Friend, 1976; Decker & Friend, 1974; Yancey *et al.*, 1979; Montesano, 1980) presumably because these regions are sites of close membrane apposition thus increasing the probability of connexons interacting (Yancey *et al.*, 1979).

The rate of turnover of gap junctions in normal tissues is not known. However, they can be removed from the cell surface at a reasonable rate as shown by the regenerating weanling rat liver where in a four hour period their abundance falls from 2% of the area of the lateral plasma membrane of hepatocytes to 0.05%.

Two mechanisms of junctional breakdown have been proposed and both seem to operate in normal tissues. The first mechanism involves internalisation of whole gap junctions into the cytoplasm of one cell to form what have been called 'annular gap junctions' (Larsen, 1977). This process has been followed in detail in rabbit granulosa cells. Here there is morphological evidence for the involvement of actin filaments in the invagination process (Larsen, Tung, Murray & Swanson, 1979) and for the fusion of the resulting annular junctions with lysosomes (Larsen & Nan, 1978). The other mechanism involves simply the dispersal of the connexons into the plane of the membrane – a reversal of the formation process. This appears to operate in the regenerating liver (Yancey *et al.*, 1979) and also during metamorphosis in the moth and in this latter example there may be re-utilisation of the connexons as judged from freeze-fracture analysis (Lane & Swales, 1980; also see the chapter by Lane). The relative usage of these two mechanisms of junctional breakdown is not known but there are now quite a number of reported instances of cytoplasmic annular gap junctions (Larsen, 1977), though formation of annular gap junctions can be promoted under annoxic conditions (Yancey *et al.*, 1979).

SUMMARY

Gap junctions are quite simple cell surface specialisations which appear to form a network of channels between the cytoplasms of neighbouring cells. These channels allow for the intercytoplasmic movement of possibly all water-soluble cellular components up to molecular weight 1000 or so. Their apparent ubiquity in metazoan animals leads to the speculation that this form of intercellular communication must be fundamental to the maintenance of the multicelled state. In this regard it is interesting to note that there is a similar distribution of the plant counterpart, the plasmodesmata (see the chapter by Robards).

Membrane channels are not peculiar to multicelled animals but are indeed universal features (Klingenberg, 1981). It would thus seem that the gap junction is a member of a whole family of membrane proteins and perhaps it was derived from one of these proteins early on in the evolution of multicelled animals.

One aspect that is coming to the forefront is the apparent 'closing off' of the junctional channels. Recent electrophysiological evidence suggests they are particularly responsive to small changes in intracellular pH (Spray *et al.*, 1981; chapter by Peterson, Findlay, Daoud & Collins, this volume). Obviously if cells could finely regulate their junctional permeability it would add a new dimension to how their physiological role is viewed. One of the problems this raises is whether gap junctions that are seen in tissues by morphological studies are at that point of time allowing cells to communicate (cf. *Hydra*). This means that future investigations should study junctional communication by both morphological and functional analysis.

Acknowledgements

This work was supported by a grant from the Cancer Research Campaign. The author is grateful to Dr John D. Pitts for ideas and critically reading the manuscript.

REFERENCES

ALBERTINI, D. F., FAWCETT, D. W. & OLDS, D. J. (1975). Morphological variations in gap junctions of ovarian granulosa cells. *Tissue and Cell*, **7**, 389–405.

ASADA, Y. & BENNETT, M. V. L. (1971). Experimental alteration of coupling resistance at an electrotonic synapse. *Journal of Cell Biology*, **49**, 159–72.

AZARNIA, R. & LOEWENSTEIN, W. R. (1977). Intercellular communication and tissue growth. VIII A genetic analysis of junctional communication and cancerous growth. *Journal of Membrane Biology*, **34**, 1–38.

BALDWIN, K. M. (1979). Cardiac gap junction configuration after uncoupling treatment as a function of time. *Journal of Cell Biology*, **82**, 66–75.

BARR, L., BERGER, W. & DEWEY, M. M. (1968). Electrical transmission at the nexus between smooth muscle cells. *Journal of General Physiology*, **31**, 347–68.

BARR, L., DEWEY, M. M. & BERGER, W. (1965). Propagation of action potentials and the structure of the nexus in cardiac muscle. *Journal of General Physiology*, **48**, 797–823.

BENEDETTI, E. L., DUNIA, I. & BLOEMENDAHL, H. (1974). Development of junctions during differentiation of lens fibers. *Proceedings of the National Academy of Sciences, USA*, **71**, 5073–7.

BENNETT, M. V. L. (1973). Function of electrotonic junctions in embryonic and adult tissues. *Federation Proceedings*, **32**, 65–75.

BENNETT, M. V. L. & GOODENOUGH, D. A. (1978). Gap junctions, electrotonic coupling and intercellular communication. *Neurosciences Research Progress Bulletin*, **16**, 373–486.

BENNETT, M. V. L., NAKAJIMA, Y. & PAPPAS, G. D. (1967a). Physiology and ultrastructure of electrotonic junctions. I Supramedullary neurons. *Journal of Neurophysiology*, **30**, 161–79.

BENNETT, M. V. L., NAKAJIMA, Y. & PAPPAS, G. D. (1967b). Physiology and ultrastructure of electrotonic junctions. II Giant electromotor neurons of *Malapterurus electricus*. *Journal of Neurophysiology*, **30**, 209–35.

BENNETT, M. V. L., PAPPAS, G. D., ALJURE, E. & NAKAJIMA, Y. (1967). Physiology and ultrastructure of electrotonic junctions. II Spinal and medullary electromotor nuclei in mormyrid fish. *Journal of Neurophysiology*, **30**, 180–208.

BENNETT, M. V. L., PAPPAS, G. D., GIMENEZ, M. & NAKAJIMA, Y. (1967). Physiology and ultrastructure of electrotonic junctions. IV Medullary electromotor nuclei in gymnotid fish. *Journal of Neurophysiology*, **30**, 236–300.

BENNETT, M. V. L., SPIRA, M. E. & SPRAY, D. C. (1978). Permeability of gap junctions between embryonic cells of *Fundulus*. A reevaluation. *Developmental Biology*, **65**, 114–25.

BENNETT, M. V. L. & TRINKAUS, J. P. (1970). Electrical coupling between embryonic cells by way of extracellular space and specialized junctions. *Journal of Cell Biology*, **44**, 592–610.

BRANTON, D., BULLIVANT, S., GILULA, N. B., KARNOVSKY, M. J., MOOR, H., MÜHLETHALER, K., NORTHCOTE, D. H., PACKER, L., SATIR, B., SATIR, P., SPETH, V., STAEHELIN, L. A., STEERE, R. & WEINSTEIN, R. S. (1975). Freeze-etching nomenclature. *Science*, **190**, 54–56.

BRINK, P. R. & DEWEY, M. M. (1978). Nexal membrane permeability to anions. *Journal of General Physiology*, **72**, 67–86.

BRINK, P. R. & DEWEY, M. M. (1980). Evidence for fixed charge in the nexus. *Nature, London*, **285**, 101–2.

BRINK, P. R., KENSLER, R. W. & DEWEY, M. M. (1979). The effect of lanthanum on the nexus of the anterior byssus retractor muscle of *Mytilus edulis L. American Journal of Anatomy*, **154**, 11–26.

CAMPBELL, F. R. (1980). Gap junctions between cells of bone marrow: an ultrastructural study using tannic acid. *Anatomical Record*, **196**, 101–11.

CANTINO, D. & MUGAINI, E. (1975). The structural basis for electrotonic coupling in the avian ciliary ganglion. A study with thin sectioning and freeze-fracture. *Journal of Neurocytology*, **4**, 505–36.

CASPAR, D. L. D., GOODENOUGH, D. A., MAKOWSKI, L. & PHILLIPS, W. C. (1977). Gap junction structures. I Correlated electron microscopy and X-ray diffraction. *Journal of Cell Biology*, **74**, 605–28.

CAVENEY, S. & PODGORSKI, C. (1975). Intercellular communication in a positional field. Ultrastructural correlates and tracer analysis of communication between insect epidermal cells. *Tissue and Cell*, **195**, 539–74.

COGGESHALL, R. E. (1974). Gap junctions between identified glial cells in the leech. *Journal of Neurobiology*, **5**, 463–7.

COX, R. P., KRAUSS, M. R., BALIS, M. E. & DANCIS, J. (1970). Evidence for transfer of enzyme product as the basis of metabolic cooperation between tissue culture fibroblasts of Lesch-Nyhan disease and normal cells. *Proceedings of the National Academy of Sciences, USA*, **67**, 1573–9.

COX, R. P., KRAUSS, M. R., BALIS, M. E. & DANCIS, J. (1974). Metabolic cooperation in cell culture: studies of the mechanism of cell interaction. *Journal of Cellular Physiology*, **84**, 237–52.

DAHL, G. & BERGER, W. (1978). Nexus formation in the myometrium during parturition and induced by estrogen. *Cell Biology International Reports*, **2**, 381–6.

DECKER, R. S. (1976). Hormonal regulation of gap junction differentiation. *Journal of Cell Biology*, **69**, 669–85.

DECKER, R. S. & FRIEND, D. S. (1974). Assembly of gap junctions during amphibian neuralation. *Journal of Cell Biology*, **62**, 32–47.

DE LAAT, S. W., TERTOOLEN, L. G. J. & GRIMMELIKHUIJZEN, C. J. P. (1980). No junctional communication between epithelial cells in *Hydra*. *Nature, London*, **288**, 711–13.

DE MELLO, W. C. (1975). Uncoupling of heart cells produced by intracellular sodium injection. *Experientia*, **31**, 460–2.

DEWEY, M. M. & BARR, L. (1964). A study of the structure and distribution of the nexus. *Journal of Cell Biology*, **23**, 553–86.

DICAPRIO, R. A., FRENCH, A. S. & SANDERS, E. J. (1976). On the mechanism of electrical coupling between cells of early *Xenopus* embryos. *Journal of Membrane Biology*, **27**, 393–408.

DREIFFUS, J. J., GIRARDIER, L. & FORSMANN, W. G. (1966). Etude de la propagation de l'excitation dans le ventricle de rat au moyen de solution hypertonique. *Pflügers Archiv für die Gesamte Physiologie*, **292**, 13–33.

DUNIA, I., NICOLAS, J. F., JAKOB, H., BENEDETTI, E. L. & JACOB, F. (1979). Junctional modulation in mouse embryonal carcinoma cells by Fab fragments of rabbit anti-embryonal carcinoma cell serum. *Proceedings of the National Academy of Sciences, USA*, **76**, 3387–91.

EDDLESTONE, C. T. & ROJAS, E. (1980). Evidence of electrical coupling between mouse pancreatic B cells. *Journal of Physiology*, **303**, 76P–77P.

ELIAS, P. & FRIEND, D. S. (1976). Vitamin A induced mucous metaplasia: an *in vitro* system for modulating tight and gap junction differentiation. *Journal of Cell Biology*, **68**, 173–88.

EPSTEIN, M. L. & GILULA, N. B. (1977). A study of communication specificity between cells in culture. *Journal of Cell Biology*, **75**, 769–87.

FALLON, J. F. & KELLEY, R. O. (1977). Ultrastructural analysis of the apical ectoderm ridge during vertebrate limb morphogenesis. II Gap junctions as distinctive ridge structures common to birds and mammals. *Journal of Embryology and Experimental Morphology*, **41**, 223–32.

FILSHIE, B. K. & FLOWER, N. (1977). Junctional structure in *Hydra*. *Journal of Cell Science*, **23**, 151–72.

FINBOW, M. E. (1979). Permeability of intercellular junctions formed between animal cells. Ph.D. thesis, University of Glasgow.

FINBOW, M. E. & PITTS, J. D. (1981). Permeability of junctions between animal cells: Intercellular exchange of various metabolites and a vitamin-derived cofactor. *Experimental Cell Research*, **131**, 1–13.

FINBOW, M. E. & YANCEY, S. B. (1981). The roles of intercellular junctions. In *The Cell Surface, Volume IV of Biochemistry of Cellular Regulation*, ed. P. Knox, chapter 8. Florida: CRC Press.

FINBOW, M E., YANCEY, S. B., JOHNSON, R. G. & REVEL J.-P. (1980). Independent lines of evidence suggesting a major gap junctional protein with a molecular weight of 26 000. *Proceedings of the National Academy of Sciences, USA*, **77**, 970–4.

FISHER, R. S., ERLIJ, D. & HELMAN, S. I. (1980). Intracellular voltage of isolated epithelia of frog skin-apical and basolateral cell puncture. *Journal of General Physiology*, **76**, 447–53.

FLAGG-NEWTON, J. L. & LOEWENSTEIN, W. R. (1979). Experimental depression of junctional membrane permeability in mammalian cell culture – a study with tracer molecules in the 300 to 800 dalton range. *Journal of Membrane Biology*, **50**, 65–100.

FLAGG-NEWTON, J. L., SIMPSON, I. & LOEWENSTEIN, W. R. (1979). Permeability of the cell-to-cell membrane channels in mammalian cell junctions. *Science*, **205**, 404–7.

FLAXMAN, B. A. (1972). Growth *in vitro* and induction of differentiation in cells of basal cell cancer. *Cancer Research*, **32**, 462–9.

FLAXMAN, B. A. & CAVATO, F. V. (1973). Low resistance junctions in epithelial outgrowths from normal and cancerous epidermis *in vitro*. *Journal of Cell Biology*, **58**, 219–23.

FRY, G. N., DEVINE, C. E. & BURNSTOCK, G. (1977). Freeze-fracture studies of nexuses between smooth muscle cells. *Journal of Cell Biology*, **72**, 26–34.

FURSHPAN, E. J. (1964). Electrical transmission at an excitory synapse in a vertebrate brain. *Science*, **144**, 878–80.

FURSHPAN, E. J. & POTTER, D. D. (1959). Transmission at the giant motor synapses of the crayfish. *Journal of Physiology*, **145**, 289–325.

FURSHPAN, E. J. & POTTER, D. D. (1968). Low resistance junctions between cells in embryos and tissue culture. *Current Topics in Developmental Biology*, **3**, 95–127.

FURUKAWA, T., FUKAMI, Y. & ASADA, Y. (1963). A third type of inhibition in the Mauthner cell of goldfish. *Journal of Neurophysiology*, **26**, 759–74.

GABELLA, G. & BLUNDELL, D. (1979). Nexuses between smooth muscle cells of the guinea pig ileum. *Journal of Cell Biology*, **82**, 239–47.

GAUNT, S. J. & SUBAK-SHARPE, J. H. (1979). Selectivity in metabolic cooperation between cultured mammalian cells. *Experimental Cell Research*, **120**, 307–20.

GILULA, N. B., EPSTEIN, M. L. & BEERS, W. H. (1978). Cell-to-cell communication and ovulation: a study of the cumulus-oocyte complex. *Journal of Cell Biology*, **78**, 58–75.

GILULA, N. B., REEVES, D. R. & STEINBACH, A. (1972). Metabolic coupling, ionic coupling and cell contacts. *Nature, London*, **235**, 262–5.

GINZBERG, R. D. & GILULA, N. B. (1979). Modulation of cell junctions during differentiation of the chick otocyst sensory epithelium. *Developmental Biology*, **68**, 110–29.

GOLD, G. H. (1979). Photoreceptor coupling in retina of the toad *Bufo marinus*. II Physiology. *Journal of Neurophysiology*, **42**, 311–28.

GOLD, G. H. & DOWLING, J. E. (1979). Photoreceptor coupling in retina of toad *Bufo marinus*. I Anatomy. *Journal of Neurophysiology*, **42**, 292–310.

GOODENOUGH, D. A. (1976). *In vitro* formation of gap junction vesicles. *Journal of Cell Biology*, **68**, 220–31.

GOODENOUGH, D. A., DICK, J. S. B. & LYONS, J. E. (1980). Lens metabolic cooperation – a study of lens transport and permeability visualized with freeze substitution autoradiography and electron microscropy. *Journal of Cell Biology*, **86**, 576–89.

GOODENOUGH, D. A. & GILULA, N. B. (1974). The splitting of hepatocyte gap junctions and zonula occludens with hypertonic disaccharides. *Journal of Cell Biology*, **61**, 575–90.

GOODENOUGH, D. A. & REVEL, J.-P. (1970). A fine structural analysis of inter-cellular junctions in the mouse liver. *Journal of Cell Biology*, **45**, 272–90.

GRAF, J. & PETERSON, O. H. (1978). Cell membrane potential and resistance in liver. *Journal of Physiology*, **284**, 105–26.

GRIEPP, E. B. & BERNFIELD, M. R. (1978). Acquisition of synchrony between embryonic heart cell aggregates and layers. *Experimental Cell Research*, **113**, 263–72.

GRIEPP, E. B., PEACOCK, J. H., BERNFIELD, M. R. & REVEL, J.-P. (1978). Morphological and functional correlates of synchronous beating embryonic heart cell aggregates and layers. *Experimental Cell Research*, **113**, 273–82.

HAGIWARA, S. & MORITA, H. (1962). Electrotonic transmission between two nerve cells in the leech ganglion. *Journal of Neurophysiology*, **25**, 721–31.

HAMMER, M. G. & SHERIDAN, J. D. (1978). Electrical coupling and dye transfer between acinar cells in the rat salivary gland. *Journal of Physiology*, **275**, 495–505.

HAND, A. R. & GOBEL, S. (1972). The structural organization of the septate and gap junctions of *Hydra*. *Journal of Cell Biology*, **52**, 397–408.

HANNA, R. B. KEETER, J. S. & PAPPAS, G. D. (1978). The fine structure of a rectifying electrotonic synapse. *Journal of Cell Biology*, **79**, 764–73.

HANNA, R. B., MODEL, P. G., SPRAY, D. C., BENNETT, M. V. L. & HARRIS, A. L. (1980). Gap junctions in early amphibian embryos. *American Journal of Anatomy*, **158**, 111–14.

HENDERSON, D., EIBL, H. & WEBER, K. (1979). Structure and biochemistry of mouse hepatic gap junctions. *Journal of Molecular Biology*, **132**, 193–218.

HERMANN, A., RIESKE, G., KREUTZBERG, G. W. & LUX, H. D. (1975). Transjunc-tional flux of radioactive precursors across electrotonic synapse between lateral giant axons of the crayfish. *Brain Research*, **95**, 125–31.

HERTZBERG, E. L. (1980). Biochemical and immunological approaches to the study of gap junctional communication. *In Vitro*, **16**, 1057–67.

HERTZBERG, E. L. & GILULA, N. B. (1979). Isolation and characterisation of gap junctions from rat liver. *Journal of Biological Chemistry*, **254**, 2138–47.

HOGAN, J. C. & TRINKAUS, J. P. (1977). Intercellular junctions, intramembranous particles and cytoskeletal elements of deep cells of the *Fundulus* gastrula. *Journal of Embryology and Experimental Morphology*, **40**, 125–41.

HOOPER, M. L. & PARRY, J. (1980). Incidence of gap junctions and microvilli in variant cell lines with altered capacity for metabolic cooperation. *Experimental Cell Research*, **128**, 461–6.

HULL, B. E. & STAEHELIN, L. A. (1979). The terminal web – a reevaluation of its structure and function. *Journal of Cell Biology*, **81**, 67–82.

HÜLSER, D. F. & DEMPSEY, A. (1973). Gap and low resistance junctions between cells in culture. *Zeitschrift für Naturforschung, Band C*, **28**, 603–6.

IWATSUKI, N. & PETERSON, O. H. (1978). Electrical coupling and uncoupling of exocrine acinar cells. *Journal of Cell Biology*, **79**, 533–45.

IWATSUKI, N. & PETERSON, O. H. (1979). Direct visualization of cell to cell coupling: transfer of fluorescent probes in living mammalian pancreatic acini. *Pflügers Archiv*, **380**, 277–81.

JAMAKOSMANOVIC, A. & LOEWENSTEIN, W. R. (1968). Intercellular communication and tissue growth. III Thyroid cancer. *Journal of Cell Biology*, **38**, 556–61.

JOHNSON, R. G., HAMMER, M., SHERIDAN, J. D. & REVEL, J.-P. (1974). Gap junction formation between reaggregated Novikoff hepatoma cells. *Proceedings of the National Academy of Sciences, USA*, **71**, 4536–40.

JOHNSON, R. G. & SHERIDAN, J. D. (1971). Junctions between cancer cells in culture: ultrastructure and permeability. *Science*, **174**, 717–19.

JOSEPH, T., SLACK, C. & GOULD, R. P. (1973). Gap junctions and electrotonic coupling in foetal rabbit adrenal cortical cells. *Journal of Embryology and Experimental Morphology*, **29**, 681–96.

JOSEPHSON, R. K. & SCHWAB, W. E. (1979). Electrical properties of an excitable epithelium. *Journal of General Physiology*, **74**, 213–36.

KACZMAREK, L. K., FINBOW, M. E., REVEL, J.-P. & STRUMWASSER, F. (1979). The morphology and coupling of *Aplysia* bag cells within the abdominal ganglion and in cell culture. *Journal of Neurobiology*, **10**, 535–50.

KALDERON, N., EPSTEIN, M. L. & GILULA, N. B. (1977). Cell-to-cell communication and myogenesis. *Journal of Cell Biology*, **75**, 788–806.

KATER, S. B. & GALVIN, N. J. (1978). Physiological and morphological evidence for coupling in mouse salivary gland acinar cells. *Journal of Cell Biology*, **79**, 20–6.

KEETER, J. S., PAPPAS, G. D. & MODEL, P. G. (1975). Inter- and intramyotomal gap junctions in the axolotl embryo. *Developmental Biology*, **45**, 21–33.

KELLEY, R. O., VOGEL, K. G., CRISSMAN, H. A., LUJAN, C. J. & SKIPPER, B. E. (1979). Development of the aging cell surface: reduction of gap junction-mediated metabolic cooperation with progressive subcultivation of human embryo fibroblasts (IMR-90). *Experimental Cell Research*, **119**, 127–43.

KENSLER, R. W., BRINK, P. & DEWEY, M. M. (1977). Nexus of frog ventricle. *Journal of Cell Biology*, **73**, 768–81.

KENSLER, R. W., BRINK, P. R. & DEWEY, M. M. (1979). The septum of the lateral axon of the earthworm: a thin section and freeze-fracture study. *Journal of Neurocytology*, **8**, 565–90.

KING, M. G. & SPENCER, A. N. (1979). Gap and septate junctions in the excitable endoderm of *Polyorchis penicillatus* (Hydrozoa *Anthomedusae*). *Journal of Cell Science*, **36**, 391–400.

KLINGENBERG, M. (1981). Membrane protein oligomeric structure and transport function. *Nature, London*, **290**, 449–54.

KOHEN, E. & KOHEN, C. (1977). Rapid automated multichannel microspectrofluorometry – a new method for studies on the cell-to-cell transfer of molecules. *Experimental Cell Research*, **107**, 261–8.

KOHEN, E., KOHEN, C., THORELL, B., MINTZ, D. H. & RABINOVITCH, A. (1979). Intercellular communication in pancreatic islet monolayer culture: a microfluorometric study. *Science*, **204**, 862–5.

KUFFLER, S. W. & POTTER, D. D. (1964). Glia in the leech central nervous system: physiological properties and neuron-glia relationships. *Journal of Neurophysiology*, **27**, 290–320.

KÜHN, K. & REALE, E. (1975). Junctional complexes of the tubular cells in the human kidney as revealed with freeze-fracture. *Cell and Tissue Research*, **160**, 193–205.

KÜHN, K., REALE, E. & WERMBTER, G. (1975). The glomeruli of the human and rat kidney studied by freeze-fracturing. *Cell and Tissue Research*, **160**, 177–91.

LANE, N. J. & SWALES, L. S. (1980). Dispersal of junctional particles, not internalization, during *in vivo* disappearance of gap junctions. *Cell*, **19**, 579–86.

LARSEN, W. J. (1977). Structural diversity of gap junctions: a review. *Tissue and Cell*, **9**, 373–94.

LARSEN, W. J., AZARNIA, R. & LOEWENSTEIN, W. R. (1977). Intercellular communication and tissue growth. IX Junctional membrane structure of hybrids between communication-competent and communication-incompetent cells. *Journal of Membrane Biology*, **34**, 39–54.

LARSEN, W. J. & NAN, H. (1978). Origin and fate of cytoplasmic gap junctional vesicles in rabbit granulosa cells. *Tissue and Cell*, **10**, 585–98.

LARSEN, W. J., TUNG, H.-N., MURRAY, S. A. & SWANSON, C. A. (1979). Evidence for the participation of actin filaments and bristle coats in the internalization of gap junction membrane. *Journal of Cell Biology*, **83**, 576–87.

LAWRENCE, T. S., BEERS, W. H. & GILULA, N. B. (1978). Transmission of hormonal stimulation by cell-to-cell communication. *Nature, London*, **272**, 501–6.

LEDBETTER, M. L. S. & LUBIN, M. (1979). Transfer of potassium. A new method of cell-cell coupling. *Journal of Cell Biology*, **80**, 150–65.

LO, C. W. & GILULA, N. B. (1979a). Gap junctional communication in the pre-implantation mouse embryo. *Cell*, **18**, 399–409.

LO, C. W. & GILULA, N. B. (1979b). Gap junctional communication in the post-implantation mouse embryo. *Cell*, **18**, 411–22.

LO, C. W. & GILULA, N. B. (1980a). PCC4azal teratocarcinoma stem cell differentiation in culture. II Morphological characterization. *Developmental Biology*, **75**, 93–111.

LO, C. W. & GILULA, N. B. (1980b). PCC4azal teratocarcinoma stem cell differentiation in culture. III Cell-to-cell communication properties. *Developmental Biology*, **75**, 112–20.

LOEWENSTEIN, W. R. (1979). Junctional intercellular communication and the control of growth. *Biochimica et Biophysica Acta*, **560**, 1–65.

LOEWENSTEIN, W. R., KANNO, Y. & SOCOLAR, S. J. (1978). Quantum jumps of conductance during formation of membrane channels at cell-cell junction. *Nature, London*, **274**, 133–6.

LOEWENSTEIN, W. R. & PENN, R. A. (1967). Intercellular communication and tissue growth. II Tissue regeneration. *Journal of Cell Biology*, **33**, 235–42.

LOEWENSTEIN, W. R., SOCOLAR, S. J., HIGASHINO, S., KANNO, Y. & DAVIDSON, N. (1965). Intercellular communication: renal, urinary bladder, sensory and salivary glands. *Science*, **149**, 295–8.

LUCIANO, L., THIELE, J. & REALE, E. (1979). Development of follicles and of occluding junctions between the follicular cells of the thyroid gland. *Journal of Ultrastructural Research*, **66**, 164–81.

MACKIE, G. D. (1976). Propagated spikes and secretion in a Coelenterate glandular epithelium. *Journal of General Physiology*, **68**, 313–25.

McNUTT, N. S. & WEINSTEIN, R. S. (1970). The ultrastructure of the nexus: a correlated thin-section and freeze-fracture study. *Journal of Cell Biology*, **47**, 666–88.

MAKOWSKI, L., CASPAR, D. L. D., PHILLIPS, W. C. & GOODENOUGH, D. A. (1977). Gap junction structures. II Analysis of the X-ray diffraction data. *Journal of Cell Biology*, **74**, 629–45.

MARTIN, A. R. & PILAR, G. (1964). An analysis of electrical coupling at synapses in the avian ciliary ganglion. *Journal of Physiology*, **171**, 454–75.

MATTER, A. (1973). A morphometric study on the nexus of rat cardiac muscle. *Journal of Cell Biology*, **56**, 690–6.

MEDA, P., PERRELET, A. & ORCI, L. (1979). Increase of gap junctions between pancreatic B-cells during stimulation of insulin secretion. *Journal of Cell Biology*, **82**, 441–8.

MONTESANO, R. (1980). Intramembranous events accompanying junction formation in a liver cell line. *Anatomical Record*, **198**, 403–14.

MONTESANO, R. & PHILIPPEAUX, M. M. (1981). Membrane particle arrays in SV-40 transformed 3T3 cells. *Journal of Cell Science*, **47**, 311–30.

MOOR, R. M., SMITH, M. W. & DAWSON, R. M. C. (1980). Measurement of intercellular coupling between oocyte and cumulus cells using intracellular markers. *Experimental Cell Research*, **126**, 15–29.

NADOL, J. B., MULROY, M. J., GOODENOUGH, D. A. & WEISS, T. F. (1976). Tight and gap junctions in a vertebrate inner ear. *American Journal of Anatomy*, **147**, 281–302.

NE'EMAN, Z., SPIRA, M. E. & BENNETT, M. V. L. (1980). Formation of gap and tight junctions between reaggregated blastomeres of killifish, *Fundulus*. *American Journal of Anatomy*, **156**, 251–62.

NELSON, P. G. & PEACOCK, J. H (1972). Transmission of an active electrical response between fibroblasts (L cells) in culture. *Journal of General Physiology*, **62**, 25–36.

NICOLAS, J.-F., KEMLER, R. & JACOB, F. (1981). Effects of anti-embryonal carcinoma serum on aggregation and metabolic cooperation between teratocarcinoma cells. *Developmental Biology*, **81**, 127–32.

PAPPAS, G. D., ASADA, Y. & BENNETT, M. V. L. (1971). Morphological correlates of increased coupling resistance at an electrotonic synapse. *Journal of Cell Biology*, **49**, 173–88.

PAPPAS, G. D. & BENNETT, M. V. L. (1966). Specialised sites involved in electrical transmission between neurons. *Annals of the New York Academy of Sciences*, **137**, 495–508.

PAYTON, B. W., BENNETT, M. V. L. & PAPPAS, G. D. (1969). Permeability and structure of junctional membranes at an electrotonic synapse. *Science*, **166**, 1641–3.

PEDERSON, D. C., JOHNSON, R. G. & SHERIDAN, J. D. (1980). The development of metabolite transfer between reaggregating Novikoff hepatoma cells. *Experimental Cell Research*, **127**, 159–77.

PENN, R. A. (1966). Ionic communication between liver cells. *Journal of Cell Biology*, **29**, 171–4.

PERRACHIA, C. & DULHUNTY, A. F. (1976). Low resistance junctions in crayfish:

structural changes with functional uncoupling. *Journal of Cell Biology*, **70**, 419–39.

PFENNINGER, K. H. & ROVAINEN, C. M. (1974). Stimulation and calcium dependence of vesicle attachment sites in the pre-synaptic membrane: a freeze cleave study of the lamprey spinal cord. *Brain Research*, **72**, 1–23.

PICKETT, P. B., PITELKA, D. R., HAMAMOTO, S. T. & MISFELDT, D. S. (1975). Occluding junctions and cell behaviour in primary culture and neoplastic mammary gland cells. *Journal of Cell Biology*, **66**, 316–32.

PITTS, J. D. (1971). Molecular exchange and growth control in tissue culture. In *Ciba Foundation Symposium on Growth Control*, ed. G. E. W. Wolstenholme & J. Knight, pp. 89–105. London: Churchill-Livingstone.

PITTS, J. D. & BÜRK, R. R. (1976). Specificity of junctional communication between animal cells. *Nature, London*, **264**, 762–4.

PITTS, J. D. & SIMMS, J. W. (1977). Permeability of junctions between animal cells: intercellular transfer of nucleotides but not macromolecules. *Experimental Cell Research*, **104**, 153–63.

PORVAZNIK, M., JOHNSON, R. G. & SHERIDAN, J. D. (1976). Intercellular junctions and other cell surface differentiations of H4-11E hepatoma cells *in vitro*. *Journal of Ultrastructural Research*, **55**, 343–59.

RAVIOLA, E., GOODENOUGH, D. A. & RAVIOLA, G. (1980). Structure of rapidly frozen gap junctions. *Journal of Cell Biology*, **87**, 273–9.

REVEL, J.-P. & BROWN, S. (1976). Cell junctions in development with particular reference to the neural tube. *Cold Spring Harbor Symposium on Quantitative Biology*, **40**, 433–55.

REVEL, J.-P. & KARNOVSKY, M. J. (1967). Hexagonal arrays of subunits in intercellular junctions of the mouse heart and liver. *Journal of Cell Biology*, **33**, C7–C12.

REVEL, J.-P., YANCEY, S. B., MEYER, D. J. & FINBOW, M. E. (1980a). Behaviour of gap junctions during liver regeneration. In *Communications of Liver Cells*, ed. H. Popper, L. Bianchi, F. Gudat & W. Reutter, pp. 163–76. MTP Press.

REVEL, J.-P., YANCEY, S. B., MEYER, D. J. & NICHOLSON, B. (1980b). Cell junctions and intercellular communication. *In Vitro*, **16**, 1010–17.

REVEL, J.-P., YEE, A. G. & HUDSPETH, A. J. (1971). Gap junctions between electrotonically coupled cells in tissue culture and in brown fat. *Proceedings of the National Academy of Sciences, USA*, **68**, 2924–7.

RIESKE, E., SCHUBERT, P. & KREUTZBERG, G. W. (1975). Transfer of radioactive material between electrically coupled neurons of the leech central nervous system. *Brain Research*, **84**, 365–82.

ROBERTSON, J. D. (1963). The occurrence of a subunit pattern in the unit membranes of club endings in Mauthner cell synapses in goldfish brains. *Journal of Cell Biology*, **19**, 201–21.

ROSE, B. (1971). Intercellular communication and some structural aspects of membrane junctions in a simple cell system. *Journal of Membrane Biology*, **5**, 1–19.

ROSE, B. & LOEWENSTEIN, W. R. (1971). Junctional membrane permeability. *Journal of Membrane Biology*, **5**, 20–50.

ROSE, B. & LOEWENSTEIN, W. R. (1976). Permeability of a cell junction and the local cytoplasmic free ionized calcium concentration: a study with aequorin. *Journal of Membrane Biology*, **28**, 87–119.

ROVAINEN, C. M. (1974). Synaptic interactions of reticulospinal neurons and nerve cells in the spinal cord of the sea lamprey. *Journal of Comparative Neurobiology*, **154**, 207–24.

SANDERS, E. J. & DiCAPRIO, R. A. (1976). Intercellular junctions in the *Xenopus* embryo prior to gastrulation. *Journal of Experimental Zoology*, **197**, 415–21.

SCHNEIDER-PICARD, G., CARPENTIER, J. L. & ORCI, L. (1980). Quantitative evaluation of gap junctions during development of the brown adipose tissue. *Journal of Lipid Research*, **21**, 600–5.

SCHOENWOLF, G. C. & KELLEY, R. O. (1980). Characterisation of intercellular junctions in the caudal portion of the developing neural tube of chick embryo. *American Journal of Anatomy*, **158**, 29–41.

SHAKLAI, M. & TAUASSOLI, M. (1979). Cellular relationships in the rat bone marrow studied by freeze-fracture and lanthanum impregnation thin-section electron microscopy. *Journal of Ultrastructural Research*, **69**, 343–61.

SHEN, S. S., HAMAMOTO, S. T. & PITELKA, D. R. (1976). Electrophysiological study of coupling between cultured cells of the mouse mammary gland in five distinct physiological states. *Journal of Membrane Biology*, **29**, 373–82.

SHERIDAN, J. D. (1968). Electrophysiological evidence for low resistance junctions in the early chick embryo. *Journal of Cell Biology*, **37**, 650–9.

SHERIDAN, J. D. (1971a). Dye movement and low-resistance junctions between reaggregated embryonic cells. *Developmental Biology*, **26**, 627–36.

SHERIDAN, J. D. (1971b). Electrical coupling between fat cells in newt fat body and mouse brown fat. *Journal of Cell Biology*, **50**, 795–803.

SHERIDAN, J. D. (1973). Functional evaluation of low resistance junctions: influence of cell shape and size. *American Zoology*, **13**, 1119–28.

SHERIDAN, J. D. (1976). Cell coupling and cell communication during embryogenesis. In *The Cell Surface in Animal Embryogenesis*, ed. G. Poste & G. L. Nicholson, pp. 409–47. New York: Elsevier.

SHERIDAN, J. D. (1978). Junctional formation and experimental modification. In *Intercellular Junctions and Synapses*, ed. J. Feldman, N. B. Gilula & J. D. Pitts, pp. 37–59. London: Chapman-Hall.

SHERIDAN, J. D., FINBOW, M. E. & PITTS, J. D. (1979). Metabolic interactions between animal cells through permeable intercellular junctions. *Experimental Cell Research*, **123**, 111–17.

SHERIDAN, J. D., HAMMER-WILSON, M., PREUS, D. & JOHNSON, R. G. (1978). Quantitative analysis of low-resistance junctions between cultured cells and correlation with gap-junctional area. *Journal of Cell Biology*, **76**, 532–44.

SHIBATA, Y., NAKATA, K. & PAGE, E. (1980). Ultrastructural changes during development of gap junctions in rabbit left ventricular myocardial cells. *Journal of Ultrastructural Research*, **71**, 258–71.

SIMIONESCU, M., SIMIONESCU, N. & PALADE, G. E. (1975). Segmental differentiations of cell junctions in the vascular endothelium – the microvasculature. *Journal of Cell Biology*, **67**, 863–85.

SIMPSON, I., ROSE, B. & LOEWENSTEIN, W. R. (1977). Size limit of molecules permeating the junctional membrane channels. *Science*, **195**, 294–6.

SLACK, C., MORGAN, R. H. M. & HOOPER, M. L. (1978). Isolation of metabolic cooperative-defective variants from mouse embryonal carcinoma cells. *Experimental Cell Research*, 117, 195–205.

SPRAY, D. C., HARRIS, A. L. & BENNETT, M. V. L. (1981). Gap junctional conductance is a simple function of intracellular pH. *Science*, 211, 712–15.

SUBAK-SHARPE, J. H., BÜRK, R. R. & PITTS, J. D. (1969). Metabolic cooperation between biochemically marked cells in tissue culture. *Journal of Cell Science*, 4, 353–67.

THIELE, J. & REALE, E. (1976). Freeze-fracture study of the junctional complexes of human and rabbit follicles. *Cell Tissue Research*, 168, 133–40.

TWAROG, B. M., DEWEY, M. M. & HIDAKA, T. (1973). The structure of *Mytilus* smooth muscle and the electrical contacts of the resting muscle. *Journal of General Physiology*, 61, 207–21

UNWIN, P. N. T. & ZAMPIGHI, G. (1980). Structure of the junctions between communicating cells. *Nature, London*, 283, 545–9.

WADE, J. B. (1978). Membrane structural specialization of the toad urinary bladder. III Location, structure and vasopressin dependence of intramembrane particles. *Journal of Membrane Biology*, 40, 281–96.

WEISS, T. F., MULROY, M. J. & ALFMANN, D. W. (1974). Intracellular responses to acoustic clicks in the inner ear of the alligator lizard. *Journal of the Acoustic Society of America*, 55, 606–19.

WOOD, R. L. (1977). The cell junctions of *Hydra* as viewed by freeze-fracture replication. *Journal of Ultrastructural Research*, 58, 299–315.

WRIGHT, E. D., GOLDFARB, P. S. G. & SUBAK-SHARPE, J. H. (1976). Isolation of variant cells with defective metabolic cooperation (mec⁻) from polyoma transformed syrian hamster cells. *Experimental Cell Research*, 103, 63–77.

YANCEY, S. B., EASTER, D. & REVEL, J.-P. (1979). Cytological changes in gap junctions during liver regeneration. *Journal of Ultrastructural Research*, 67, 229–42.

YEE, A. G. & REVEL, J.-P. (1978). Loss and reappearance of gap junctions in regenerating liver. *Journal of Cell Biology*, 78, 554–64.

Extracellular molecular messengers at the cell and tissue level: hypotheses and speculations

GRAHAM J. DOCKRAY* AND COLIN R. HOPKINS†

Departments of Physiology,* and Histology and Cell Biology (Medical),†
University of Liverpool, Liverpool, UK

INTRODUCTION

Modern ideas on the mechanisms of extracellular chemical communication can be traced to the early years of the century. In particular, the notion of blood-borne chemical messengers (hormones) was introduced by Bayliss and Starling in 1902 to describe the control of the exocrine pancreas by the intestinal hormone secretin and, at about the same time, Elliot postulated that the similarity between the actions of adrenal extracts and sympathetic nerve stimulation could be explained by the release from sympathetic nerve endings of adrenaline-like substances (Bayliss & Starling, 1902; Elliot, 1904). During the first half of the century however, concepts of hormonal control and chemical neuro-transmission developed more or less independently, and it has been only during the last two decades, as events at the cellular level have become more clearly defined, that the essential similarities between these and other regulatory systems which employ extracellular messenger molecules have become fully appreciated. In this chapter we shall try to identify the characteristics of these systems which together describe the basic cellular organization and associated regulatory mechanisms involved in the transfer of information by extracellular messengers.

INFORMATION EXCHANGE BETWEEN CELLS

All cells have the capacity to monitor changes in their immediate environment and to respond appropriately. At the level of the single cell the monitoring of environmental signals is illustrated by the chemotactic responses of unicellular organisms. In multicellular organisms free cells are surrounded by a more structured, signal-bearing environment but they too display individual orientative behaviour.

However, in multicellular organisms with functionally different cell types there is a need for the constituent cells to communicate with each other in order to co-ordinate their activities.

Information exchange between cells is achieved by two types of mechanism, (1) direct cell–cell contact which involves either gap junctions and diffusible internal messengers (see Finbow, this volume), or fixed, external signals and (2) indirect communication in which the signal molecule travels through the extracellular space.

Gap junctions are well suited to the synchronization of activity of groups of functionally identical cells e.g. cardiac and smooth muscle cells, liver parenchymal cells. They also allow for an increased maximal response to an external stimulus, for example as in the exocrine pancreas where the stimulation of a single acinar cell may lead to the activation of its neighbours (Petersen, 1980). It is of interest that gap junctions can apparently exist between functionally different cells e.g. A and B cells of the pancreatic islets (Orci et al., 1975). Their possible role in such situations is uncertain, although conceivably they might mediate information exchange ensuring that antagonistic cells function as a unit. In general, however, it would seem that extracellular signals play the dominant regulatory role in functionally heterogeneous cell groupings.

Fixed external signals on the cell surface are able to mediate interactions between immediately adjacent cells which may or may not be functionally different. This form of communication allows two cells to recognize each other, and to form appropriate associations. It is likely that fixed signals are of special importance in systems characterized by free cells such as the T-cells of the immune system, or the migrating cell masses of early development. In the solid tissues of multicellular organisms there is an obvious need for distant cells to communicate, and this requirement is met by freely diffusible messengers. As pointed out below the mechanisms concerned with transporting macromolecules to the cell surface are essentially the same as those responsible for exporting them to the extracellular space. It is therefore conceivable that fixed signal molecules were the evolutionary progenitors of diffusible messengers. Also in the context of evolutionary developments, it seems reasonable to suppose that extracellular control systems were a necessary precondition for the differentiation of tissues containing more than one cell type. This supposition implies that the emergence of a group of functionally different cells embodies two complementary processes: autonomy from surrounding cells, and an ability to respond to an independent set of extrinsic controls. These aspects of development can be illustrated by the evolution of the exocrine pancreas. A scattered population of zymogen cells along the intestinal mucosa, is generally believed to have preceded the development of a separate pancreatic gland. In the ancestral condition the zymogen cells probably responded directly to stimulation by nutrients in the gut lumen. The establishment of neuronal and hormonal mechanisms to control enzyme secretion allowed a more sophisticated co-ordination between feeding and enzyme secretion, and was a clear requirement for the evolution of a compact organ spatially removed from the intestine (Dockray, 1979).

Diffusible extracellular messengers are frequently promiscuous; that is to say they may be used to carry different messages in a variety of different systems. This is strikingly demonstrated by the now widely appreciated observation that many active peptides produced in gut, adrenal or pituitary endocrine cells, also occur in brain where they are thought to have a neuro-regulatory function separate from their peripheral hormonal roles (Dockray & Gregory, 1980). During the course of evolution a messenger molecule may establish new functions, particularly if its original role has become redundant, e.g. prolactin is an osmoregulatory hormone in fish, and a mammatrophin in mammals. This suggests an economical use of the genetic information required for the action of extracellular messengers. Similarities in amino acid sequence between a variety of peptide hormones, growth factors and putative neuro-regulatory molecules also raise the possibility that during evolution a common pool of regulatory molecules has been drawn on by a variety of different systems (see for example Blundell & Humbel, 1980). As discussed below, information transfer is essentially a bi-molecular event, because in addition to the signal molecule it requires a receptor site on the target cell. Since the evolution of these systems depends on mutations of both signal molecule and receptor it is perhaps not surprising to find that messenger molecules (and presumably their receptors) are frequently well conserved.

Many information-carrying molecules are hydrophilic so that recognition devices (receptors) are of necessity often located at the cell surface. The extracellular messengers that a cell is capable of recognizing are determined by its complement of surface receptor populations. In this sense a panel of receptors serves to define the possible range of functions of the cell, and during morphogenesis it can be considered the primary biochemical correlate of cell differentiation. However, it is worth emphasizing that while the existence of receptors on target cells is a necessary requirement for information exchange, it is not itself sufficient. Functional transducing systems are also needed. Several types of so-called non-functional receptors are now known, the occupancy of which is not linked to any demonstrable biological response. In some instances, such as the asialo-glycoprotein-binding receptors on liver parenchymal cells, these receptors are probably concerned with clearance of the ligand from the circulation (Ashwell & Morell, 1974). In other cases apparent receptor-binding may in fact be due to some other binding function such as degradation by membrane-bound enzymes (Terris & Steiner, 1975). There remain still further examples, such as the epidermal growth factor receptors of ovarian granulosa cells, which survive but lose their ability to stimulate the cells to divide after luteinization (Vlodaavsky *et al.*, 1978). Conceivably as the cells become luteinized their receptors take on these new responsibilities and become linked to another kind of effector system.

For practical purposes it is convenient to divide extracellular communication systems into three compartments, or divisions (Fig. 1): the production compartment or cell of origin; the delivery compartment, through which the messenger must pass, and the third compartment – the recognition, receptor or target compartment. These divisions may be regarded as constituting the primary unit of information exchange, although in reality there is an additional requirement for some type of feedback to control the overall activity of the system. In each division it is possible to identify mechanisms for the control of information transfer and for ensuring the specificity of the message.

ORGANIZATION OF THE PRODUCTION COMPARTMENT

Secretory pathway

The insertion of protein into cell membranes is a crucial event in nearly all secretory pathways involving hydrophilic messenger molecules. In many cases the protein (or a peptide derivative) is itself destined to become the extracellular messenger; in other cases the protein may be a biosynthetic enzyme or a constituent of uptake pumps for concentrating small messengers e.g. amines into vesicles. The molecular mechanisms controlling secretion of hydrophobic substances such as steroids, are rather different, and closely linked to the regulation of their

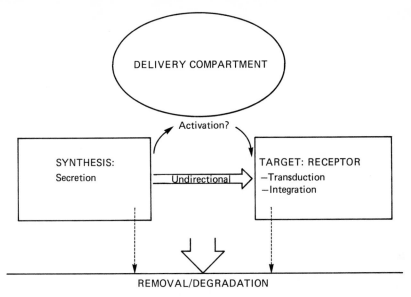

Fig. 1. Schematic representation of the organization of systems using extracellular messengers. Note that these systems can be as diverse as classical endocrine systems, in which the delivery compartment is provided by the circulation, and nervous systems, in which the delivery compartment is limited to a synaptic cleft.

synthesis. The basic mechanisms for insertion of proteins into membranes are well established (Blobel *et al.*, 1979). A signal sequence which is rich in hydrophobic amino acids, and which is generally located at the N-terminus of the peptide chain, serves to promote the insertion of the protein into the membrane of the endoplasmic reticulum. The chain may remain embedded in the membrane or alternatively pass through into the lumen. The information responsible for determining the progress of the growing peptide chain through the membrane is embodied in the nucleotide code of the messenger RNA and it is probable that similar kinds of instructions which predispose the molecule to further specific processing steps are also included at this initial stage. Further processing steps may involve amidation, glycosylation, sulphation, acylation and, as discussed below, cleavage of the chain. The least elaborate and perhaps most primitive form of secretory pathway for the peptide may require only the transport of the membrane-bound protein to the cell surface, and cleavage there to release a fragment into the extracellular space. Such a mechanism is obviously not suitable for bulk delivery of large quantities of protein, and is probably not readily modulated except by controlling protein synthesis. However, it may well be appropriate for the constitutive production of low concentrations of informational molecules (see below).

Processing the messenger

The messenger molecules secreted by a cell are, of course, the final product of the enzymic modifications to which they are exposed. Changes in populations of biosynthetic enzymes to yield a variety of related but different active products is a point of control that has been exploited by numerous systems e.g. steroid, catecholamine, prostaglandin-producing cells. Recently the importance of cleavage in the processing of peptides has received considerable attention, since it has become clear that in different cell types a common precursor can be cleaved and processed to give rise to final products with very different biological activities. It has long been realised that hydrolases capable of selectively degrading peptides exist within special intracellular compartments; i.e. the lysosomes, but only recently has it been shown that the selective packaging of these hydrolases probably depends upon their possession of a signal sugar (mannose-6-phosphate) (Sly & Stahl, 1978; Fischer *et al.*, 1980). Without this sugar these hydrolases are secreted, probably having travelled along the same intracellular pathway as the secretory peptide; conceivably they participate in its processing.

The endogenous opiate peptides illustrate the range of products which can arise. There are at least two main groups. One group includes β-endorphin and is derived from a precursor of 31 000 molecular weight which also contains the sequence of ACTH, β MSH, and a third, γ MSH (Eipper & Mains, 1978; Nakanishi *et al.*, 1979). In different pituitary cells, and in brain, the main secretory product can be ACTH, β-endorphin, or MSHs. Both MSH and β-endorphin may have acyl groups at their N-termini; it is known that acylation of the N-terminal residue of β-endorphin abolishes biological activity (Zakarian & Smyth, 1979). Apparently pituitary β-endorphin is acylated but the brain peptide is not. This type of post-translational modification therefore serves to determine the biological activity of the peptide. Although the sequence of Met-enkephalin is contained in β-endorphin there is little evidence to suggest a biosynthetic relationship between the two peptides. Instead there is evidence from a number of studies that at least in adrenal medulla, and perhaps in brain too, both Leu^5-enkephalin and Met^5-enkephalin are derived from the same precursor of 50 000 molecular weight (Lewis *et al.*, 1980). A number of variants containing the enkephalin sequence are also produced; these range in size from peptides of 22 000 molecular weight to hexapeptides e.g. Met^5-Arg^6-enkephalin, Met^5-Lys^6-enkephalin and heptapeptides e.g. Met^5-Arg^6-Phe^7-enkephalin (see Brownstein, 1980). It is likely, although not yet firmly established, that there are cellular differences in the biosynthetic processing pathways leading to the production of different amounts of the various enkephalins. Since there are probably two, and perhaps three types, of opiate receptors (Kosterlitz & Paterson, 1980), and since the various forms of enkephalin may differ in their relative affinities for these receptors, the mechanisms controlling production and identity of particular ligands are clearly an important site of control.

Storage

The storage of messengers in granules or vesicles is a common feature of many different communicator cells. The particular characteristic of this pattern of organization is that it is well suited to the prompt release of relatively large quantities of messenger (Hopkins, 1979). The storage-secretion pathway can conveniently be considered as a specialization of the general method for insertion of proteins into cell membranes. The available evidence points to the fact that within a particular cell there is a single pathway for storage and release of proteins. There is little or no evidence for different messengers being packaged by a particular cell into different granules. However, a reservation needs to be applied to certain neurons where there is morphological evidence for the existence of two or more populations of secretory granules, e.g. small agranular vesicles and large granular vesicles (see for example, Cook & Burnstock, 1976; Gabella, 1979). The content of different vesicles has yet to be firmly established, but recent studies on neuro-peptides are probably relevant in this context. There is now rapidly accumulating evidence to indicate that a variety of neuro-peptides may co-exist within a particular neuron with conventional transmitters e.g. somatostatin and noradrenaline, substance P and serotonin (Hökfelt *et al.*, 1980). These relationships are not rigid since in different neurons a particular conventional transmitter can exist with various peptides e.g. noradrenaline together with somatostatin, pancreatic polypeptide-like immunoreactivity or enkephalins. It is worth emphasizing that the capacity for protein synthesis is generally restricted to the neuronal cell body, so that neuro-peptides must be packaged into granules and transported intra-axonally to nerve endings for release. In contrast, amines and small transmitters like acetylcholine, are synthesized at nerve endings, and quite possibly are located in vesicles that can be re-cycled. Thus neuro-peptides may constitute a readily depletable pool of messenger, whereas conventional transmitters are readily replenished. If the two transmitters interact at their target, early stimulation of a neuron will release a more potent combination of messengers than later stimuli. In at least one case there is evidence for interaction of co-existing molecules at post-synaptic sites. Thus both acetylcholine and VIP exist together in post-ganglionic parasympathetic neurons serving the salivary gland. These two agents are able to potentiate each other's action in causing vasodilation of the gland (Lundberg *et al.*, 1980).

Although the pathway for production of secretory granules is most likely invariant in a given cell, this does not necessarily imply that identical products must always be secreted. Many of the processing steps for production of active peptides from large precursors may not have been completed by the time the secretory granule is formed. Thus, if newly formed granules are secreted in preference to mature granules there will be released incompletely processed material. Even when two products are stored in the same granule in constant proportions there need not be a fixed ratio delivered to the targets. A good example is provided by the pituitary gonadotropins FSH and LH which are probably stored in a reasonably constant ratio in the same granules. The metabolic clearance rate of FSH is considerably less than that of LH and in a steady state equilibrium between secretion and clearance from the plasma, the plasma ratios would be constant. However, it is known that the secretion of FSH and LH is in fact pulsatile, and as a consequence the relative concentrations of the two gonadotropins in plasma can be modulated; appropriate short intervals between pulses maintain the proportion of the two gonadotropins at an approximately constant level, longer intervals which give the LH concentrations time to decay provide for a relative elevation of FSH levels (Knobel, 1980).

Cell types

Endocrine cells, neuroendocrine cells and neurons are highly specialized for the production and secretion of chemical messengers (Fig. 2). Typically they have a high order of intracellular organization and they often form specialized groupings. However the constituent elements of the intracellular organization (i.e. the organelles) required for the synthesis and secretion in these cells are a feature of most somatic cells and it would not be surprising therefore if the production of extra-cellular messenger molecules was found to be a property of many less specialized cell types. There is now evidence that this is the case. Thus small amounts of immunoreactive insulin can be found in most, or all, cells (Rosenzweig *et al.*, 1980). This material is unlikely to be derived from insulin bound to receptors (either surface or internalized) since it can be demonstrated in cells maintained in insulin-free culture. Other regulatory molecules, e.g. fibroblast growth factor, epidermal growth factor and nerve growth factor, may arise from similar sources, with virtually ubiquitous distribution (Mobely *et al.*, 1977). In each case it seems plausible to suppose that there is relatively little modulation of the rates of secretion of these molecules. Moreover, in each case the cell of origin is not specialized for production, storage and secretion of messengers in the way that typifies conventional glandular cells. Perhaps this type of the control system should be called panocrine (pan, everywhere).

Tissue organization

The spatial relationships between the producer compartment and the target compartment vary considerably. At one extreme are the classical endocrine systems in which the hormone is delivered to its target in the systemic circulation, and consequently all tissues in the body may be exposed to a particular messenger. In these systems the relative spatial positions of the hormone-producing cell and its target are virtually irrelevant, and specificity is determined by the receptor populations of the target cell. In contrast, at the other extreme are messengers that may mediate a variety of responses in different systems, so that speci-ficity in each system depends on the relative spatial positions of the producer and target cells providing a context for the message. The most obvious example of the latter arrangement is the nervous system where a relatively small number of neuro-transmitters can be employed because the pattern of connections between the participating cells is all-important in determining the function of the system.

Similar considerations can be applied equally well to other 'solid' tissues in which there are groups of functionally distinct cells. For instance in the gut there are endocrine-like cells in the mucosa that possess basal axon-like processes which project to other cells in the mucosa (Fig. 2). These cells are specialized for communication in a way

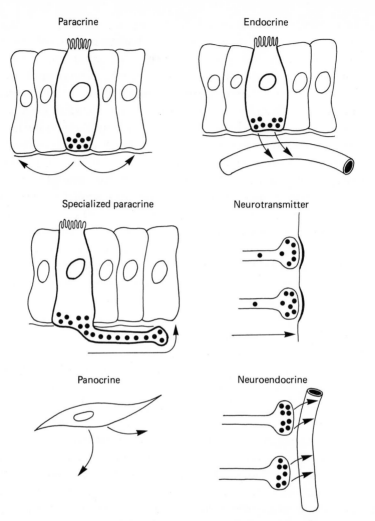

Fig. 2. Types of systems in which extracellular messengers are employed. It should be emphasized that the same extracellular signal molecule might function in any or all of these systems. See text for further information.

that suggests they achieve a degree of specificity with their targets that is normally associated with neuronal systems. This mode of control is a clear refinement of the paracrine regulation first postulated by Feyrter to describe the role of scattered endocrine-like cells (notably in the gut), from which an active factor might diffuse through the extracellular space to neighbouring target cells, rather than be transported in the blood (Feyrter, 1954). A good example of specialized paracrine control is the somatostatin cell of the gut mucosa. In the pyloric antral mucosa the basal processes of this cell project to nearby gastrin cells (Larsson *et al.*, 1979). The release of gastrin is known to be inhibited by exogenous somatostatin. In addition, somatostatin has numerous other actions, including the suppression of hormones such as insulin, glucagon and growth hormone. It is reasonable to suppose that the structural relationships between the gastrin and somatostatin cells in antral mucosa allow locally high concentrations of somatostatin to occur in the immediate vicinity of gastrin cells and so inhibit gastrin release while having little effect on circulating concentrations and so sparing other endocrine systems from inhibition. The importance of the topographic location of cells in non-neuronal regulatory functions is further exemplified by somatostatin cells in the islets of Langerhans. In many species, e.g. rat, glucagon (and/or pancreatic polypeptide) cells are arranged around the periphery of the islet; immediately within this layer lie somatostatin cells and the insulin-secreting B cells are grouped in the central portion (Orci & Unger, 1976). The blood supply to the islet breaks into capillaries at the periphery of each islet and only forms into venules at the centre of the islet (Fujita, Yanatoir & Murakami, 1976). However it is of particular significance that in a few species, e.g. horse, the insulin-secreting cells are arranged around the periphery of the islet, and that in these species arterioles go to the centre of the islet before breaking into capillaries that reform into venules at the periphery. Evidently, then, in both cases the local organization of cells within each islet ensures that insulin cells are bathed in a solution rich in glucagon and somatostatin. Glucagon enhances insulin secretion and, as already mentioned, somatostatin inhibits it. The net hormonal secretion (insulin and glucagon) of the pancreas may therefore be the result of integration within individual islets of several types of information including neuronal, endocrine, i.e. blood-borne factors such as gut hormones, and local or paracrine factors.

DELIVERY CHANNELS

The organization of the delivery channel constitutes the most striking variable when comparing extracellular control systems. In the case of endocrine systems the channel may include virtually the entire body fluid (for hydrophobic substances that freely penetrate cell membranes), or at least the extracellular fluid. In the case of the nervous system the channel can be as circumscribed as a synaptic cleft. The mechanisms which determine the concentrations of messenger in the delivery compartment are clearly of primary importance in regulating activity at the receptor. Secretory rates from the cells of origin are obviously important. In addition several other factors are worth identifying. The volume of the compartment is a limiting factor in determining the concentration of messenger after secretion. By implication the barriers which serve to define the compartment need to be identified. Other chemical constituents of the compartment which may interact with the messenger either enhancing its activity or decreasing it (by degradation, or passively by binding e.g. carrier proteins for steroids) need to be identified. Finally mechanisms of clearance or removal from the compartment have to be considered. The latter can include reuptake mechanisms e.g. for catecholamines, filtration systems e.g. liver and kidneys, and free or bound degrading enzymes.

The limiting barriers

The outer limits which serve to define a particular channel may also ensure specificity either by keeping a particular messenger within a channel and preventing penetration elsewhere, or conversely by preventing exogenous active substances gaining access to the channel. A good example is the blood/brain barrier which resides at the level of the tight junctions that seal the endothelial lining of brain capillaries. In general the barrier keeps circulating messengers out of the brain, and brain messengers out of the circulation (except in specific circumstances e.g. in the hypothalamus). The channel may also serve to deliver appropriate concentrations of messenger without dilution e.g. portal blood systems such as transport hypothalamic peptides to the pituitary, or gut peptides to the liver. Insofar as there is any advantage in the organization of these portal systems it is presumably to ensure delivery of peptide in appropriately high concentrations, and to prevent the messenger getting to other receptor sites.

Modifications and metabolism during delivery

It is worth noting that it may often be quite difficult to establish the chemical nature of the molecular messenger once it enters the delivery channel and so identify precisely the physiologically relevant form of the ligand binding to the receptor. Within the plasma there clearly exist systems for production or activation of some messengers e.g. angiotensin and bradykinin. There may also be similar post-secretory processing events involved in the activation of neuro-regulators or locally active substances that do not enter the plasma – but they are obviously difficult to study. However, the mechanisms of re-uptake and metabolism of classical neuro-transmitters are well established (see for example, Iversen, 1973). Almost certainly there are comparable mechanisms for the degradation of peptides which act as neuro-regulators and local signals, although again the details of these are poorly understood. Nevertheless, it is probable that the half-times for removal of neuro-transmitter and locally active substances are extremely short, of the order of seconds, or fractions of a second. In contrast, it seems that plasma half-lives of 1–10 min are common for peptide hormones, and may be up to several hours for the larger protein and glycoprotein hormones e.g. growth hormone, luteinizing hormone (see Bennett & McMartin, 1979). For some peptides degradative enzyme systems exist in the plasma (e.g. bradykinin and angiotensin II) but in general this form of degradation is not the primary *in vivo* removal mechanism, instead degradation by mechanisms associated with the vascular wall are common e.g. in lung, kidney, and liver. The mechanisms involved are poorly characterized although the activity of angiotensin-converting enzyme (responsible for converting angiotensin I to angiotensin II and for the inactivation of bradykinin) in the lung has been well defined and localized to the caveloae on the luminal surfaces of capillary endothelia (Ryan *et al.*, 1975). In some cases the mechanisms may be highly specific. For example the liver clears small biologically active fragments of gastrin e.g. C-terminal tetra- and penta-peptides, more effectively than heptadecapeptides (Struntz, Walsh & Grossman, 1978). It is tempting to speculate that the enzymes involved in degrading neuropeptides and locally active substances are associated with the external surface of plasma membranes surrounding the delivery channel in these systems. Glomerular filtration in the kidney is generally important in the clearance of circulating peptides unless they are bound to a larger protein carrier. After filtration molecules may be excreted unchanged (melanostatin) degraded to products which are excreted (oxytocin, vasopressin) or reabsorbed and degraded within the proximal tubule of the nephron (glucagon, insulin, growth hormone, conticotrophin analogues), (Bennett & McMartin, 1979).

REGULATION AT THE LEVEL OF THE TARGET CELL

Target cell stimulation is initiated by hormone binding to its receptors.

Where a hormone–receptor interaction is essentially an equilibrium reaction, termination will be a consequence of decreasing hormone levels in the extracellular environment causing a shift in equilibrium to the left, and consequent dissociation of the receptor-bound hormone. In many systems, however, it is clear that termination of the stimulus does not rely exclusively upon the dissociation of the hormone ligand (Catt *et al.*, 1979). In these circumstances termination depends upon specific mechanisms for uncoupling the receptor-effector mechanism. At the present time the best defined alteration is due to the desensitization of adenyl cyclase which has been shown to occur in response to β adrenergic stimulation (Rodbell, 1980). However, tachyphyllaxis and other acute conditions in which target cells continue to bind hormone while becoming refractory to further stimulation may also illustrate this form of regulation.

Other regulatory events at the level of the target cell receptors have been described and widely discussed. They include changes in the affinity of the receptor for the hormone; the so called 'negative cooperativity' displayed by insulin receptors being the best known example (de Meyts, 1976). The most clearly demonstrated form of regulation at the target cell level is, however, change in receptor number. Thus in a wide variety of systems it has been shown that the rapid reduction in hormone binding which often occurs as a consequence of hormone becoming associated with its receptors is frequently due to the endocytic activity removing the receptor into the cell. The majority of receptors removed in this way may still be occupied by hormone but it is not yet clear if such internalized hormone–receptor complexes can continue to be involved in information transfer (i.e. can continue signalling) (Hopkins, 1980). Studies on an increasing variety of ligands show that the subsequent fate of internalized receptors can vary considerably and it is clear that while some kinds of receptor are rapidly recycled back to the cell surface others, along with their bound ligands, are degraded within the cytoplasmic lysosomal system. For receptors which bind hormones the available evidence suggests that the internalized receptors are neither recycled nor readily replaced so that with their removal the cell is no longer able to respond to circulating hormone (Catt *et al.*, 1979).

This process is often called 'down regulation'. It is dose dependent and may be a useful device preventing over-stimulation of the target cell. Of perhaps wider significance is the possibility that the internalization of occupied receptors also plays a role in systems which depend

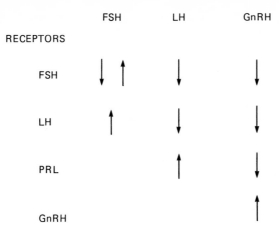

Fig. 3. Influence of gonadotropin on ovarian granulosa cell responsiveness to homologous and heterologous gonadotropins. ↑ indicates increased and ↓ indicates decreased responsiveness. LH, luteinizing hormone; FSH, follicle-stimulating hormones; PRL, prolactin; GnRH, gonadotropin-releasing hormone.

upon short or intermittent phases of stimulation. The best evidence for this has been obtained from studies on the gonads where in the ovary for example the binding of a gonadotropin has been shown not only to induce the removal of its own receptors but also to induce the appearance of a new, separate population of receptors for a second gonadotropin (Rao *et al.*, 1977; Richards, 1979). Fig. 3 indicates some of the interrelationships which have been described; more information is required but it is already clear how this form of receptor regulation could contribute to the cyclic programme of follicular differentiation.

REFERENCES

ASHWELL, G. & MORELL, A. G. (1974). The role of surface carbohydrates in the hepatic recognition and transport of circulating glycoproteins. *Advances in Enzymology,* **41**, 99–128.

BAYLISS, W. M. & STARLING, E. (1902). Mechanism of pancreatic secretion. *Journal of Physiology,* **28**, 325–53.

BENNETT, H. P. J. & McMARTIN, C. (1979). Peptide hormones and their analogues: distribution, clearance from the circulation, and inactivation in vivo. *Pharmacological Reviews,* **30**, 247–92.

BLOBEL, F., WALTER, P., CHANGO, C. N., GOLDMAN, B. M., ERICKSON, A. H. & LINGAPPA, V. R. (1979). Translocation of proteins across membranes: the signal hypothesis and beyond. In *Secretory Mechanisms, Symposia of the Society for Experimental Biology 33,* ed. C. R. Hopkins & C. J. Duncan, pp. 9–36. Cambridge University Press.

BLUNDELL, T. L. & HUMBEL, R. E. (1980). Hormone families: pancreatic hormones and homologous growth factors. *Nature*, **287**, 781–7.

BROWNSTEIN, M. J. (1980). Opioid peptides: search for the precursors. *Nature*, **287**, 678–9.

CATT, K. J., MARMOD, J. A., AGUILERA, G. & DUFAU, M. L. (1979). Hormonal regulation of peptide receptors and target cell responses. *Nature*, **280**, 109–16.

COOK, R. D. & BURNSTOCK, G. (1976). The ultrastructure of Auerbach's plexus in the guinea pig. I. Neuronal elements. II. Non-neuronal elements. *Journal of Neurocytology*, **5**, 171–206.

DE MEYTS, P. (1976). Cooperative properties of hormone receptors in cell membranes. *Journal of Supramolecular Structure*, **4**, 241–58.

DOCKRAY, G. J. (1979). Evolutionary relationships of the gut hormones. *Federation Proceedings*, **38**, 2295–301.

DOCKRAY, G. J. & GREGORY, R. A. (1980). Relations between neuropeptides and gut hormones. *Proceedings of the Royal Society, London, Series B*, **210**, 151–64.

EIPPER, B. A. & MAINS, R. E. (1978). Analysis of the common precursor to corticotropin and endorphin. *Journal of Biological Chemistry*, **253**, 5732–44.

ELLIOT, T. R. (1904). The action of adrenaline. *Journal of Physiology*, **32**, 401–67.

FEYRTER, F. (1954). *Uber die peripheren endokrinen (parakrinen) drusen des menschen.* Vienna. Maudrich.

FISCHER, H. D., GONZALEZ-NORIEGA, A., SLY, W. S. & MORRE, J. D. (1980) Phosphomannosyl-enzyme receptors in rat liver. Subcellular distribution and role in intracellular transport of lysosomal enzymes. *Journal of Biological Chemistry*, **255**, 9608–15.

FUJITA, T., YANATOIR, Y. & MURAKAMI, T. (1976). Insulo-acinar axis, its vascular basis and its functional and morphological changes caused by CCK-PZ and caerulein. In *Endocrine Gut and Pancreas*, ed. T. Fujita, pp. 347–58. Amsterdam: Elsevier.

GABELLA, G. (1979). Innervation of the gastrointestinal tract. *International Reviews of Cytology*, **59**, 129–93.

HÖKFELT, T., JOHANSSON, O., LJUNGDAHL, Å., LUNDBERG, J. M. & SCHULTZBERG, M. (1980). Peptidergic neurones. *Nature*, **284**, 515–21.

HOPKINS, C. R. (1979). The secretory pathway in outline. In *Secretory Mechanisms, Symposia of the Society for Experimental Biology 33*, ed. C. R. Hopkins & C. J. Duncan, pp. 1–12. Cambridge University Press.

HOPKINS, C. R. (1980). Epidermal growth factor and mitogenesis. *Nature*, **286**, 110–12.

IVERSEN, L. L. (1973). Catecholamine uptake processes. *British Medical Bulletin*, **29**, 130–5.

KNOBEL, E. (1980). *Recent Progress in Hormone Research*, **36**, 53–74.

KOSTERLITZ, H. W. & PATERSON, S. J. (1980). Characterization of opioid receptors in nervous tissue. *Proceedings of the Royal Society, London, Series B*, **210**, 113–22.

LARSSON, L.-I., GOLTERMANN, N., MAGISTRIS, L. DE, REHFELD, J. F. & SCHWARTZ, T. W. (1979). Somatostatin cell processes as pathways for paracrine secretion. *Science*, **205**, 1393–5.

LEWIS, R. V., STERN, A. S., KIMURA, S., ROSSIER, J., STEIN, S. & UDENFRIEND, S.

(1980). An about 50,000-Dalton protein in adrenal medulla: a common precursor of (Met)- and (Leu)-enkephalin. *Science*, **208**, 1459–61.

LUNDBERG, J. M., ANGGARD, A., FAHRENKRUG, J., HÖKFELT, T. & MUTT, V. (1980). Vasoactive intestinal polypeptide in cholinergic neurones of exocrine glands: Functional significance of coexisting transmitters for vasodilation and secretion. *Proceedings of the National Academy of Sciences, USA*, **77**, 1651–5.

MAY, J. V., MCCARTHY, K., REICHART, L. E. & SCHOMBERG, D. W. (1980). Follicle stimulating hormone mediated induction of functional luteinizing hormone human chorionic gonadotrophin receptors during monolayer culture of porcine granulosa cells. *Endocrinology*, **107**, 1041–7.

MOBELY, W. C., SERVER, A. C., ISHI, D., RIOPELLI, G. J. & SHOTTER, E. M. (1977). Nerve growth factor. *New England Journal of Medicine*, **297**, 1096–104.

NAKANISHI, S., INOUE, A., KITA, T., NAKAMURA, M., CHANG, A. C. Y., COHEN, S. N. & NUMA, S. (1979). Nucleotide sequence of cloned cDNA for bovine corticotropin-β-lipotropin precursor. *Nature*, **278**, 423–7.

ORCI, L., MALAISSE-LAGAE, E., RAVAZZOLA, M., ROUILLER, D., RENOLD, A. E., PERRELET, A. & UNGER, R. (1975). A morphological basis for intercellular communication between α- and β- cells in the endocrine pancreas. *Journal of Clinical Investigation*, **56**, 1066–70.

ORCI, L. & UNGER, R. H. (1976). Functional subdivision of islets of Langerhans and possible role of D cells. *Lancet*, ii, 1243–4.

PETERSEN, O. H. (1980). *The Electrophysiology of Gland Cells*. London: Academic Press.

RAO, M. C., RICHARDS, J. S., MIDGELY, A. R. & REICHERT, L. E. (1977). Regulation of gonadotropin receptors by luteinizing hormone in granulosa cells. *Endocrinology*, **101**, 512–23.

RICHARDS, J. C. (1979). Hormonal regulation of hormone receptors in ovarian follicular development. In *Ovarian Follicular Development and Function*, ed. A. R. Midgley & W. A. Sadler, pp. 132–9. New York: Raven Press.

RODBELL, M. (1980). The role of hormone receptors and GTP regulatory protein in membrane transduction. *Nature*, **284**, 17–22.

ROSENZWEIG, J. L., HAVRANKOVA, J., LESHIAK, M. A., BROWNSTEIN, M. & ROTH, J. (1980). Insulin is ubiquitous in extrapancreatic tissues of rats and humans. *Proceedings of the National Academy of Sciences, USA*, **77**, 572–6.

ROSSIER, J., AUDIGIER, Y., LING, N., CROS, J. & UDENFRIEND, S. (1980). Met-enkephalin-Arg^6-Phe^7, present in high amounts in brain of rat, cattle and man, is an opioid agonist. *Nature*, **288**, 88–90.

RYAN, J. W., RYAN, V. S., SHULTZ, D. R., WHITAKER, C., CHUNG, A. & DORER, F. E. (1975). Subcellular localisation of pulmonary angiotensin-converting enzyme. *Biochemical Journal*, **146**, 497–9.

SLY, W. S. & STAHL, P. (1978). Receptor mediated uptake of lysosomal enzymes. In *Transport of Macromolecules in Cellular System*, ed. S. Silverstein, pp. 229–44. Dahler Konderenzen.

STRUNTZ, U. T., WALSH, J. H. & GROSSMAN, M. I. (1978). Removal of gastrin by various organs in dogs. *Gastroenterology*, **74**, 32–3.

TERRIS, S. & STEINER, D. F. (1975). Binding and degradation of ^{125}I-insulin by rat hepatocytes. *Journal of Biological Chemistry*, **250**, 8389–98.

VLODAAVSKY, I., BROWN, K. & GOSPODAROWICZ, D. (1978). A comparison of the binding of epidermal growth factor to granulosa cells and luteal cells in tissue culture. *Journal of Biological Chemistry,* **253**, 3744–8.

ZAKARIAN, S. & SMYTH, D. (1979). Distribution of active and inactive forms of endorphins in rat pituitary and brain. *Proceedings of the National Academy of Sciences, USA,* **76**, 5972–6.

Cell interactions in plants – a comparative survey

ANTHONY W. ROBARDS

Department of Biology, University of York

INTRODUCTION

Large multicellular organisms pose problems of intercellular communication that have been solved in different ways across the plant and animal kingdoms. In animals, it is believed that the gap junctional complex is responsible for allowing the passage of solutes from cell to cell but not the transfer of large, possibly information-carrying, molecules. This basic theme is repeated, with modifications, through the whole range of multicellular organisms and, while it is the purpose of most of this volume to consider the relatively well-studied aspects of 'integration of cells in animal tissues', it falls to the present chapter to appraise the present state of knowledge concerning cell interactions in plants.

A main difference between the physical requirements of communicating pathways in plant and animal cells is that, whereas the cell membranes of animals are commonly in close apposition, those in plants are usually separated by an intercellular matrix – the cell wall – which may be of considerable ($> 1.0 \mu$m) thickness. Thus, while gap junctions appear to be created by modifications to the structure of the membranes alone, the equivalent structure in plants is necessarily more complex.

It is slightly more than 100 years since Tangl (1879) first described protoplasmic connecting threads in plants. These structures were subsequently termed plasmodesmata (singular – 'plasmodesma') by Strasburger (1901) and the word is now used unambiguously to refer to the cytoplasmic connections that occur between plant cells.

The subject of 'Intercellular Communication in Plants: Studies on Plasmodesmata' was comprehensively reviewed in the volume edited by Gunning & Robards (1976a) and the serious student should refer to this book for the background literature in this field. For the same reason, and to conserve space in this chapter, the reader will be referred to specific chapters in Gunning & Robards (1976a) where they review a topic up to that date.

Symplastic continuity in plants

Throughout the plant kingdom, structures are found linking the cytoplasm of one cell to that of its neighbours. These symplastic connections, or plasmodesmata, imply continuity from cell to cell *within* the cell membrane and, in consequence, provide at least a theoretical opportunity for the transmission from cell to cell of ions, solutes, small molecules, electrical signals and water. The important principle here, as in animal cells, is that convection or diffusive fluxes can take place through the connections but larger constituents, that would destroy the individuality of different cells if they were to intermingle, are restrained from so mixing. Much of the present work on plasmodesmata involves attempts to demonstrate their involvement in intercellular communication in different contexts as well as to elucidate the detailed structure of the connections themselves.

Functional phenomenology. It has to be stated straightaway that efforts to demonstrate the involvement of plasmodesmata in various intercellular reactions have almost without exception provided only indirect, circumstantial support for such functions. Nevertheless, the accumulated circumstantial evidence is now extremely strong and it is doubtful if there can be any serious suggestion that plasmodesmata do not fulfil a variety of functions in intercellular communication.

If plasmodesmata do provide open continuity from one cell to another, then such a junction should have a lower resistance to the passage of an electrical current than would a junction formed by a pair of intact membranes. This has, indeed, been found to be the case by a number of workers but, while the resistance is much lower than would be expected from that of a pair of membranes, it is usually still higher than would arise from totally open pores. The conclusion arrived at is that the plasmodesmata represent low-resistance connections from cell to cell but are not simply open cytoplasmic tubes (Goodwin, 1976).

The demonstration of intercellular fluxes and flows via plasmodesmata is beset with the problems of being able to confirm that the movement must be through the intercellular connections and not across the membranes from cell to cell. Thus, it has usually been assumed that such transport must be through the plasmodesmata when the intervening cell wall can be shown to be impregnated with an impermeable substance such as lignin or suberin, or when the fluxes and flows measured experimentally are so high that it seems improbable that they could take place across a normal membrane. Using such a rationale, Robards & Clarkson (1976) demonstrated that plasmodesmata probably carried the fluxes and flows across the endodermis of plant roots (where the endodermal cell walls possess an apparently impermeable barrier); Gunning (1976a) showed that the very high rates of transport through secretory glandular hairs could not occur by passage across cell membranes which would not have a sufficiently high solute permeability to allow the rates measured; Osmond & Smith (1976) illustrated the requirement for plasmodesmata in the transport of metabolites to and fro across the mestome sheath of leaves of C_4 plants; and Walker (1976) showed that plasmodesmata were responsible for the transport of solutes across the nodes of the alga *Chara*. More recently, Drake, Carr & Anderson (1978) showed that severe plasmolysis of oat coleoptile parenchyma cells may either break plasmodesmatal strands or may leave the protoplasts connected via thin strands of cytoplasm. After deplasmolysing, recovery was not immediate but electrical continuity was restored within a few hours. This suggests that plasmolysis may sufficiently disrupt the plasmodesmata (possibly by callose formation) to inhibit their normal function but not to the extent that they cannot recover when normal conditions are restored. Further experiments by Drake & Carr (1978, 1979) suggested that, though plasmodesmata may not be necessary for auxin transport through coleoptile parenchyma cells, gibberellin may well be transported symplastically. In the latter case it was shown that sodium azide reduced longitudinal transport and this has led to the suggestion that the sodium azide may have a direct effect on the plasmodesmata themselves – a hypothesis that has received further support from the work of Drake (1979) who showed that both sodium azide and potassium cyanide reduced the electrical coupling of oat coleoptile parenchyma cells.

Whereas the evidence for the transport of electrical signals, growth substances, ions, low molecular weight solutes and water across plasmodesmata is relatively indirect (although none-the-less weighty when accumulated together), the movement of virus particles from cell to cell has been clearly demonstrated to take place through the intercellular connections (Gibbs, 1976). Electron micrographs clearly show virus particles within plasmodesmata and this, together with the time course of viral infection of cells across an organ, leaves little doubt that this is the pathway for the spread of infection. On the one hand, this is good evidence that plasmodesmata do provide a functional connection from cell to cell; on the other, the large size of the viral particles in relation to the pore dimensions of normal plasmodesmata implies that the virus must be capable of inducing a pathologically enlarged channel between cells. Thus, interpretation of such data must take account of such modifications.

Structural possibilities. The latest view of gap junctions in animal cells is that they are protein-bound pores, each providing a continuous narrow channel of about 1.0 nm diameter between adjacent cells. It appears that the pores may exercise some control on intercellular communication by opening and closing. Thus, although the pore is extremely small and exercises an 'ultrafiltration' effect between cells, the bulk cytoplasm of neighbouring cells is in direct contact across the junction. The situation is both different and more variable in plants. Whereas, in animal cells, the two membranes of adjacent cells lie closely together in the area of gap junctions, this is usually impossible in plants owing to the presence of the cell wall. Consequently, some specialisation of the cell membrane is necessary if continuity is to be achieved. The simplest form of open connection would be an open tube from cell to cell but the further possibility exists that some structural component of the cytoplasm could also be contained within such a channel. In reality, plasmodesmatal structure is found to range from the simplest of open tubes to extremely complex, and varied, intercellular connections.

PLASMODESMATA

The introduction above has served to show that plant cells do communicate with each other in a variety of different ways and that structures are present that, at least from *a priori* considerations, could be responsible for such interactions. Much, therefore, hinges on the structure and distribution of these structures – the plasmodesmata.

Origin of plasmodesmata

From the earliest electron micrographs of dividing plant cells it was suggested that plasmodesmata are formed by the entrapment of a strand of endoplasmic reticulum (ER) within the vesicles that fuse together to form the cell plate (Jones, 1976). For most higher plant systems studied there appears to be little reason to change this view. In fusing together, the membranes of the vesicles that form the cell plate also provide the plasmalemma of the new partition (Fig. 1). The pores traversed by the strands of endoplasmic reticulum are hence lined by plasmalemma – one feature of plasmodesmata that appears to be general throughout the plant kingdom.

This method of formation of plasmodesmata raises a number of interesting points and possibilities. Firstly, are the strands of ER arranged across the cell plate in a predetermined manner, either in terms of pattern or number? All too little is known about this aspect at the present time. The continuity of the strand of ER from cell to cell provides the potential for not one, but two, symplastic pathways: via the internal cavity of the ER; and via the 'cytoplasmic annulus' between the strand of ER and the inside of the plasmalemma-lined tube through the wall. Further, it should not be overlooked that the plasmalemma of cells linked to each other by plasmodesmata is also in continuity and, in consequence, the cell membrane of the young plant will be virtually one continuous sheet.

In lower plants a variety of different structures can be seen, in some cases there being no traversing strand of ER at all (Marchant, 1976). The different origins and forms of plasmodesmata throughout the plant kingdom do suggest strongly that they evolved more than once during the course of evolution and that, while subject to the same fundamental evolutionary pressure for formation concomitant with the development of multicellular plants, the subsequent pattern of evolution has not been totally parallel.

Even among higher plants plasmodesmata appear not only to be formed during cytokinesis. There is good evidence that they may be formed secondarily by fusion of the plasmalemma through adjacent walls. In this way, the 'dilution' of plasmodesmata within a cell interface can be avoided (as in the gross enlargement of nematode-induced giant cells); plasmodesmata may be formed across the 'non-division' walls of graft hybrids; and they may be formed between the haustoria and host cell membranes of parasitic higher plants (Jones, 1976).

Structure of plasmodesmata

As will be seen from the above section, the basic structure of a newly formed, simple higher plant plasmodesma is that of a strand of ER

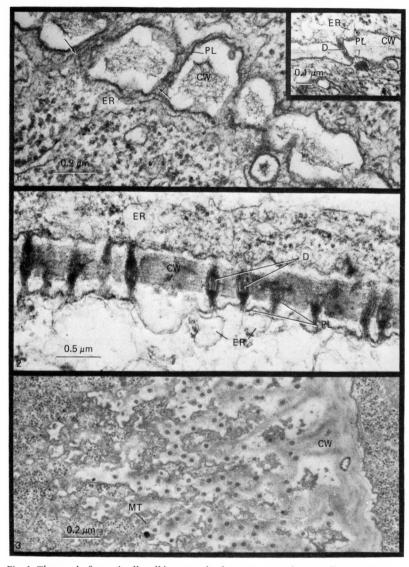

Fig. 1. The newly formed cell wall between barley root parenchyma cells. Membranous strands can be seen traversing the wall but the plasmalemma-lined tubes are not yet tightly binding the central strand. *Inset*: A slightly later stage in the development of a plasmodesma, in this case traversing the wall between an endodermal cell and a cortical cell. CW = cell wall; Pl = plasmalemma; ER = presumptive endoplasmic reticulum; D = desmotubule. Arrows indicate membranous strands (presumably ER) traversing the new intercellular partition.

running from cell to cell through a plasmalemma-lined tube. There is not space in this chapter to consider structural variations in detail except to note that the plasmodesmata of algae and fungi show quite a wide range of variation and that a central strand of ER (or any other cytoplasmic membrane) is by no means a universal feature in these groups (Marchant, 1976).

As the young cell wall develops, so the plasmalemma-lined tube appears to become more tightly appressed around the traversing strand of ER. Eventually, this strand is no longer recognisable as a membranous tube and, with a transverse diameter of about 20 nm, looks more like a tube with a wall composed of particulate subunits (see Table 1 for some comparative dimensions). This traversing strand is termed the 'desmotubule' (Robards, 1968) which distinguishes it uniquely from other components (Figs. 2, 3 & 4). It has been suggested that the desmotubule has a wall structure of 13 subunits (Robards, 1971; see also Fig. 7 in this chapter) although Olesen (1979), using fixatives incorporating tannic acid, has suggested a nine-subunit structure (Fig. 8). Whatever the structure of the desmotubule (which appears to have become modified from the original ER by the pressures as it is trapped in the cell plate), micrographs *do* indicate that it remains in continuity with ER in the cells on either side of the junction. There is the added, and potentially important, complexity that the desmotubule may contain a central rod running through it, but this also requires further study to establish whether it is a real structure or is an artifact of the preparation techniques. This desmotubule, then, runs between cells within the tube lined by plasmalemma. In many situations, the desmotubule is 'pinched' by a neck-like constriction at either end of the pore. If this seal is, indeed, a tight one, then it would have the implication that open continuity from cell to cell could *only* occur through the cavity of the desmotubule. However, the seal is certainly not tight in all cases and so the opportunity exists for communication through two separate channels.

External to the plasmalemma, some authors have demonstrated

Fig. 2. Longitudinal section of fully differentiated plasmodesmata between cells in a barley root. In some instances strands of presumptive ER can be seen approaching the mouths of the plasmodesmata. The plasmalemma-lined tubes through the wall, and the desmotubules, are clearly visible.

Fig. 3. Transverse view of barley root plasmodesmata in a cell wall that has been sectioned obliquely. Thus fragments of wall and cytoplasm can be seen intermingling. MT = microtubule.

Table 1. *Some comparative dimensions of plasmodesmata*

	Source material			
	Azolla	*Hordeum* endodermis		*Abutilon*
Parameter	(Young root cortex)	(4 mm from tip)	(120 mm from tip)	(Distal wall of nectary trichome stalk cell)
Diameter at inner face of plasma membrane[a]	25	33	44	29
Outer diameter of desmotubule	16	20	20	16
Inner diameter of desmotubule	7	9	10	10

Data taken from Robards (1976). All dimensions in nm.
[a] Measured at the middle of plasmodesmata, hence neck constrictions not taken into account.

the presence of callose surrounding the plasmodesmata. Unfortunately, it is very difficult to be sure that this is not a traumatic response to sampling and fixation. Olesen (1980) has reported the presence of callose (aniline blue positive material) associated with plasmodesmata at all stages of development in *Salsola* mesophyll– bundle sheath interfaces, but only with plasmodesmata in young, meristematic regions in *Zea*. Such results stress the requirement for still better structural information about plasmodesmata and raise doubts as to the efficacy of conventional fixatives in preserving the *in vivo* structure of such structures. Rapid freezing methods would undoubtedly have much more to offer if: (a) freezing could be so effective that representative samples of tissue could be immobilised without damage; and, (b) suitable techniques could be used to analyse the structure of the cryopreserved connections. At the moment, both of these aspects raise problems although cryoultramicrotomy of rapidly frozen plasmodesmata has shown them to have broadly the same structure and dimensions as those studied by other methods (Vian & Rougier, 1974). Freeze-etching should provide a means of studying plasmodesmata (Figs. 5 & 6) but, unfortunately, the fracture plane through the membranes is such that few further details of the desmotubule–plasmalemma complex are revealed (Willison, 1976).

This, together with the fact that plasmodesmata are not regularly arranged, means that the elegant techniques of freeze-fracture and image analysis that have been applied to gap junctions in animals cannot be so readily deployed in plants.

The possibility that plasmodesmata act as 'valves' has been one that various authors have commented on from time to time. Olesen (1979) has described 'sphincters' in tannic acid fixed plasmodesmata that he equates with possible valves. The middle layer of the plasmalemma in the neck region is composed of closely packed globular units which may correspond to the particle clusters observed at the plasmodesmatal entrance in freeze-fracture preparations (Willison, 1976). In *Zea*, no sphincters were seen associated with the neck region but they *were* found in both *Epilobium* roots and *Salsola* leaves (Olesen, 1979).

Variations in plasmodesmatal structure

Among higher plants, variations in plasmodesmatal structure involve structural changes as described above but also the possibility for anastomosing desmotubules so that a number of desmotubules on one side of the pore fuse together to form a single desmotubule on the other side – a situation commonly found between phloem sieve elements and companion cells. Some plasmodesmata develop a 'median nodule' where a central cavity within the wall becomes enlarged. The enlarged central parts of the desmotubules may also fuse in this area. Again, within the limits of this chapter it is not possible to dwell on all structural variations (see Robards, 1976).

Enzymes and plasmodesmata

Juniper (1977) has emphasised the role of plasmodesmatal dilution in differentiation and has also pointed to the manner in which plasmodesmata may act as sites of preferential wall lysis so that, through the variations in plasmodesmatal frequency that arise during development, they may lead to asymmetrical penetration of the wall by lytic enzymes. Other authors have also given considerable attention to the association of enzymes with plasmodesmata (van Steveninck, 1976; van Steveninck & Iwahori, 1977) but the situation remains far from clear and the unreliability of cytochemical enzyme localisation techniques poses constant difficulties in the interpretation of such data. Nevertheless, continuing reports of plasmodesmata-located enzyme activities – such as the association of peroxidases with plasmodesmata in abscising flower pedicels of *Nicotiana* demonstrated by Henry

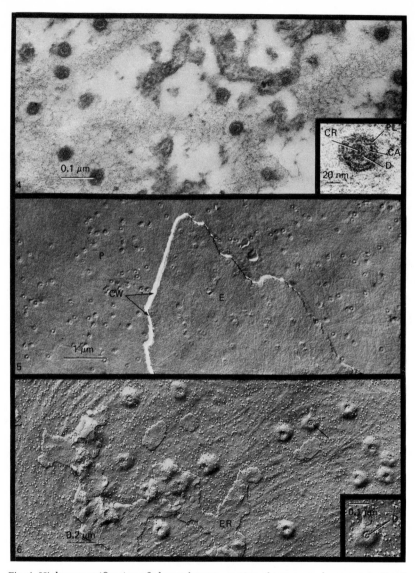

Fig. 4. Higher magnification of plasmodesmata sectioned transversely as seen in Fig. 3. Inset is at still higher magnification, showing a plasmodesma sectioned approximately in the mid-line of the wall and demonstrating the plasmalemma, cytoplasmic annulus (CA) and the desmotubule with apparent subunit structure. It remains to be established whether the central rod (CR) is a real structure or an artifact of preparation.

Fig. 5. Freeze-fracture replica from marrow root cortical cells. The micrograph illustrates the E- and P-faces of the plasma membranes of adjacent cells and demonstrates the high frequency of more or less randomly distributed plasmodesmata. P = P-face; E = E-face; CW = position of cell wall between adjacent membranes.

(1979) – do suggest that these intercellular channels may also serve a function in facilitating enzymatic modifications of cell wall structure.

Distribution of plasmodesmata

If it is assumed that plasmodesmata are, indeed, the channels for intercellular communication, then the distribution of plasmodesmata among different cell types and cell interfaces, as well as variations in frequency of plasmodesmata across different walls, should reflect something of the capacity of the particular junction for intercellular communication. Thus, at one level, we can study the distribution of plasmodesmata through the plant kingdom while, at another, we can look at variations in frequency across the same interface in different species or between different cell types of the same species.

Plasmodesmata through the plant kingdom. As has already been mentioned, plasmodesmata show differences of form among the lower plants and are not present in all multicellular algae. However, among higher plants it does appear to be valid to state that plasmodesmata occur between all neighbouring, *young* cells. Subsequently, as the plant matures, plasmodesmata may become occluded or have their frequency diluted as the wall expands. The exceptions to the general occurrence of plasmodesmata between living cells of higher plants appear to be significant where they exist. For example, the reproductive cells of plants are rarely connected by plasmodesmata, and connections are not usually seen between the gametophyte and sporophyte generations. Apart from such specialised exceptions – and those cells that *become* isolated as they develop – plasmodesmata are more distinguished by their ubiquity than by restrictions on their distribution.

Frequencies of plasmodesmata between cells. Most plasmodesmata appear to be formed during the process of cytokinesis. Consequently, a given number of plasmodesmata having been incorporated into a particular wall, the frequency of plasmodesmata expressed on a unit

Fig. 6. High-magnification view of plasmodesmata as seen in the E-face of a marrow root cortical cell. The protrusion forming the collar around each plasmodesma is evident (see Fig. 7) and, in some instances (arrowed and inset) the desmotubule with an apparent pore in its middle is clearly seen. In some areas, membrane remains attached over the plasmodesmata. This is interpreted as ER where the attachment to plasmodesmata has caused the fracture plane to jump from plasma membrane to ER membrane.

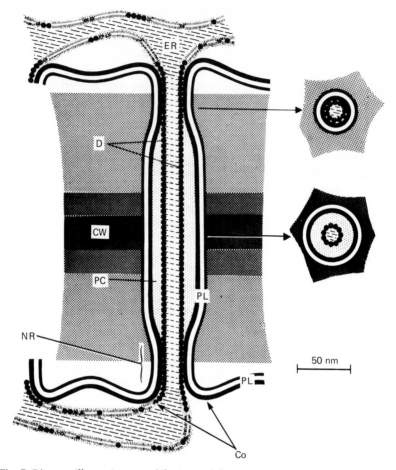

Fig. 7. Diagram illustrating a model proposed for a simple plasmodesma by Robards (1971). The desmotubule is shown in open continuity with the cisternal space of the ER. Detailed reasons for the suggestion of this model can be found in Robards (1971) and Robards (1976). ER = edoplasmic reticulum; D = desmotubule; CW = cell wall; PC = plasmodesmatal cavity; Pl = plasmalemma; Co = collar; NR = neck region. If the neck region does not form a tight seal on the desmotubule, then a clear pathway – the cytoplasmic annulus – will exist from cell to cell.

area basis will either remain the same, or will be reduced by wall expansion or total loss of plasmodesmata. The frequency would only increase if plasmodesmata were to be formed secondarily and this does not *normally* occur. A wide range of frequencies is cited in Robards (1976): young walls between meristematic cells of some ferns can have 140 or more plasmodesmata per μm^2 whereas older cell walls may have much lower frequencies (< 1.0 plasmodesma μm^{-2}). Frequencies of between 0.1 and 10 plasmodesmata μm^{-2} are common for walls of mature parenchyma cells in a wide range of species and organs. Although such frequencies do not sound very significant in themselves, they do mean that even the smallest meristematic cells in plants will be connected to their neighbours by between 1000 and 10 000 plasmodesmata. Taken in conjunction with the possible pore sizes, it is then possible to evaluate the total interface between a pair of adjacent cells with respect to its porosity. Under most combinations of pore size and plasmodesmatal frequency, the area of wall occupied by pores is unlikely to exceed 1.0% and will usually be considerably less than this.

A further feature relevant to the frequencies of plasmodesmata is their actual arrangement *within* a wall. In many situations the plasmodesmata are randomly distributed over the wall area but, in others, they are confined to pit fields. The total area fraction occupied by pores may be similar in each case: in one situation the plasmodesmata are all distributed while, in the other, the same number may be aggregated together into pit fields. Whether the plasmodesmata are so arranged during cell plate formation, or whether an originally higher frequency becomes reduced by loss of plasmodesmata from non-pit field areas remains to be demonstrated but, in either case, the pre-determination of the distribution is an intrinsic characteristic of the control mechanism within the particular cell or pair of cells.

Secondary formation, dilution and loss of plasmodesmata. So far as is known, secondary formation of plasmodesmata occurs only in response to abnormal conditions (e.g. nematode infestation of roots – Jones, 1976) or where adjacent cells have come together other than through a normal process of cytokinesis. Plasmodesmata across such 'non-division' walls are of great interest because they reflect the possibility that the plant may, at a relatively late stage in the development of a tissue, be able to modify the capacity for intercellular transport. However, relatively little is known about such possibilities during *normal* plant cell development and the subject will not be explored further here.

Of greater interest than secondary formation of plasmodesmata in the developmental history of most normal plant cell types are the effects of dilution of plasmodesmatal numbers from cell to cell. Thus, as a cell wall increases in surface area, the frequency will be steadily reduced until the mature cell may have only a fraction of the original frequency across the same wall. In many situations where cells elongate, the end walls retain relatively high frequencies while the elongating walls ultimately have far lower ones. This subject has been studied in some detail by Gunning (1978) who worked on plasmodesmata in the cell walls of developing roots of *Azolla* (a small aquatic fern). Gunning concluded that there was no secondary formation of plasmodesmata but that there could be loss of plasmodesmata from the walls of xylem and phloem elements during differentiation as well as through separation of cells from each other. Eventually, the apical cell (which produces all the cells of the *Azolla* root) fails to maintain the same number of plasmodesmata in each successive cell plate. Consequently the root apex becomes more and more isolated symplastically: a feature that could be responsible for the determinate growth pattern of the root.

The extreme stage of plasmodesmatal frequency dilution is, of course, total loss. This occurs in a number of ways, sometimes where no alternative is possible; in others, where some specific control mechanism appears to ordain such development. As an example of the first type, as vascular cells develop, xylem vessel elements may differentiate next to xylem parenchyma: the cell contents of the vessel element are lost and, at the same time, symplastic continuity is also lost. Similarly, plasmodesmata are lost where intercellular spaces form between the walls of previously closely adjacent cells. However, there are other situations where adjacent cells appear to retain living contents and yet lose symplastic continuity. Stomatal guard cells are an example: earlier reports cited widely different incidences of plasmodesmata in guard cell walls. It now appears that plasmodesmata are present in the young guard cell wall but are lost as the cell develops (Carr, 1976a; Willmer & Sexton, 1979). Other, similar, instances could be cited (see Carr, 1976a for a full review of this subject). The physiological significance of such events requires further evaluation but it cannot be doubted that, once again, the pattern of development is critical in helping to determine the specific functioning of the particular cell. Indeed, total symplastic isolation of a cell itself brings about a tendency to *de-differentiation*, and the implication is that normal symplastic continuity usually allows the transmission of 'messages' that hold such a process in check (Gunning & Robards, 1976b).

Dynamics

In the past, plasmodesmata have been thought of rather as passive structures present within the cell wall between adjacent cells. Indeed, for many years it is doubtful whether they were considered as significant *functional* features at all (Carr, 1976b). The disposition, pore dimensions and frequencies of plasmodesmata have now been relatively well studied in relation to intercellular communication although, as emphasised above, the accumulated evidence for plasmodesmatal function remains disappointingly indirect. Even the latest work, relating the area fraction of a wall to the fluxes and flows that traverse that interface, take no account of the possible dynamic characteristics of the intercellular connections. This point was emphasised by Gunning (1976b) when he stated that 'electron microscopists should be looking for the ultrastructural equivalents of sphincters'.

Plasmodesmata as valves? Olesen (1979) believes that he has found the structural equivalents of sphincters (Fig. 8). However, the difficulty of demonstrating the relationship between ultrastructural changes and modified physiological capacity is going to be a formidable one. Blake (1978) has considered the hydrodynamics of plasmodesmata from a theoretical standpoint and has investigated the effects of plasmodesmatal geometry in relation to either a Newtonian or power-law fluid, and also considered either a permeable or flexible desmotubule membrane. Although the difficulties of such an approach often reside in the poor data currently available for *precise* plasmodesmatal geometry and dimensions, it is interesting that Blake concluded that 'the single most dominant feature affecting volume flow rates is the constriction (the neck) at each end of the central cavity. By varying this neck gap opening, the plasmodesma may be able to exercise control over the relative flow of liquid through itself and the cell wall.' Gunning & Robards (1976c) also made some calculations that emphasised the effects of neck constrictions on diffusive permeability and hydraulic conductivity of the cytoplasmic annulus. Such thoughts must stimulate further work to establish whether plasmodesmata can indeed exercise such control over intercellular communication.

INTERCELLULAR COMMUNICATION THROUGH PLASMODESMATA

Fluxes and flows

Among the best data currently to hand that support the contention that plasmodesmata do provide the channels of intercellular communication, are the results of experiments where actual fluxes and flows

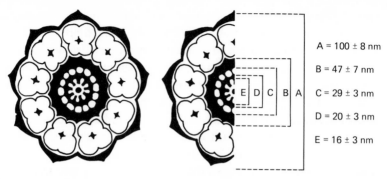

Fig. 8. Diagrammatic representation of transverse sections of the neck region of plasmodesmata after fixation in tannic acid. Reproduced, with permission, from Oleson (1979). The dimensions referred to are the means of 20 sets of measurements made on rotational image reinforcement diagrams. Because of the presence of opaque material between the desmotubule and plasmalemma and the thick, opaque covering of the outer leaflet of the latter, it is not possible to make a direct comparison with dimensions obtained from fixations not incorporating tannic acid. A = outer diameter of the ring structure; B = outer diameter of the opaque outer leaflet of the plasmalemma; C and D = outer and inner diameters of the 'bright' middle layer of the plasmalemma; and E = outer diameter of the desmotubule.

have been measured and then related to the dimensions and frequencies of plasmodesmata across specific cell interfaces. Inherent in such estimations is the assumption that the observed movement *must* proceed through the plasmodesmata because the transmembrane/cell wall pathway is either blocked or would be hopelessly inadequate. Although such determinations are limited, they have already provided some interesting results.

Quantitatively, it has been possible to obtain data on fluxes and flows through plasmodesmata from (among others) mesophyll cells and bundle sheath cells of C_3 plants, as well as sieve element/companion cell connections and *Abutilon* nectaries (Gunning, 1976a); bundle sheath cells of C_4 plants (Osmond & Smith, 1976); *Chara* nodes (Walker, 1976); and *Pisum* root nodules, *Hordeum* endodermis and *Cucurbita* endodermis (Robards & Clarkson, 1976). In some cases (the *Abutilon* nectary, for example – Gunning & Hughes, 1976) it has to be assumed that the flow must be through the plasmodesmata because the membrane permeability that would be required to transport the observed amount of sucrose would be far in excess (by 3–4 orders of magnitude) of any known membrane permeability. (A well quoted limitation to plant membrane permeability is that given by MacRobbie [1971] who cited a probable maximum figure of between 10^{-8} and 10^{-7} moles m^{-2} s^{-1}. It is probably not reasonable to assume that marginally greater fluxes could not pass across. membranes under some circumstances – indeed, there is evidence that they do. However, intercellular fluxes greatly in excess of these rates are unlikely to be able to follow the transmembrane route.) In other cases (the bundle sheath and the root endodermis) the assumption is made that the suberin lamellae within the wall totally preclude transport along that pathway. Consequently, it is possible to make some determinations and to ask some questions.

Firstly, estimations of the effective plasmodesmatal pore size and the viscosity of the pore fluid constantly impede precise calculations of the pressure drop that might be needed to sustain an observed volume flow; this, in turn, renders the calculation of the hydraulic conductivity of the plasmodesmatal pathway subject to rather large errors. In spite of this, in all except the most pessimistic assumptions of pore dimensions and viscosities, the pressure drops that would be needed could easily be envisaged to be within the plant's capability. When calculating solute fluxes from cell to cell, the difficulties arising from not knowing the precise pore dimensions again become paramount. These difficulties can, to some extent, be circumvented by expressing fluxes and flows, not in relation to the area fraction of wall across which they are presumed to pass (i.e. involving knowledge of pore dimensions), but on a 'per plasmodesma' basis. As it is considerably easier to obtain reliable estimates of plasmodesmatal frequency than of the actual dimensions of the conducting channels, citation of either a volume flow or a solute flux per plasmodesma serves well for the comparison of data from different sources. In fact, when this approach is adopted, some interesting facts emerge (Tables 2 & 3). Both in terms of volume flow and solute flux, the data from widely different sources show a striking 'per plasmodesma' similarity even despite up to 12-fold differences in plasmodesmatal frequencies. Such evidence strongly suggests that plasmodesmata have a relatively uniform capacity for sustaining fluxes and flows and that different rates of transport across different interfaces are brought about by the insertion (or dilution, or loss) of an appropriate number of plasmodesmata. Further, it can be shown that, while in some cases the plasmodesmata appear to do little more than replace the hydraulic conductivity or permeability of the

Table 2. *Some comparative data on estimated volume
flows through plasmodesmata*

Location	Plasmodesmatal frequency (pda μm^{-2})	Observed volume flow (m^3 pda^{-1} s^{-1})	Superiority factor
Abutilon, nectary	12.6	2.1×10^{-20}	230–1400[a]
Hordeum, endodermis	1.05	2.4×10^{-20}	5.2–307[a]
Cucurbita, endodermis	6.2	1.5×10^{-21}	30.4–1813[a]

The original data for this Table, and for Table 3, are provided in Gunning &
Robards (1976b, c). For clarity, the specific details of the cells and walls
have been omitted here. The 'superiority factor' is the factor by which the
plasmodesmatal volume flow per unit area of cell junction exceeds the flow
that would occur across two successive plasmalemmas of unit area, each
with a hydraulic conductivity of 2×10^{-8} m s^{-1} bar^{-1}.
[a]The two values listed are the desmotubular and cytoplasmic pathways
respectively. Dimensions are as given in Table 1 (for mature *Hordeum*) and
values of 0.016 poise (*Abutilon*) and 0.01 poise (roots) are used for viscosities. End constrictions of plasmodesmata and central rods within desmotubules have not been taken into account in these calculations.

pathway across two plasma membranes in series (as in the endodermal
cells), in others the plasmodesmata represent a highly preferred pathway in comparison with membranes (as in the volume flow through the
Abutilon nectary hairs). It should also be appreciated that plasmodesmata may not, in themselves, be the major limitation to intercellular
transport. As has been shown by Walker (1976) for the *Chara* node, the
major limitation to intercellular transport is more likely to be the rate
of cytoplasmic streaming (and hence movement of solutes up to the cell
junction) than the physical limitations of the plasmodesmata themselves.

Other opportunities for intercellular communication

From what has gone before, it will be evident that plasmodesmata
could serve, not only for the conduction of fluxes and flows from cell to
cell, or for the passage of electrical signals through the symplastic
channels, but the possibility also exists for the movement of *membrane-
bound* molecules by translational diffusion within the plasmalemma or
desmotubule/ER complex. Such transport, constrained to two dimensions, could be highly effective (Eididin, 1974) although at the moment
there is no evidence that it takes place.

Table 3. *Some comparative data on estimated solute fluxes through plasmodesmata*

Location	Plasmodesmatal frequency (pda μm^{-2})	Observed solute flux (mole pda^{-1} s^{-1})	Superiority factor
C$_3$ plant, mesophyll	3	2.7×10^{-19}	6.5–210
C$_4$ plant, bundle sheath	15	5×10^{-19} (*Amaranthus*)	164–1640
		10×10^{-19} (*Zea*)	296–2960
C$_3$ plant, bundle sheath	7.8	2.9×10^{-19}	23–230
Sieve element, companion cell	6	8.2×10^{-19}	48–480
Chara, node	4–5	9.5×10^{-19}	430–4300
Hordeum, endodermis	1.05	3.6×10^{-21} (PO$_4$)	0.04–0.01
Cucurbita, endodermis	6.2	2.2×10^{-20} (K$^+$)	1.4–14

The 'superiority factor' here is the factor by which the plasmodesmatal flux per unit area of cell junction exceeds the flux that would occur if the solutes passed through two successive plasmalemmas of unit area, each capable of carrying a flux of $10^{-7} - 10^{-8}$ mole m^{-2} s^{-1} (see MacRobbie, 1971).

CONCLUSIONS

In this brief review of plasmodesmata I have attempted to outline the important features of structure and function as they appear at the present time. Even in doing this, I have only scratched the surface of this enormous and complex topic. It is extraordinary that, almost a century after Pfeffer (1897) wrote 'General physiological considerations of the establishment and maintenance of correlative harmony, in short for the chain of stimulus transmission, render a continuity of the living substance so essential that it would be necessary to propose it, even if it were not already discovered' (translated by D. J. Carr, see Gunning & Robards, 1976a), plasmodesmata remain so unwarrantedly neglected.

Some fundamentally important questions remain to be answered. For example, why are plasmodesmata so complicated? Direct cytoplasmic continuity, even through small channels, is clearly not the solution adopted by higher plants in most situations and yet we have little, if

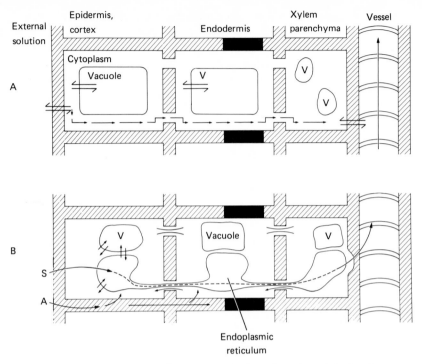

Fig. 9. A. Diagram illustrating a conventional, physiological model of a plant root showing transport between cells via open channels. The Casparian band in the anticlinal endodermal walls is shown in black.

   B. Additional opportunities for symplastic (S) or apoplastic (A) transport are suggested by the incorporation of an additional compartment (ER?) which may be continuous from cell to cell via plasmodesmata – a possibility supported from the available ultrastructural evidence. Fig. 9 is reproduced, with permission, from Robards & Clarkson (1976).

any, idea why *two* possible symplastic pathways (the desmotubule and the cytoplasmic annulus) are available in addition to the theoretically possible conduction route along the membranes themselves. If, as seems probable, the cisternal space of the 'ER' within cells can be continuous via the desmotubular pore from cell to cell, then the 'ER' under such circumstances would become a specialised compartment of the symplast (Fig. 9). The implications of this have hardly been considered by most plant physiologists.

   The possibility that plasmodesmata can exert a valve-like effect is an interesting idea for further exploration. Under many circumstances in plants it seems that bidirectional movement can take place across the same cell junction. However, there is no reason at the moment to assume that a *single* conduction channel could not accommodate movements of solutes in two opposite directions, or even movement of solutes over short distances *against* a volume flow. Only if volume flows in opposite directions were found across the same junction would it be necessary to contemplate the need for separate channels – and such a situation has not, so far, been discovered. Nevertheless, the potential capacity of plasmodesmata to act as valves may help to facilitate transport in opposite directions and may also help to explain other phenomena such as the large differences in turgor pressure that can sometimes be found in adjacent – and symplastically connected – cells.

At a time when work on junctional complexes in animal cells is reflected in a proliferation of related publications, it is paradoxical that the significant publications on plasmodesmata over the past half-decade can be numbered by not much more than the fingers on a pair of hands! The ways forward are difficult, but some of the experiments and observations that need to be made are clear. The integration of structural, physiological and biophysical data will be necessary if a significant step forward is to take place. Perhaps this contribution will stimulate others to take up their tools to apply them to this challenging – and important – field of interest.

Acknowledgements

I am most grateful to Dr R. Bolduc (Research Branch, Agriculture Canada, Quebec) who produced the freeze-fracture replicas of marrow roots while working in my laboratory. This work was supported in part by a grant from the Agricultural Research Council.

REFERENCES

BLAKE, J. R. (1978). On the hydrodynamics of plasmodesmata. *Journal of Theoretical Biology*, **74**, 33–47.

CARR, D. J. (1976a). Plasmodesmata in growth and development. In *Intercellular Communication in Plants: Studies on Plasmodesmata*, ed. B. E. S. Gunning & A. W. Robards, pp. 243–89. Berlin: Springer-Verlag.

CARR, D. J. (1976b). Historical perspectives on plasmodesmata. In *Intercellular Communication in Plants: Studies on Plasmodesmata*, ed. B. E. S. Gunning & A. W. Robards, pp. 291–5. Berlin: Springer-Verlag.

DRAKE, G. A. (1979). Electrical coupling, potentials, and resistances in oat coleoptiles: effects of azide and cyanide. *Journal of Experimental Botany*, **30**, 719–25.

DRAKE, G. A. & CARR, D. J. (1978). Plasmodesmata, tropisms and auxin transport. *Journal of Experimental Botany*, **29**, 1309–18.

DRAKE, G. A. & CARR, D. J. (1979). Symplastic transport of gibberellins: evidence from flux and inhibitor studies. *Journal of Experimental Botany*, **30**, 439–47.

DRAKE, G. A., CARR, D. J. & ANDERSON, W. P. (1978). Plasmolysis, plasmodesmata, and the electrical coupling of oat coleoptile cells. *Journal of Experimental Botany*, **29**, 1205–14.

EIDIDIN, M. (1974). Two-dimensional diffusion in membranes. *Symposia of the Society for Experimental Biology*, **28**, 1–14.

GIBBS, A. J. (1976). Viruses and plasmodesmata. In *Intercellular Communication in Plants: Studies on Plasmodesmata*, ed. B. E. S. Gunning & A. W. Robards, pp. 149–64. Berlin: Springer-Verlag.

GOODWIN, P. B. (1976). Physiological and electrophysiological evidence for intercellular communication in plant symplasts. In *Intercellular Communication in Plants: Studies on Plasmodesmata*, ed. B. E. S. Gunning & A. W. Robards, pp. 121–9. Berlin: Springer-Verlag.

GUNNING, B. E. S. (1976a). The role of plasmodesmata in short distance transport to and from the phloem. In *Intercellular Communication in Plants: Studies on Plasmodesmata*, ed. B. E. S. Gunning & A. W. Robards, pp. 203–27. Berlin: Springer-Verlag.

GUNNING, B. E. S. (1976b). Introduction to plasmodesmata. In *Intercellular Communication in Plants: Studies on Plasmodesmata*, ed. B. E. S. Gunning & A. W. Robards, pp. 1–13. Berlin: Springer-Verlag.

GUNNING, B. E. S. (1978). Age-related and origin-related control of the numbers of plasmodesmata in cells walls of developing *Azolla* roots. *Planta*, **143**, 181–90.

GUNNING, B. E. S. & HUGHES, J. E. (1976). Quantitative assessment of symplastic transport of pre-nectar into the trichomes of *Abutilon* nectaries. *Australian Journal of Plant Physiology*, **3**, 619–37.

GUNNING, B. E. S. & ROBARDS, A. W. (eds.) (1976a). *Intercellular Communication in Plants: Studies on Plasmodesmata*. Berlin: Springer-Verlag.

GUNNING, B. E. S. & ROBARDS, A. W. (1976b). Plasmodesmata: current knowledge and outstanding problems. In *Intercellular Communication in Plants: Studies on Plasmodesmata*, ed. B. E. S. Gunning & A. W. Robards, pp. 297–311. Berlin: Springer-Verlag.

GUNNING, B. E. S. & ROBARDS, A. W. (1976c). Plasmodesmata and symplastic transport. In *Transport and Transfer Processes in Plants*, ed. I. F. Wardlaw & J. B. Passioura, pp. 15–41. New York: Academic Press.

HENRY, E. W. (1979). Peroxidases in tobacco abscission zone tissue. VI. Ultra-structural localisation in plasmodesmata during ethylene-induced abscission. *Cytologia*, **44**, 135–52.

JONES, M. G. K. (1976). The origin and development of plasmodesmata. In *Intercellular Communication in Plants: Studies on Plasmodesmata*, ed. B. E. S. Gunning & A. W. Robards, pp. 81–105. Berlin: Springer-Verlag.

JUNIPER, B. E. (1977). Some speculations on the possible roles of the plasmodesmata in the control of differentiation. *Journal of Theoretical Biology*, **66**, 583–92.

MACROBBIE, E. A. C. (1971). Phloem translocation, facts and mechanisms: a comparative survey. *Biological Reviews*, **46**, 429–81.

MARCHANT, H. J. (1976). Plasmodesmata in algae and fungi. In *Intercellular Communication in Plants: Studies on Plasmodesmata*, ed. B. E. S. Gunning & A. W. Robards, pp. 59–80. Berlin: Springer-Verlag.

OLESEN, P. (1979). The neck constriction in plasmodesmata: evidence for a peripheral sphincter-like structure revealed by fixation with tannic acid. *Planta*, **144**, 349–58.

OLESEN, P. (1980). Fluorescence and electron microscopical cytochemistry of plasmodesmata neck regions. In *Abstracts of the VIth International Congress on Histochemistry and Cytochemistry, Brighton 1980*, p. 291. Oxford: Royal Microscopical Society.

OSMOND, C. B. & SMITH, F. A. (1976). Symplastic transport of metabolites during C_4-photosynthesis. In *Intercellular Communication in Plants: Studies on Plasmodesmata*, ed. B. E. S. Gunning & A. W. Robards, pp. 229–41. Berlin: Springer-Verlag.

PFEFFER, W. (1897). *Pflanzenphysiologie*, 2nd. edn. Leipzig: Engelmann.

ROBARDS, A. W. (1968). Desmotubule – a plasmodesmatal substructure. *Nature*, **218**, 784.

ROBARDS, A. W. (1971). The ultrastructure of plasmodesmata. *Protoplasma*, **72**, 315–23.

ROBARDS, A. W. (1976). Plasmodesmata in higher plants. In *Intercellular Communication in Plants: Studies on Plasmodesmata*, ed. B. E. S. Gunning & A. W. Robards, pp. 15–57. Berlin: Springer-Verlag.

ROBARDS, A. W. & CLARKSON, D. T. (1976). The role of plasmodesmata in the transport of water and nutrients across roots. In *Intercellular Communication in Plants: Studies on Plasmodesmata*, ed. B. E. S. Gunning & A. W. Robards, pp. 181–201. Berlin: Springer-Verlag.

STEVENINCK, R. F. M. VAN (1976). Cytochemical evidence for ion transport through plasmodesmata. In *Intercellular Communication in Plants: Studies on Plasmodesmata*, ed. B. E. S. Gunning & A. W. Robards, pp. 131–47. Berlin: Springer-Verlag.

STEVENINCK, R. F. M. VAN & IWAHORI, S. (1977). Do plasmodesmata possess ATPase activity? *Colloques Internationaux du Centre National de la Recherche Scientifique*, **258**, 499–506.

STRASBURGER, E. (1901). Ueber Plasmaverbindungen pflanzlicher Zellen. *Jahrbuch für Wissenschaftliche Botanik*, **36**, 493–610.

TANGL, E. (1879). Ueber offene Communicationen zwischen den Zellen des Endosperms einiger Samen. *Jahrbuch für Wissenschaftliche Botanik*, **12**, 170–190.

VIAN, B. & ROUGIER, M. (1974). Ultrastructure des plasmodesmes après cryoultramicrotomie. *Journal de Microscopie*, **20**, 307–12.

WALKER, N. A. (1976). Transport of solutes through the plasmodesmata of *Chara* nodes. In *Intercellular Communication in Plants: Studies on Plasmodesmata*, ed. B. E. S. Gunning & A. W. Robards, pp. 165–79. Berlin: Springer-Verlag.

WILLISON, J. H. M. (1976). Plasmodesmata: a freeze-fracture view. *Canadian Journal of Botany*, **54**, 2842–7.

WILLMER, C. M. & SEXTON, R. (1979). Stomata and plasmodesmata. *Protoplasma*, **100**, 113–24.

Studies on invertebrate cell interactions – advantages and limitations of the morphological approach

NANCY J. LANE

ARC Unit of Invertebrate Chemistry and Physiology, Department of Zoology, Downing Street, Cambridge CB2 3EJ, UK

Fine structural studies have been enormously important in establishing basic cellular arrangements and patterns in tissues; from such associations functional interrelationships may be surmised and ultimately tested by physiological, biochemical or other experimental criteria. Since the advent of the electron microscope in the 1950s, many new ultrastructural techniques have been developed, each of which has enabled the cell biologist to unravel yet another aspect of cellular structure of which some have had implications for potential cell-to-cell interactions. Inevitably the morphology of the cell membrane and its modifications have been the chief sphere of concern in this regard, since cell interactions must involve the surface plasmalemma. Cell interactions relating to the plasma membrane have been considered to fall into three main categories of membrane-mediated events (Wolpert, 1977); these include situations where membranes act as mechanical sensors, enabling cells to respond to such stimuli as different degrees of adhesiveness, or as channels, whereby signals may be exchanged from one cell to another as occurs via the gap junctions, or finally as transducers, as would be the result of a stimulus setting off a membrane permeability change or synthesising secondary messengers such as cyclic AMP (cAMP) by the activation of adenyl cyclase. In addition, cells may interact via other more highly organised kinds of adhesiveness which may take the form of desmosomal or hemi-desmosomal membrane modifications or by other quite different sorts of junctional specialisations such as tight junctions. In all cases, morphological studies are required to elucidate the distinguishing characteristics and distribution of these cell-to-cell associations.

Conventional thin-sections are the means whereby initial morphological studies are made, since by such investigations the general histological features of tissues can be determined as well as the way the component cells are organised with respect to one another. Using one of a range of electron-opaque tracers, added to the chemical fixative, the extracellular spaces between cells can be analysed and mapped. When such tracers are added to physiological saline, the extent to which their entry between cells is restricted can also be established; in this way permeability barriers in tissues may be characterised and the nature of the occluding junctions giving rise to them determined. Low molecular weight fluorescent tracers may be injected into cells and any intercellular coupling monitored light microscopically. Electron-opaque substances may similarly be injected, and, depending on their molecular weight, either reveal the ultrastructural distribution of the injected cell with respect to other cells, or indicate the cell-to-cell migration of the tracer via any communicating junctions present. Similarly, autoradiographical studies permit the visualisation of the transport from one cell to another of radioactively labelled metabolites that are small enough to migrate by gap junctional channels. Cyto-chemical and antibody-binding techniques permit the localisation of either enzymatic activities which may be transducers, like adenyl cyclase, or of specific cell-surface sites such as receptor or contact sites. Freeze-fracture studies have recently also made possible an under-standing of the arrangement of intramembranous cellular components, and in particular, the three-dimensional configuration of those parts of the cell membrane which form intercellular associations by means of junctional structures. With this procedure as with all the fine structural methodology, there is a spectrum of technical variables, such as the kind of cryoprotectant used and rapid versus conventional freezing, which may modify the way a tissue appears in response to treatment and hence to the scope of the data produced. It is in the interpretation of this structural data, gleaned from the range of available techniques, and the way the morphological features are deemed to suggest func-tional significance, that leads to speculations as to the physiological interactions between cells. These, in some systems, have proved to be very fruitful, whereas in other situations, the evidence is inadequate and correlated biochemical or physiological studies are required to move beyond the limitations imposed by the purely morphological criteria. A good case in point is the dilemma which arises when more than one structure is to be found in the same location, and it is not clear which is responsible for an observed physiological phenomenon; in-sect salivary glands where both septate and gap junctions are present serve as an example; the problem of which junctional structure was responsible for the cell-to-cell communication, which was observed electrophysiologically as early as 1964 (Loewenstein & Kanno), could not at first be established unequivocally by fine structural means alone (Rose, 1971).

In the case of this cell-to-cell coupling, subsequent electrophysiological and morphological studies on vertebrate tissues have revealed communicating systems in which only one junctional type, the gap junction, is present, (for example, Revel, Yee & Hudspeth, 1971; Pinto da Silva & Gilula, 1972; Gilula, Reeves & Steinbach, 1972; Pinto da Silva & Martinéz-Palomo, 1975). However, invertebrate tissues have also proved invaluable in these studies such as in the low-resistance electrotonic junctions that occur between the giant axons of earthworms (Kensler, Brink & Dewey, 1979; Brink & Dewey, 1980); and crayfish (Peracchia, 1973a, b; Peracchia & Dulhunty, 1976; Hanna, Keeter & Pappas, 1978; Zampighi, Ramón & Durán, 1978). In these cases, the analysis of the kind and location of intercellular junctions can be precisely and accurately established by fine structural studies, particularly by bringing together evidence from several approaches, such as tracer-impregnated (Fig. 1) and freeze-cleaved (Fig. 2) tissues. However, electrophysiological or permeability studies are required to prove that the cells under consideration are ionically or metabolically coupled, whereby parameters such as the modulating control mechanisms that are operative (Rose, Simpson & Loewenstein, 1977; Caveney, 1978; Peracchia, 1980) or junction asymmetry due to directional permselectivity (Flagg-Newton & Loewenstein, 1980) may be established. Moreover, biochemical analyses of isolated junctions are needed to determine the chemical nature of the purified junctional material. From these, as well as X-ray diffraction and specimen-tilting with low-dose electron microscopical studies, correlated with physiological studies, models have been constructed (Staehelin, 1974; Makowski, Caspar, Phillips & Goodenough, 1977; Unwin & Zampighi, 1980). For most of these latter studies vertebrate materials have been used, as bulk samples are more readily obtained from them than they are from invertebrate sources.

The tissues from invertebrate organisms, however, exhibit a number of cellular interactions which differ from those of vertebrates and which may therefore reveal unexpected information. For example, vertebrate tissues tend to possess a rather restricted number of cell-to-cell associations which occur fairly ubiquitously to produce certain

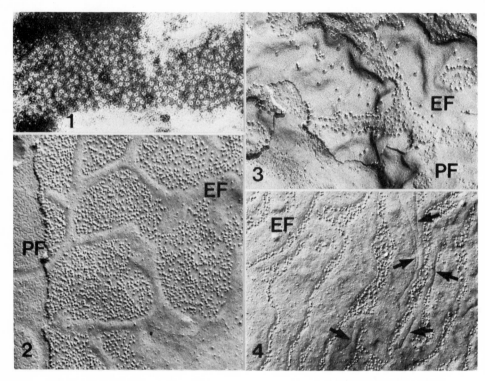

Fig. 1. Arthropod gap junction infiltrated with lanthanum to reveal the dense extra-cellular space surrounding the junctional particles that span the intercellular cleft. Note the opaque central channel in the particles which are arranged in rather loosely packed clusters in these glial cell membranes. × 142 000.

Fig. 2. Freeze-fractured replica of an arthropod gap junction demonstrating the characteristic pattern of cleaving with the 13 nm junctional particle in the E face (EF) and their complementary pits on the P face (PF). Note that, although some particles are arrayed in aggregates, those on the right-hand side appear to be undergoing a degree of translational migration away from, or to, the maculae. × 43 100.

Fig. 3. Glial gap junctions from the nervous system of an arachnid, in this case a scorpion, exhibiting the loose arrangement of EF gap junctional particles, even in the adult state. This may represent junctional turnover or could reflect the degree of cell-to-cell coupling. Note the suggestion of two different categories of junctional particles as regards size which may merely reflect the extent to which they are able to protrude beyond the surrounding lipid. × 50 000.

Fig. 4. Linear gap junctions from the outer glial layer ensheathing the CNS of a garden centipede. The junctions are transformed from macular to linear configurations over the membrane face. Note that the EF rows of particles may run into grooves (arrows); these are complementary to PF ridges which together form a discontinuous system of tight junctional punctate appositions between adjacent cells. × 32 000.

effects. Hence, desmosomes tend to maintain the structural integrity of vertebrate tissues, while tight junctions produce seals wherever permeability barriers are required and gap junctions form the basis of cell-to-cell communication as regards the intercellular exchange of ions and small molecules. In the tissues of invertebrates, there appear to be several junctional types which could act as adhesive devices and several which might be occluding in action. Low-resistance pathways between cells appear to be based, however, as in vertebrates, on gap junctional structures, except that in certain groups such as the arthropods, they exhibit distinct differences from those of vertebrates, undetectable in thin-sections, but visualisable by freeze-fracturing.

The gap junctions of arthropods have long been established as inverted, or B-type (Flower, 1972) because their component junctional particles in both fixed and unfixed tissues, cleave preferentially onto the E face (extracellular half-membrane leaflet). This contrasts with the situation in most vertebrate tissues, where the reverse is true. This arthropod characteristic, not initially of great significance, has emerged as a feature of considerable convenience in the analysis of certain aspects of cellular organisation. This fracturing pattern, together with a particle diameter in the range of 12 to 14 nm, considerably larger than the other junctional particles or other populations of intramembranous particles (IMPs), means that the gap junctional particles in arthropods are always readily identifiable and distinguishable from any other class of E face IMP. Any changes in their distribution that may occur, therefore, are immediately apparent, a situation which does not hold for vertebrate systems, where the gap junctional particles are generally indistinguishable from the other P face (cytoplasmic half-membrane leaflet) IMPs present.

In insect tissues, those intercellular junctions that are found in holometabolous organisms, undergo profound changes during the metamorphosis that transforms larval into adult systems. The cells that are organised and integrated into perfectly functioning larval tissues undergo considerable pupal rearrangments to give rise to the very different adult tissues, and these changes are preceded by cell separation and dissociation which must then be followed by cellular re-association. Two systems that have been analysed in this regard are the insect central nervous system (CNS) (Lane & Swales, 1978b, 1980) and the digestive tract (Lane, Swales & Lee, 1980).

In the ganglia of arthropod CNS, glial cells are found isolating, protecting and possibly nourishing (Radojcic & Pentreath, 1979; Lane,

1981d) the nerve cells and their processes. They are of two types, outer and inner, the former being modified to form an encompassing perineurial sheath. These glial systems become disarrayed upon the onset of pupation, and as the cells separate so their junctional complexes become non-functional, that is, the junctions disappear as judged by thin-section criteria. By freeze-fracture criteria, however, intramembranous modifications occur that are not normally evident in vertebrate material when cells dissociate, although recently there has been some support for these phenomena in regenerating rat liver (Yancy, Meyer & Revel, 1980) or anoxic sheep ovarian follicles (Lee, Cran & Lane, unpublished observations).

The macular larval gap junctions are not maintained then, during pupal development, for almost immediately the maculae break down by a 'streaming' out of junctional particles in linear arrays from the plaque periphery (Figs. 2–4 in Lane & Swales, 1980). A particle-free aisle, running along an intramembranous groove, lies in the centre of the aligned arrays. As development proceeds, these aisles widen and the gap junctional particles become dispersed over the intramembranous E face (Lane & Swales, 1980) rather than being disposed of by 'internalisation' (see Larsen, 1977). Diapausing control pupae suggest that the intramembranous groove along which the particles become aligned may be a cellular response to hormonal stimulation since glial cells in diapause remain associated together and their gap junctions retain the macular larval configuration. Moreover, such organisms are known to enter diapause due to an absence of ecdysone (Williams, 1952) and the hormonal regulation of intercellular communication has been demonstrated for other insect tissues (Caveney, 1978). Interestingly, recent cytochemical studies suggest that the junction-bearing glial cell membranes may act as transducers since adenyl cyclase activity has been found associated specifically with them (Lane & Swales, unpublished observations) and the actual effector molecules at pupation may be peptide in nature, triggering off the production of cAMP. Here then is an example of integrated cellular interaction and response to an event, possibly hormonal, which can be observed by morphological criteria. After the migration of nerve cells to the positions they will assume for adult life, the glial cells again take up their intimate spatial association with the neurones by ensheathing them as they did in the larval CNS. In addition, they appear to recouple with one another via gap junctions which reform by intramembranous particle migration into linear arrays and finally macular clusters (Lane

& Swales, 1978b, 1980); the junctional particles may be the original larval ones undergoing reutilisation, since no interiorisation of gap junctions is seen in the early pupal stages.

The evidence that arthropod glial cells are coupled functionally is more difficult to establish than it is to see that they are structurally associated by macular gap junctions. In an attempt to correlate structure with function recent work on tracer injection at the electron microscopical level shows that horseradish peroxidase (HRP) can be injected into a single glial cell whose distribution can then be precisely charted; such a large molecule (40 000 MW) cannot move through the gap junctional particle channels and hence remains within one cell (Schofield, Swales & Treherne, in preparation, quoted in Treherne & Schofield, 1981). However, there is preliminary evidence that smaller molecular weight electron dense tracers may move from one glial cell to another, presumably via gap junctions, in mature adults (Swales, Lane & Schofield, unpublished results) and experiments are currently underway in an attempt to extend these to insects in the pupal state. Of course, the cell-to-cell transport of tracers cannot reveal whether physiologically important molecules are also exchanged but it indicates the potential for such exchange.

The functional integration of many cell-to-cell responses other than glial ones has been thought to be by gap junctional communication as well, but a shortcoming with all such studies is that although freeze-fracture may tell one whether gap junctions are present it does not reveal whether ions and molecules are actively being exchanged via these structures. The assumption that cells are coupled when they are linked by closely-packed, macular gap junctions has recently had to undergo revision due to the ultra-rapid freezing methods that have become available by 'smash' or impact freezing (for example, Heuser *et al.*, 1979), or jet freezing (Müller, Meister & Moor 1980; Pscheid, Schudt & Plattner, 1981). These and studies on the effects of Ca^{2+} and pH, have suggested that the closeness of the packing of the component gap junctional particles may be an indication of the degree of cell coupling (Peracchia, 1977, 1978, 1980; Raviola, Goodenough & Raviola, 1980), the more closely-packed, the more uncoupled. This being the case and the fact that the gap junctions of arthropods have always been very pleomorphic and loosely packed (Figs. 2, 3 and 4), would suggest that the latter are more closely physiologically coupled than the tightly packed ones found occurring in vertebrate tissues. This could of course, alternatively, reflect a continuous turnover in the gap junc-

tional particle population as has already been observed in some arthropod tissues (Lane, 1978b). However, it now appears that within the
same vertebrate tissue, gap junctions, analysed after rapid freezing,
may be in different stages of packing (Raviola *et al.*, 1980) and certainly
this is also the case for insect tissue (Figs. 2–4) (Lane, 1981d) even when
frozen conventionally. The morphological differences observed in the
gap junctions of tissues from different adult arthropods, wherein loose
macular plaques (Fig. 2) may be found, or loose linear arrays (Figs. 3
& 4), are not easily interpreted physiologically. It seems very reasonable, though, that this packing should be functionally significant
and it will be interesting to see what rapid freezing reveals about
arthropod gap junctional structure. Unfortunately none of this morphological information will actually bear on the extent of coupling or
the nature of the hypothetically exchanged ions or molecules. Even in
the insect salivary gland, where Loewenstein and colleagues (1978)
have demonstrated light microscopically the exchange of fluorescently
labelled molecules of different sizes between cells, the morphology of
the communicating junctions is not fully understood with regard to the
state of the junctional packing given different degrees of coupling.

One interesting aspect of the presence of gap junctions between
the modified glial cells which form the morphological basis of the
physiological blood–brain barrier in arthropods, is their co-existence
with tight junctions (Fig. 5). In certain arachnids (Lane & Chandler,
1980) these two junctional types can be seen to become assembled into
mature structures at about the same time (Lane, 1981b) (Fig. 6). However here, in contrast with vertebrate tissues, the two junctions are
composed of different sized intramembranous particles which fracture
onto distinct fracture faces. The 13 nm gap junctional IMPs are E face
and the 8–10 nm tight junctional ones are P face; during junctional
development the two types are quite distinct for this reason (Lane,
1980, 1981b) (Fig. 6). In such tissues, it is clear that no *one* precursor
particle could give rise to both junctional types, a theory put forward
for vertebrate tissues where both junctions are composed of 8–10 nm
IMPs on the P face (Montesano, Friend, Perrelet & Orci, 1975; Elias &
Friend, 1976; Porvaznik, Johnson & Sheridan, 1979; Wynograd &
Nicolas, 1980).

Once formed, the tight junctions produce very prominent networks
of ridges (Fig. 7) which nearly always fracture on the P face in arthropod tissues (Lane, 1981c). The presence of tight junctions in
vertebrates is usually associated with a restriction to the passage of

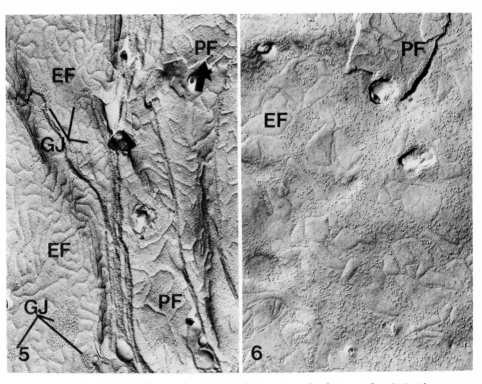

Fig. 5. Circumferential band of arachnid tight junctions displaying P face (PF) ridges and E face (EF) grooves with intercalated gap junctional (GJ) plaque. These occluding structures, in adult spiders, occur between the outer glial cells and form the basis of a blood–brain barrier (see Lane & Chandler, 1980). Note the coincidence, over face transitions, of PF ridges and EF grooves (arrow) indicating their complementary nature. × 24 500.

Fig. 6. Outer glial cells ensheathing the CNS in an immature spider in which the junctional structures are still in the process of assembly. The tight junctions appear as incomplete PF ridges composed of 8–10 nm particles or EF grooves, while the gap junctional, 13 nm EF particles, are clustering together between the grooves, in the process of aggregating. The two junctional types are therefore readily distinguishable and are unlikely to arise from a single common precursor particle (see Lane, 1981b). × 33 000.

ions and molecules across the intercellular cleft between the lateral borders of the cells thus associated (Claude & Goodenough, 1973) and this is also the case for arthropods (Fig. 7, inset). Changes in the nature of this seal may be followed during insect pupal metamorphosis, again, by differences in the degree of impedance to tracer entry (Lane & Swales, 1978b). Hence the accessibility of the tracer alone can give a physiological clue to the absence or presence of occluding junctions. Conventional or tracer-incubated thin-sections and freeze-cleaved replicas, then, indicate the features of this kind of cell-to-cell association whereby, for example, the component cells in the outer tissue layer of the CNS are integrated into a functional unit. There are, in arthropod CNS, tight junctions which are indecipherable from one another in thin-section, but quite distinct by freeze-cleave criteria (Figs. 7, 8 and 9) (Lane, 1981c), since the latter reveals the three-dimensional arrangements that are undetectable by any other procedure. In spiders (Fig. 7) (Lane & Chandler, 1980) and scorpions (Lane, Harrison & Bowerman, 1981) (Fig. 8) the tight junctions are composed of anastomosing rows of particles fused into ridges which display complementary grooves, although in scorpions (Fig. 8) they are far less continuous in nature. In centipedes (Fig. 4) and insects (Fig. 9), the ridges are discontinuous, non-anastomosing and relatively infrequent (Lane, Skaer & Swales, 1977; Lane & Swales, 1978a, b, 1979; Lane & Skaer, 1980; Lane, 1981c). Moreover, in spiders and insects the ridges fracture onto the PF while in scorpions they cleave preferentially onto the E face.

These tight junctions then, within the arthropod CNS, are structurally very diverse but play, it is believed, a similar physiological role in each system in which they occur. Peripheral glial cells in each of these groups have evolved, therefore, to form an integrated system which is the morphological basis of a permeability or blood–brain barrier, but their membranes have become modified in rather different ways. This is in some respects similar to the situation in different vertebrate systems, such as epithelia compared with myelin, where tight junctions, which form the basis of intercellular seals, are to be found exhibiting varying degrees of complexity (Claude & Goodenough, 1973; Shinowara, Beutel & Revel, 1980). It is interesting to discover, though, that the arthropods exhibit a spectrum of structural diversity comparable to the vertebrates. The simple junctions of insect tissues, however, seem at one extreme pole in contrast with the complex ones of the arachnids. Here, then, although morphological studies have indicated that junctions are present and reveal their distribution and structural complexity, further electrophysiological studies are required, on the CNS of arachnids, for example, to determine the differences they exhibit in comparison with the better physiologically-defined glial cell barriers of the insects (Treherne & Pichon, 1972). Clearly also, isolation of these junctions followed by chemical and immunological analysis and characterisation would be invaluable in

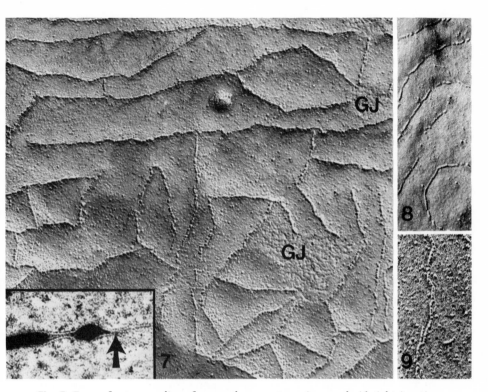

Fig. 7. Freeze-fracture replica of a complex, anastomosing arachnid tight junction which is composed of PF ridges made up of fused 8–10 nm particles. The strands or ridges form a continuum with a complementary network in the adjacent cell to form punctate appositions so that a permeability barrier is produced. Incubation with tracers (in inset) shows exogenous horseradish peroxidase being stopped by an occluding punctate junction (arrow). Intercalated gap junctional pits (GJ) are also evident. × 56 500; inset, × 170 000.

Fig. 8. Replica of discontinuous tight junctional ridges in the glial cells of scorpion CNS. Here the ridges, which show a variable degree of fusion between adjacent particles, fracture onto the E face with complementary PF grooves. × 63 000.

Fig. 9. Tight junctional ridge from the outer glial cells of the insect CNS, revealing the simply aligned PF particles of which they are comprised. × 70 000.

our understanding of the degree of their homology; their differences in appearance could reflect subtle chemical differences rather than any basic divergence, although the evolutionary link between arachnids and insects seems not to be direct (Lane, 1981c).

Further to this problem concerning the function of a structure described only as to its morphology, additional enigmas are confronted when, between the cells of different tissues, a variety of junctions occur, each of which exhibits a similar circumferential distribution and, to a varying extent, a similar capacity to exclude certain exogenous molecules. These enigmas relate, not to differing varieties of the same intercellular junctional type, such as discussed above for the tight junction, but to rather distinct junctions. For example, structures known as septate junctions (Wood, 1959; Noirot-Timothée & Noirot, 1980) which are not, with a very few exceptions (see Lane & Skaer, 1980) found in vertebrates, are seen to link adjacent epithelial cells in many systems, including those along the length of the intestinal tract, in many invertebrate tissues. For example, in the gut in arthropods it is found that, in the oesophagus and rectum, the lateral borders of the component cells are always linked by pleated septate junctions, while in the mid-gut, they are inevitably associated by smooth septate (or 'continuous') junctions (Noirot & Noirot-Timothée, 1967; Flower & Filshie, 1975; Lane, 1978a; Lane & Skaer, 1980; Skaer, Lane & Lee, 1980). These two kinds of septate junction exhibit certain similarities in that they both have a constant 15–20 nm intercellular cleft and both possess septa. Beyond this, however, the differences begin and tracer-infiltration and freeze-fracture reveal that pleated septate junctions have undulating intercellular septa and rows of separate intramembranous particles (Fig. 10) while smooth septate junctions exhibit non-ribbon-like intercellular septa separated by rows of columns, with intramembranous particle rows frequently fused laterally into semi-solid ridges (Fig. 11). Moreover, in different groups, for example the centipedes and *Limulus*, the features of these septate junctions may be rather different (Lane, unpublished data; Lane & Harrison, 1978). These structures have repeatedly been considered to be the basis of invertebrate permeability barriers due to evidence based on tracer exclusion or blistering under hypertonic stress (Satir & Gilula, 1973; Noirot & Noirot-Timothée, 1976; Mills, Lord & Di Bona, 1976; Filshie & Flower, 1977; Wood, 1977; Noirot-Timothée, Smith, Cayer & Noirot, 1978; Green, Bergquist & Bullivant, 1979, 1980; Wood & Kuda, 1980) and indeed have been thought to be the invertebrate equivalent of the vertebrate tight junctions, representing their evolutionary forerunners (Green, 1978). It is now clear that true tight junctions do exist in the invertebrates (Lane, Skaer & Swales, 1977; Toshimori, Iwashita & Ōura, 1979; Lane & Chandler, 1980; Lane & Harrison, 1980; Lane & Skaer, 1980; Lane, 1981c) as well as septate junctions, and so it may be that a spectrum of very different structures could be involved in handling a range of rather comparable physiological problems. The degree to

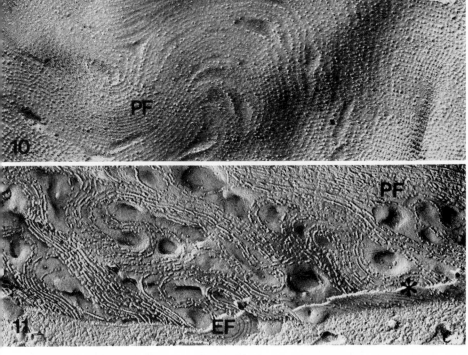

Fig. 10. Pleated septate junctions from the testis of a cockroach. Freeze-fracturing reveals undulating stacked rows of intramembranous P face (PF) particles clearly separated from one another, with stacked rows of complementary pits, on the E face, the two faces lying 15–20 nm apart. The intercellular septa seen in thin-sections (as in Fig. 16) might insert into the membrane via these particles. × 59 000.

Fig. 11. Smooth septate, or continuous, junctions from the midgut of a moth. Freeze-cleaving shows the parallel rows of ridges of laterally fused intramembranous particles these fracture onto the EF in unfixed tissue and onto the PF in fixed tissue, as shown here. The septa as seen in thin-sections, but not the intercellular columns, may be attached to these intramembranous structures. The faces of the two cells are separated by a distinct 15–20 nm intercellular cleft (*). × 52 000.

which these various junctions are capable of producing physiologically measurable barriers is not yet clear except in the case of the CNS of insects (Treherne & Pichon, 1972) and crustaceans, (Abbott, Pichon & Lane, 1977). In terms of tracer exclusion by septate junctions in the insect gut, only preliminary results are as yet available (Skaer, Lane & Lee, 1980) and these need to be extended to determine at what molecular weight exogenous substances are prohibited intercellular passage and what charge and potential binding capacity the matrix of the intercellular clefts might carry. Suggestions have been made that there are charged groups in this matrix (Dallai, 1970; Skaer, Harrison & Lee, 1979) which could bind ions and molecules and so the matrix, rather than the septa, could inhibit the free inward intercellular diffusion of substances.

The differences in septate junction structure along the gut length have been claimed to be related simply to the embryonic origin of the tissues, the pleated septate junctions always occurring in non-regenerating, ectodermal tissues and the smooth septate junctions, always in regenerating, endodermal tissues (Noirot & Noirot-Timothée, 1967), the two being analogous in function. There are, however, various points of controversy here (see Lane & Skaer, 1980, page 72) which include the existence of both smooth and pleated septate junctions in such tissues as Malpighian tubules (Dallai, 1976; Skaer *et al.*, 1979) which are known not to undergo turnover and which are thought to have an ectodermal embryonic origin. Hence the cells lining the gut in arthropods are clearly integrated to form an effective functional unit, but the subtle differences in the precise roles played by the various intercellular junctional structures are still obscure.

The smooth septate junctions of the midgut may undergo turnover in adult tissues (Lane, 1979a). Moreover, in pupal tissues of metamorphosing insects such as the moth *Manduca sexta*, very striking changes occur in the gut tract. The structure becomes transformed from a straight-sided larval tube, heavily tracheated, to an adult structure with a distended portion together with undulating tubular regions, with altogether much thinner walls (Lane, Swales & Lee, 1980) as might befit the changed diet from voracious herbivorous larva to nectar-imbibing adult. At the same time, during pupation, the cells become more loosely associated so that the lateral borders are very much less junction-rich and the tissue very fragile; the junctions therefore probably maintain the physical integrity of the gut wall in addition to whatever barrier function they may effect. During these stages the characteristic freeze-fracture appearance of the smooth septate junctions become disrupted; the ridges become broken down into fragments and particle dispersal occurs over the membrane face. Towards the end of pupal life the smooth septate junctions become reassembled again (Fig. 12) into the typical adult structures. It is possible that some of the junctional particles could be being reutilised since no signs of junctional internalisation (Larsen 1977) are apparent. Here the cells are being reintegrated from a disrupted system into a

Fig. 12. Smooth septate junctions in the midgut of a metamorphosing moth, demonstrating a stage in the reassembly of the junctional particles (arrows) into ridges after their initial disruption at the onset of pupation. × 30 300.

functional, junction-bearing epithelial layer and it seems that the response of the gut to the advent of pupation may be a hormonally stimulated event. Diapausing controls enter pupation but their gut remains larval; the onset of pupation is stimulated by ecdysone which seems not to be released in diapausing pupae (Williams, 1952; Lane & Swales, 1980). In addition, other factors may be involved, for it has been shown in flea midgut that after ingestion of a blood meal, the columnar epithelium becomes squamous, entailing, perhaps, breakdown and reassociation of the smooth septate junctions (Houk, 1977). Morphological studies are here again revealing changes in cellular interrelationships, but physiological investigations are clearly required to elucidate the ambiguities remaining.

One further point which bears mentioning is that along the digestive tract of insects, two other modifications of the septate junction may occur, the scalariform junction (Fain-Maurel & Cassier, 1972; Noirot-Timothée, Noirot, Smith & Cayer, 1979; Lane, 1979b) and the reticular septate junction (Lane, 1979b; Flower & Walker, 1979). Both these seem to have quite distinct physiological roles to play but may in some cases be unique to adult rather than larval tissues and are fairly specialised in their location (Lane & Skaer, 1980). Here then we have cells of the digestive tract integrated via a range of different intercellular junctions into an efficient functional entity, each junctional modification, it must be presumed, especially suited to the job it performs in that particular part of the gut tract and in the overall scheme of the organism.

Another tissue in insects which presents somewhat comparable junctional complexity is the compound eye of dipteran flies. Here the glial cells surrounding the axons exhibit both interglial, and axo-glial junctions which may be very similar; for example, retinular and septate junctions (Figs. 13–16) (Carlson & Chi, 1979; Chi & Carlson, 1980a, 1981; Chi, Carlson & Ste Marie, 1979; Lane, 1979c; Lane & Skaer, 1980) as well as marginal glial junctions (Chi & Carlson, 1980b) all have several features in common. They all exhibit a 15–20 nm cleft, possess some intercellular cross-striations and have been thought by some investigators to exert a degree of restriction to the free diffusion of ions and molecules. The retinular junctions (Fig. 13 & 15) give way to septate junctions (Fig. 16) along one and the same cleft (Fig. 14). Lanthanum-staining shows that both junctions may permit the free intercellular passage of ions in certain circumstances (Figs. 15 & 16) but reveals that the former possesses intercellular columns, and the latter, intercellular septa. These structures, as well as more conventional junctional interactions such as gap junctions (Fig. 15) (Ribi, 1977) and tight junctions (Lane, 1981a), have also been found by morphological studies in the same general area of the eye but their physiological significance has not yet been elucidated in anything remotely approaching a satisfactory way. There are technical problems in even attempting to determine the function of some of these junctions and so their role remains speculative. Yet it seems very likely, given their morphological distinctions, albeit subtle ones, that they will all turn out to have a rather specific role related perhaps to their unique structure, to the embryological origin of the tissue in which they are found or to their particular position along the intercellular borders.

Morphological studies then, using a range of techniques, have produced a good deal of information as to the nature of cell-to-cell associations, which can even lead to the construction of models relating to the detailed arrangements of their intercellular membranous modifications. From these investigations, functional hypotheses can be put

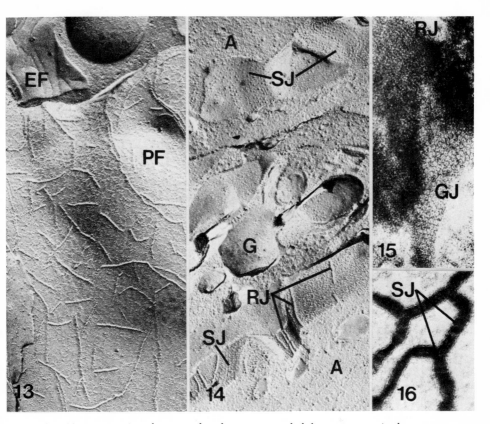

Figs. 13–16. Junctions between the photoreceptor rhabdomeres, or retinular axons (A), in the distal retina of the dipteran compound eye; glial cells (G) closely encompass these axons and some axo-glial junctions therefore also occur. Fig. 13 shows the retinular junctions with discontinuous PF ridges and EF grooves, often non-complementary, the ridges composed of fused, laterally aligned PF particles. Fig. 14 reveals the more proximal areas where the retinular junctions (RJ) give way to septate junctions (SJ) as seen by freeze-fractured or in tracer-incubated thin-sections (Fig. 16). In oblique, lanthanum-treated sections the intercellular columns of the retinular junctions (RJ) can be seen (Fig. 15), sometimes beside a gap junction (GJ). Fig. 13, × 35 000; Fig. 14, × 45 000; Fig. 15, × 126 000; Fig. 16, × 110 000.

forward, but, although preliminary studies require morphological investigations which may lay the foundation stones for many immensely important parameters of intercellular associations, ultimately physiological assessments of various kinds, as well as biochemical analyses, need to be pursued, in order to establish definitively the features and significance of the range of cell-to-cell interactions that exist in animal tissues.

Acknowledgements

I am deeply indebted to Mr William M. Lee for producing the photographic montages and to Mrs Vanessa Rule who prepared the typescript for this contribution.

REFERENCES

ABBOTT, N. J. (1979). Primitive forms of brain homeostasis. *Trends in Neurosciences*, **2**, 91–3.

ABBOTT, N. J., PICHON, Y. & LANE, N. J. (1977). Primitive forms of potassium homeostasis: observations on crustacean central nervous system with implications for vertebrate brain. *Experimental Eye Research*, **25** (suppl.), 259–71.

ABBOTT, N. J. & TREHERNE, J. E. (1977). Homeostasis of the brain microenvironment: a comparative account. In *Transport of Ions and Water in Animals*, ed. B. L. Gupta, R. B. Moreton, J. L. Oschman & B. J. Wall, pp. 481–510. London: Academic Press.

BRINK, P. R. & DEWEY, M. M. (1980). Evidence for fixed charge in the nexus. *Nature*, **285**, 101–2.

CARLSON, S. D. & CHI, C. (1979). The functional morphology of the insect photoreceptor. *Annual Review of Entomology*, **24**, 379–416.

CASPER, D. L. D., GOODENOUGH, D. A., MAKOWSKI, L. & PHILLIPS, W. C. (1977). Gap junction structures. 1. Correlated electron microscopy and X-ray diffraction. *Journal of Cell Biology*, **74**, 605–28.

CAVENEY, S. (1978). Intercellular communication in insect development is hormonally controlled. *Science*, **199**, 192–5.

CHI, C. & CARLSON, S. D. (1980a). Membrane specializations in the first optic neuropile of the housefly, *Musca domestica* L. I Junctions between neurons. *Journal of Neurocytology*, **9**, 429–49.

CHI, C. & CARLSON, S. D. (1980b). Membrane specializations in the first optic neuropile of the housefly, *Musca domestica* L. II Junctions between glial cells. *Journal of Neurocytology*, **9**, 451–69.

CHI, C. & CARLSON, S. D. (1981). Lanthanum and freeze-fracture studies on the retinular cell junction in the compound eye of the housefly. *Cell and Tissue Research*, **214**, 541–52.

CHI, C., CARLSON, S. D. & STE MARIE, R. (1979). Membrane specializations in the peripheral retina of the housefly, *Musca domestica* L. *Cell and Tissue Research*, **198**, 501–20.

CLAUDE, P. & GOODENOUGH, D. A. (1973). Fracture faces of *zonulae occludentes* from 'tight' and 'leaky' epithelia. *Journal of Cell Biology*, **58**, 390–400.

DALLAI, R. (1970). Glycoproteins in the *zonula continua* of the epithelium of the mid-gut in an insect. *Journal de Microscopie*, **9**, 277–80.

DALLAI, R. (1976). Septate and continuous junctions associated in the same epithelium. *Journal of Submicroscopic Cytology*, **8**, 163–74.

ELIAS, P. M. & FRIEND, D. S. (1976). Vitamin-A-induced mucous metaplasia. An in vitro system for modulating tight and gap junction differentiation. *Journal of Cell Biology*, **68**, 173–88.

FAIN-MAUREL, M-A. & CASSIER, P. (1972). Une nouveau type de jonctions: les jonctions scalariformes. Étude ultrastructurale et cytochimique. *Journal of Ultrastructure Research*, **39**, 222–38.

FILSHIE, B. K. & FLOWER, N. E. (1977). Junctional structures in *Hydra*. *Journal of Cell Science*, **23**, 151–72.

FLAGG-NEWTON, J. L. & LOEWENSTEIN, W. R. (1980). Asymmetrically permeable membrane channels in cell junction. *Science*, **207**, 771–3.

FLOWER, N. E. (1972). A new junctional structure in the epithelia of insects of the order Dictyoptera. *Journal of Cell Science*, **10**, 683–91.

FLOWER, N. E. & FILSHIE, B. K. (1975). Junctional structures in the midgut cells of lepidopteran caterpillars. *Journal of Cell Science*, **17**, 221–39.

FLOWER, N. E. & WALKER, G. D. (1979). Rectal papillae in *Musca domestica*: the cuticle and lateral membranes. *Journal of Cell Science*, **39**, 167–86.

GILULA, N. B. (1978). Structure of intercellular junctions. In *Intercellular Junctions and Synapses. Receptors and Recognition*, series B, vol. 2, ed. J. Feldman, N. B. Gilula & J. D. Pitts. pp. 3–22. London: Chapman and Hall.

GILULA, N. B., BRANTON, D. & SATIR, P. (1970). The septate junction: a structural basis for intercellular coupling. *Proceedings of the National Academy of Sciences, USA*, **67**, 213–20.

GILULA, N. B., REEVES, O. R. & STEINBACH, A. (1972). Metabolic coupling, ionic coupling and cell contacts. *Nature*, **235**, 262–5.

GREEN, C. R. (1978). Variations of septate junction structure in the invertebrates. In *Electron Microscopy 1978*, vol. 2, ed. J. M. Sturgess. Proceedings of the 9th International Congress on Electron Microscopy, pp. 338–9. Canada: The Imperial Press Ltd.

GREEN, C. R., BERGQUIST, P. R. & BULLIVANT, S. (1979). An anastomosing septate junction in endothelial cells of the phylum *Echinodermata*. *Journal of Ultrastructure Research*, **68**, 72–80.

GREEN, L. F. B., BERGQUIST, P. R. & BULLIVANT, S. (1980). The structure and function of the smooth septate junction in a transporting epithelium: the Malpighian tubules of the New Zealand glow-worm *Arachnocampa luminosa*. *Tissue and Cell*, **12**, 365–81.

HANNA, R. B., KEETER, J. S. & PAPPAS, G. D. (1978). The fine structure of a rectifying electrotonic synapse. *Journal of Cell Biology*, **79**, 764–73.

HEUSER, J. E., REESE, T. S., DENNIS, M. J., JAN, Y., JAN, L. & EVANS, L. (1979). Synaptic vesicle exocytosis captured by quick freezing and correlated with quantal transmitter release. *Journal of Cell Biology*, **81**, 275–300.

HOUK, E. J. (1977). Midgut ultrastructure of *Culex tarsalis* (Diptera: Culcidae) before and after a blood meal. *Tissue and Cell*, **9**, 103–18.

KENSLER, R. W., BRINK, P. R. & DEWEY, M. M. (1979). The septum of the lateral axon of the earthworm: a thin section and freeze-fracture study. *Journal of Neurocytology*, **8**, 565–90.

LANE, N. J. (1978a). Intercellular junctions and cell contacts in invertebrates. In *Electron Microscopy 1978. Vol. 3. State of the Art. Symposia*, ed. J. M. Sturgess. Proceedings of the 9th International Congress on Electron Microscopy, pp. 673–91. Canada: The Imperial Press Ltd.

LANE, N. J. (1978b). Developmental stages in the formation of inverted gap junctions during turnover in the adult horseshoe crab, *Limulus. Journal of Cell Science*, **32**, 293–305.

LANE, N. J. (1979a). Intramembranous particles in the form of bracelets or assemblies in arthropod tissues. *Tissue and Cell*, **11**, 1–18.

LANE, N. J. (1979b). Freeze-fracture and tracer studies on the intercellular junctions of insect rectal tissues. *Tissue and Cell*, **11**, 481–506.

LANE, N. J. (1979c). A new kind of tight junction-like structure in insect tissues. *Journal of Cell Biology*, **83**, 82A.

LANE, N. J. (1980). Stages in the development of co-existing tight junctional P face ridges and gap junctional E face plaques: two distinct intramembranous junctional particle populations. *Journal of Cell Biology*, **87**, 198A.

LANE, N. J. (1981a). Vertebrate-like tight junctions in the insect eye. *Experimental Cell Research*, **132**, 482–8.

LANE, N. J. (1981b). Evidence for two separate categories of junctional particle during the concurrent formation of tight and gap junctions. *Journal of Ultrastructure Research*, (in press).

LANE, N. J. (1981c). Tight junctions in arthropod tissues. *International Review of Cytology*, **73**, 243–318.

LANE, N. J. (1981d). Invertebrate neuroglia: junctional structure and development. In *Glial-Neuronal Interactions. Journal of Experimental Biology*, (in press).

LANE, N. J. & CHANDLER, H. J. (1980). Definitive evidence for the existence of tight junctions in invertebrates. *Journal of Cell Biology*, **86**, 765–74.

LANE, N. J. & HARRISON, J. B. (1978). An unusual type of continuous junction in *Limulus. Journal of Ultrastructure Research*, **64**, 85–97.

LANE, N. J. & HARRISON, J. B. (1980). An unusual form of tight junction in the nervous system of the scorpion. *European Journal of Cell Biology*, **22**, 244.

LANE, N. J., HARRISON, J. B. & BOWERMAN, R. F. (1981). A vertebrate-like blood-brain barrier, with intraganglionic blood channels and occluding junctions, in the scorpion. *Tissue and Cell*, (in press).

LANE, N. J., SKAER, H. LE B. & SWALES, L. S. (1977). Intercellular junctions in the central nervous system of insects. *Journal of Cell Science*, **26**, 175–99.

LANE, N. J. & SKAER, H. LE B. (1980). Intercellular junctions in insect tissues. *Advances in Insect Physiology*, **15**, 35–213.

LANE, N. J. & SWALES, L. S. (1978a). Changes in the blood-brain barrier of the central nervous system in the blowfly during development, with special reference to the formation and disaggregation of gap and tight junctions. I Larval development. *Developmental Biology*, **62**, 389–414.

LANE, N. J. & SWALES, L. S. (1978b). Changes in the blood-brain barrier of the central nervous system in the blowfly during development with special reference to the formation and disaggregation of gap and tight junctions. II Pupal development and adult flies. *Developmental Biology*, **62**, 415–31.

LANE, N. J. & SWALES, L. S. (1979). Intercellular junctions and the development of the blood-brain barrier in *Manduca sexta. Brain Research*, **169**, 227–45.

LANE, N. J. & SWALES, L. S. (1980). Dispersal of gap junctional particles, not internalization, during *in vivo* disappearance of gap junctions. *Cell*, **19**, 579–86.

LANE, N. J., SWALES, L. S. & LEE, W. M. (1980). Junctional dispersal and reaggregation: IMP reutilization? *Cell Biology International Reports*, **4**, 738.

LARSEN, W. J. (1977). Structural diversity of gap junctions. A review. *Tissue and Cell*, **9**, 373–94.

LOEWENSTEIN, W. R. & KANNO, Y. (1964). Studies on an epithelial (gland) cell junction. 1. Modifications of surface membrane permeability. *Journal of Cell Biology*, **22**, 565–86.

LOEWENSTEIN, W. R., KANNO, Y. & SOCOLAR, S. J. (1978). The cell-to-cell channel. *Federation Proceedings*, **37**, 2645–50.

MAKOWSKI, L., CASPAR, D. L. D., PHILLIPS, W. C. & GOODENOUGH, D. A. (1977). Gap junction structures. II Analysis of the x-ray diffraction data. *Journal of Cell Biology*, **74**, 629–45.

MILLS, J. W., LORD, B. A. P. & DI BONA, D. R. (1976). Osmotic sensitivity of septate junctions in the crayfish midgut. *Journal of Cell Biology*, **70**, 327A.

MONTESANO, R., FRIEND, D. S., PERRELET, A. & ORCI, L. (1975). *In vivo* assembly of tight junctions in fetal rat liver. *Journal of Cell Biology*, **67**, 310–19.

MÜLLER, M., MEISTER, N. & MOOR, H. (1980). Freezing in a propane jet and its application in freeze-fracturing. *Mikroskopie*, (*Wien*), **36**, 129–40.

NOIROT, C. & NOIROT-TIMOTHÉE, C. (1967). Un nouveau type de jonction inter-cellulaire (*zonula continua*) dans l'intestin moyen des insectes. *Compte rendu hebdomadaire des séances de l'Académie des sciences*, **264**, 2796–8.

NOIROT, C. & NOIROT-TIMOTHÉE, C. (1976). Fine structure of the rectum in cock-roaches (*Dictyoptera*): General organization and intercellular junctions. *Tissue and Cell*, **8**, 345–68.

NOIROT-TIMOTHÉE, C. & NOIROT, C. (1980). Septate and scalariform junctions in arthropods. *International Review of Cytology*, **63**, 97–140.

NOIROT-TIMOTHÉE, C., NOIROT, C., SMITH, D. S. & CAYER, M. L. (1979). Jonc-tions et contacts intercellulaires chez les insectes. II Jonctions scalariformes et complexes formés avec les mitochondries. Étude par coupes fines et cryofracture. *Biologie Cellulaire*, **34**, 127–36.

NOIROT-TIMOTHÉE, C., SMITH, D. S., CAYER, M. L. and NOIROT, C. (1978). Septate junctions in insects: comparison between intercellular and intramembranous structures. *Tissue and Cell*, **10**, 125–36.

PERACCHIA, C. (1973a). Low resistance junctions in crayfish. I. Two arrays of globules in junctional membranes. *Journal of Cell Biology*, **57**, 54–65.

PERACCHIA, C. (1973b). Low resistance junctions in crayfish. II Structural details and further evidence for intercellular channels by freeze-fracture and negative staining. *Journal of Cell Biology*, **57**, 66–76.

PERACCHIA, C. (1977). Gap junctions. Structural changes after uncoupling pro-cedure. *Journal of Cell Biology*, **72**, 628–64.

PERACCHIA, C. (1978). Calcium effects on gap junction structure and cell coupling. *Nature*, **271**, 669–71.

PERACCHIA, C. (1980). Structural correlates of gap junction permeation. *Interna-tional Review of Cytology*, **66**, 81–146.

PERACCHIA, C. & DULHUNTY, A. F. (1976). Low resistance junctions in crayfish. Structural changes with functional uncoupling. *Journal of Cell Biology*, **70**, 419–39.

PINTO DA SILVA, P. & GILULA, N. B. (1972). Gap functions in normal and trans-formed fibroblasts in culture. *Experimental Cell Research*, **71**, 393–401.

PINTO DA SILVA, P. & MARTINÉZ-PALOMO, A. (1975). Distribution of membrane particles and gap junctions in normal and transformed 3T3 cells studied *in situ*, in suspension, and treated with concanavalin A. *Proceedings of the National Academy of Sciences, USA*, **72**, 572–6.

PORVAZNIK, M., JOHNSON, R. G. & SHERIDAN, J. D. (1979). Tight junction development between cultured hepatoma cells: Possible stages in assembly and enhancement with dexamethasone. *Journal of Supramolecular Structure*, **10**, 13–30.

PSCHEID, P., SCHUDT, C. & PLATTNER, H. (1981). Cryofixation of monolayer cell cultures for freeze-fracturing without chemical pre-treatments. *Journal of Microscopy*, **121**, 149–67.

RADOJCIC, T. & PENTREATH, V. W. (1979). Invertebrate glia. *Progress in Neurobiology*, **12**, 115–79.

RAVIOLA, E., GOODENOUGH, D. A. & RAVIOLA, G. (1980). Structure of rapidly frozen gap junctions. *Journal of Cell Biology*, **87**, 273–9.

REVEL, J. P., YEE, A. G. & HUDSPETH, A. J. (1971). Gap junctions between electrotonically coupled cells in tissue culture and in brown fat. *Proceedings of the National Academy of Sciences, USA*, **68**, 2924–7.

RIBI, W. A. (1977). Fine structure of the first optic ganglion (lamina) of the cockroach, *Periplaneta americana*. *Tissue and Cell*, **9**, 57–72.

ROSE, B. (1971). Intercellular communication and some structural aspects of membrane junctions in a simple cell system. *Journal of Membrane Biology*, **5**, 1–19.

ROSE, B., SIMPSON, I. & LOEWENSTEIN, W. R. (1977). Calcium ion produces graded changes in permeability of membrane channels in cell junction. *Nature*, **267**, 625–7.

SATIR, P. & GILULA, N. B. (1973). The fine structure of membranes and intercellular communication in insects. *Annual Review of Entomology*, **18**, 143–660.

SHINOWARA, N. J., BEUTEL, W. B. & REVEL, J. P. (1980). Comparative analysis of junctions in the myelin sheath of central and peripheral axons of fish, amphibians and mammals: a freeze-fracture study using complementary replicas. *Journal of Neurocytology*, **9**, 15–38.

SKAER, H. LE B., HARRISON, J. B. & LEE, W. M. (1979). Topographical variations in the structure of the smooth septate junction. *Journal of Cell Science*, **37**, 373–89.

SKAER, H. LE B., LANE, N. J. & LEE, W.M. (1980). Junctional specializations of the digestive system in a range of arthropods. *European Journal of Cell Biology*, **22**, 245.

STAEHELIN, L. A. (1974). Structure and function of intercellular junctions. *International Review of Cytology*, **39**, 191–283.

TOSHIMORI, K., IWASHITA, T. & ŌURA, C. (1979). Cell junctions in the cyst envelope in the silkworm testis, *Bombyx mori* Linné. *Cell and Tissue Research*, **202**, 63–73.

TREHERNE, J. E. & PICHON, Y. (1972). The insect blood-brain barrier. In *Advances in Insect Physiology*, vol. 9, ed. J. E. Treherne, M. J. Berridge & V. B. Wigglesworth, pp. 257–313. London: Academic Press.

TREHERNE, J. E. & SCHOFIELD, P. K. (1981). Mechanisms of ionic homeostasis in an insect CNS. In *Glial Neuronal Interactions*, ed. J. E. Treherne. *Journal of Experimental Biology*, (in press).

UNWIN, P. N. T. & ZAMPIGHI, G. (1980). Structure of the junction between communicating cells. *Nature*, **283**, 545–9.

WILLIAMS, C. M. (1952). Physiology of insect diapause. 4. The brain and prothoracic glands as an endocrine system in the cecropia silkworm. *Biological Bulletin*, **103**, 120–38.

WOLPERT, L. (1977). Introductory remarks: Cell-to-cell interactions. In *International Cell Biology*, ed. B. R. Brinkley & K. R. Porter, pp. 31–5. New York: Rockefeller University Press.

WOOD, R. L. (1959). Intercellular attachment in the epithelium of *Hydra* as revealed by electron microscopy. *Journal of Biophysical and Biochemical Cytology*, **6**, 343–52.

WOOD, R. L. (1977). The cell junctions of *Hydra* as viewed by freeze-fracture replication. *Journal of Ultrastructure Research*, **58**, 299–315.

WOOD, R. L. & KUDA, A. M. (1980). Formation of junctions in regenerating *Hydra*: septate junctions. *Journal of Ultrastructure Research*, **70**, 104–17.

WYNOGRAD, O. G. & NICOLAS, G. (1980). Intercellular junctions in the adrenal medulla: a comparative freeze-fracture study. *Tissue and Cell*, **12**, 661–72.

YANCY, S. D., MEYER, D. J. & REVEL, J. P. (1980). Intercellular communication in regenerating rat liver. *European Journal of Cell Biology*, **22**, 246.

ZAMPIGHI, G., RAMÓN, F. & DURÁN, W. (1978). Fine structure of the electrotonic synapse of the lateral giant axons in a crayfish (*Procambarus clarkii*). *Tissue and Cell*, **10**, 413–26.

Functional organization of cells in exocrine gland acini

O. H. PETERSEN,* I. FINDLAY,* M. DAOUD AND
R. C. COLLINS

Department of Physiology, The University, Dundee DD1 4HN, UK

INTRODUCTION

Mammalian exocrine glands, like the pancreas, salivary glands and the lacrimal gland, are organized in functional units consisting of hundreds of cells linked together by gap junctional channels. Under resting conditions intercellular communication is extremely efficient as measured by cell-to-cell transfer of fluorescent probes or by spread of electrotonic potentials (Petersen, 1980). The pancreatic acinar tissue provides good opportunities for studying acute, marked and reversible changes in cell-to-cell communication (Iwatsuki & Petersen, 1978). Exposing an isolated tissue to solutions equilibrated with high CO_2 concentrations is a suitable way of uncoupling the communication network (Turin & Warner, 1977, 1980; Iwatsuki & Petersen, 1979). This method can be applied to study the functional consequences of interrupting intercellular communication.

CO_2-EVOKED BLOCK OF ACINAR CELL-TO-CELL COMMUNICATION

It is well documented that an elevation of P_{CO_2} causes a decrease in intracellular pH (pH_i) and that exposure of tissues to NH_3 can cause a transient increase in pH_i (Caldwell, 1958; Thomas, 1974; Boron & De Weer, 1976) (Fig. 1). Turin & Warner (1977, 1980) first demonstrated that high CO_2 levels decreased ionic communication between cells in *Xenopus* embryos. In this tissue, at least, the intracellular acidification was not accompanied by any noticeable change in the intracellular ionized Ca concentration ($[Ca^{2+}]_i$) (Rink, Tsien & Warner, 1980).

Iwatsuki & Petersen (1979) first described CO_2-evoked reversible

*Present address: The Physiological Laboratory, University of Liverpool, PO Box 147, Liverpool L69 3BX.

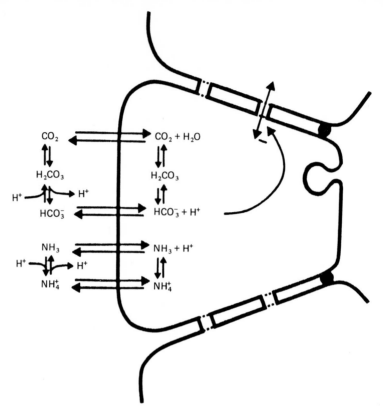

Fig. 1. Schematic diagram showing how changes in extracellular CO_2 and NH_3 concentration can influence pH_i. (From Petersen *et al.*, 1980.)

uncoupling in adult mouse pancreas and exorbital lacrimal glands. A number of controls were carried out ruling out effects of anoxia and extracellular pH. Small uncoupling effects caused by exposing tissues to solutions equilibrated with as little as 10% CO_2 could be demonstrated. We have recently shown that the uncoupling effect of 50% CO_2 can be partly counteracted by exposure to a solution containing 10 mM NH_4Cl (Findlay, Daoud, Collins & Petersen, in preparation).

THE EFFECT OF CO_2 ON PANCREATIC Ca-TRANSPORT

Lea & Ashley (1978) showed that CO_2 caused a marked increase in skeletal muscle $[Ca^{2+}]_i$ presumably due to H^+ displacement of Ca at intracellular binding sites. Since the permeability of gap junctional channels is widely believed to be regulated by changes in $[Ca^{2+}]_i$

(Loewenstein & Rose, 1978) it seemed important to investigate possible effects of CO_2 on Ca metabolism in the pancreatic acinar tissue. Fig. 2 shows the effect of briefly exposing the tissue to a Krebs solution equilibrated with 50% CO_2 (as compared to the normal 5% CO_2) on fractional ^{45}Ca efflux. This effect is compared with the well-known acetylcholine-evoked increase in ^{45}Ca efflux. CO_2 clearly evokes a very marked increase in ^{45}Ca release. This effect is not due to anoxia nor to the inevitable change in extracellular pH. In fact exposure to a solution equilibrated with 20% CO_2 with an elevated bicarbonate concentration, maintaining the normal control pH of 7.4, still caused a massive increase in ^{45}Ca outflux. This enhanced ^{45}Ca release was effectively reduced by exposure to 10 mM NH_4Cl (pH = 7.4) (Findlay et al., in preparation). Intracellular acidification therefore causes a dramatic change in cellular Ca metabolism. It seems likely that this effect is due to $H^+ - Ca^{2+}$ competition for the same intracellular binding sites. A decrease in pH_i would therefore release Ca perhaps mainly from the very extensive endoplasmic reticulum. This would result in an elevation of $[Ca^{2+}]_i$. In the pancreatic acinar tissue evidence is available showing that intracellular Ca application electrically isolates the injected cell from its neighbours (Iwatsuki & Petersen, 1977; Petersen & Iwatsuki, 1978; Petersen, 1980). The finding that CO_2 evokes ^{45}Ca-release might therefore suggest that Ca is the ultimate mediator of the control of gap junctional permeability also in situations where intracellular acidification is the primary signal. However, the data do not rule out the possibility that H^+ acts directly on the junctions and it is possible, although not yet demonstrated in the case of the pancreatic acinar cells, that a primary increase in $[Ca^{2+}]_i$ causes a decrease in pH_i. This type of interaction has been shown to occur in snail neurones (Meech & Thomas, 1977). One could therefore also postulate that Ca-evoked uncoupling is mediated by H^+.

THE EFFECT OF CO_2 ON PANCREATIC AMYLASE SECRETION

With regard to electrically non-excitable cells there is still considerable uncertainty about the functional implications of changes in cell-to-cell communication (Petersen, 1980). The major function of pancreatic acinar cells is to secrete digestive enzymes in response to appropriate stimuli and it is therefore important to monitor secretion during modulation of intercellular communication. Fig. 3a shows the effect of

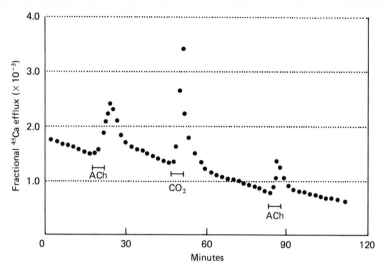

Fig. 2. The effect of acetylcholine (ACh) $(5 \times 10^{-7}$ M) and CO_2 (superfusion solution equilibrated with 50% CO_2, 50% O_2) on fractional ^{45}Ca efflux from isolated fragments of mouse pancreas.

an increase in P_{CO_2} on unstimulated amylase secretion from isolated superfused mouse pancreatic fragments. There is a marked reduction in basal secretion which is entirely reversible. An even larger and steeper CO_2-evoked depression of secretion is observed during sustained ACh-evoked secretion (Fig. 3b). During a period of sustained exposure to high CO_2 levels (equilibration with 20 or 50% CO_2) ACh evokes a much smaller increase in amylase output than under control conditions.

CONCLUSION

While the findings presented here make it clear that the CO_2-evoked cellular uncoupling of pancreatic acini is associated with decreased enzyme secretion we do not know the cause of this inhibition. It might be due to the intracellular acidification inhibiting the exocytosis process or due to an inhibiting effect of an excessively elevated $[Ca^{2+}]_i$ or to some effect caused by the closure of gap junctional channels. At this stage we lack experimental data allowing discrimination between these possibilities. There are, however, some recent interesting findings that perhaps can give us some useful hints. The exocytosis process appears not to be particularly sensitive to changes in pH$_i$ (Baker & Knight, 1980). Isolated pancreatic acinar cells secrete poorly in response to

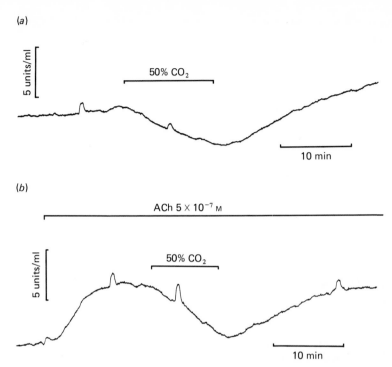

Fig. 3. The effect of exposing isolated mouse pancreatic fragments to a Krebs solution equilibrated with 50% CO_2, 50% O_2 on (a) basal amylase secretion and (b) amylase secretion in response to 5×10^{-7} M ACh.

hormonal stimulation while isolated acini secrete very much better (Peikin *et al.*, 1978). The dose–response curve for hormone-evoked (e.g. CCK or ACh) amylase secretion in isolated acini is biphasic with maximal secretion occurring at a secretagogue concentration very close to that causing a just measurable decrease in interacinar electrical communication, while higher concentrations decrease the secretory rate (Peikin *et al.*, 1978; Iwatsuki & Petersen, 1978; Petersen & Iwatsuki, 1979). Interestingly the stimulant-evoked increase in ^{45}Ca outflux does not exhibit this biphasic behaviour (Jensen, Lemp & Gardner, 1980). All these results, taken together with our own recent findings, would be compatible with the following simple hypothesis. Both enzyme secretion and cell-to-cell coupling are controlled by $[Ca^{2+}]_i$. Within the lower range of $[Ca^{2+}]_i$, increasing $[Ca^{2+}]_i$ stimulates exocytosis. Within a higher range, increasing $[Ca^{2+}]_i$ inhibits stimulated enzyme secretion and causes closure of gap junctional channels. The effects of changing pH_i on both secretion and intercellular communication is mainly mediated by changes in $[Ca^{2+}]_i$. No doubt this hypothesis is an over-simplification, but it is at least a framework for further experimentation.

Acknowledgement

Supported by a grant from the Medical Research Council (UK).

REFERENCES

BAKER, P. F. & KNIGHT, D. E. (1980). Gaining access to the site of exocytosis in bovine adrenal medullary cells. *Journal de Physiologie*, **76**, 497–504.

BORON, W. F. & DE WEER, P. (1976). Intracellular pH transients in squid giant axons caused by CO_2, NH_3 and metabolic inhibitors. *Journal of General Physiology*, **67**, 91–112.

CALDWELL, P. C. (1958). Studies on the internal pH of large muscle and nerve fibres. *Journal of Physiology*, **142**, 22–62.

IWATSUKI, N. & PETERSEN, O. H. (1977). Acetylcholine-like effects of intracellular calcium application in pancreatic acinar cells. *Nature*, **268**, 147–9.

IWATSUKI, N. & PETERSEN, O. H. (1978). Electrical coupling and uncoupling of exocrine acinar cells. *Journal of Cell Biology*, **79**, 533–45.

IWATSUKI, N. & PETERSEN, O. H. (1979). Pancreatic acinar cells: the effects of carbon dioxide, ammonium chloride and acetylcholine on intercellular communication. *Journal of Physiology*, **291**, 317–26.

JENSEN, R. T., LEMP, G. F. & GARDNER, J. D. (1980). Interaction of cholecystokinin with specific membrane receptors on pancreatic acinar cells. *Proceedings of the National Academy of Sciences, USA*, **77**, 2079–83.

LEA, T. J. & ASHLEY, C. C. (1978). Increase in free Ca^{2+} in muscle after exposure to CO_2. *Nature*, **275**, 236–8.

LOEWENSTEIN, W. R. & ROSE, B. (1978). Calcium in (junctional) intercellular communication and a thought on its behaviour in intracellular communication. *Annals of the New York Academy of Sciences*, **307**, 285–307.

MEECH, R. W. & THOMAS, R. C. (1977). The effect of calcium injection on the intracellular sodium and pH of snail neurones. *Journal of Physiology*, **265**, 867–79.

PEIKIN, S. R., ROTTMAN, A. J., BATZRI, S. & GARDNER, J. D. (1978). Kinetics of amylase release by dispersed acini prepared from guinea-pig pancreas. *American Journal of Physiology*, **235**, E743–E749.

PETERSEN, O. H. (1980). *The Electrophysiology of Gland Cells* (Monographs of the Physiological Society no. 36) pp. 1–253. London, New York: Academic Press.

PETERSEN, O. H., FINDLAY, I., MEDA, P., LAUGIER, R. & IWATSUKI, N. (1980). Control of cell to cell communication in exocrine glands by the intracellular hydrogen ion concentration. In *Hydrogen Ion Transport in Epithelia*, ed. I. Schulz, pp. 227–34. Amsterdam: Elsevier/North-Holland.

PETERSEN, O. H. & IWATSUKI, N. (1978). The role of calcium in pancreatic acinar cell stimulus-secretion coupling: an electrophysiological approach. *Annals of the New York Academy of Sciences*, **307**, 599–617.

PETERSEN, O. H. & IWATSUKI, N. (1979). Hormonal control of cell to cell coupling in the exocrine pancreas. In *Hormone Receptors in Digestion and Nutrition*, ed. G. Rosselin, P. Fromageot & S. Bonfils, pp. 191–202. Amsterdam: Elsevier/North-Holland.

RINK, T. J., TSIEN, R.-Y. & WARNER, A. E. (1980). Free calcium in *Xenopus* embryos measured with ion-selective microelectrodes. *Nature*, **283**, 658–60.

THOMAS, R. C. (1974). Intracellular pH of snail neurones measured with a new pH-sensitive glass microelectrode. *Journal of Physiology*, **238**, 159–80.

TURIN, L. & WARNER, A. E. (1977). Carbon dioxide reversibly abolishes ionic communication between cells of early amphibian embryo. *Nature*, **270**, 56–7.

TURIN, L. & WARNER, A. E. (1980). Intracellular pH in early *Xenopus* embryos: its effect on current flow between blastomeres. *Journal of Physiology*, **300**, 489–504.

Endocrine cell interactions in the islets of Langerhans

PAOLO MEDA, ALAIN PERRELET AND LELIO ORCI

Institute of Histology and Embryology, University of Geneva, Medical School, Geneva, Switzerland

INTRODUCTION

The primary role of the islets of Langerhans is to maintain blood glucose concentration within a narrow range in which the fuel requirements of the central nervous system are fulfilled and an excessive, probably deleterious, glycosylation of proteins is prevented. To achieve this, the islets precisely regulate throughout the day the mixture of hormones they secrete to compensate for the variations in the supply and/or utilization of glucose and other fuels due to meals and starvation, exercise and rest (Unger, 1981). This regulation implies that the different types of endocrine islet cells must coordinate their level of activity in relation to need. Such an integrated activity is certainly influenced by hormones and neurotransmitters reaching the islets from outside (Unger, Dobbs & Orci, 1978). However, several lines of morphological and physiological evidence suggest that the total islet output is also controlled by direct communications between homologous and heterologous endocrine cells within the islets.

The purpose of this contribution is to summarize these lines of evidence and to discuss the possible role of junctional-mediated coupling in the regulation of islet cell interactions.

STRUCTURAL INTERRELATIONSHIPS OF ISLET CELLS

In several species, studies using the indirect immunofluorescence staining technique, have revealed that the islets of Langerhans are composed of at least four different types of endocrine cells, each containing a separate hormone (Larsson, Sundler & Håkanson, 1976). The numerical proportions and the topographical relationships of these cells are characteristic for a given species. For example, in rodents, the insulin-producing B-cells represent 60–80% of the endocrine population and the somatostatin-containing D-cells represent 2–5% of the islet cells (Orci *et al.*, 1976c). The remaining cells are glucagon-containing A-cells and pancreatic polypeptide-containing PP-cells. The proportions of the two latter cell types differ in the islets located in the body or tail and in the paraduodenal region of the rat pancreas (Orci *et al.*, 1976b). In the tail, A- and PP-cells represent 15 and 0.5% of the islet cells, respectively, whereas in the paraduodenal region the proportions are reversed, A-cells representing 1% and PP-cells 14% of the islet cells (Baetens, Malaisse-Lagae, Perrelet & Orci, 1979) (Fig. 1). Morphological studies have also shown that the endocrine cells are not randomly distributed within an islet. In numerous species, B-cells are located in the centre of the islet whereas the A-, D- and PP-cells are located at its periphery (Fig. 1). As a result of this arrangement, each islet consists of a homocellular region where B-cells are surrounded almost exclusively by other B-cells and a heterocellular region where the four main types of islet cells are mixed (Orci & Unger, 1975b). The heterocellular region is also the site where afferent nerves and vessels enter the islet (Fujita, Yanatori & Murakami, 1976).

Throughout the islet, endocrine cells directly interact via specialized domains of their plasma membrane. At these sites of contact, islet cells are linked by desmosomes (Lacy & Greider, 1972), tight and gap junctions (Orci, Unger & Renold, 1973) (Figs. 2–4).

FUNCTIONAL INTERACTIONS OF ISLET CELLS

The precise arrangement and the close structural relationships of homologous and heterologous islet cells suggest that these cells might directly interact during islet functioning. This view is supported by the findings that insulin, glucagon, somatostatin and pancreatic polypeptide, at concentrations thought to be in the physiological range, can directly modify the secretion of at least one of the other islet hormones. Thus glucagon stimulates both insulin (Samols, Marri & Marks, 1965) and somatostatin release (Patton *et al.*, 1977) while somatostatin inhibits both insulin (Alberti *et al.*, 1973) and glucagon secretion (Koerker *et al.*, 1974). Insulin inhibits glucagon secretion (Samols, Tyler & Marks, 1972) while its effects on somatostatin secretion are still unclear (Samols & Harrison, 1976; Honey & Weir, 1979). Pancreatic polypeptide has been reported to inhibit somatostatin (Arimura *et al.*, 1979) and insulin release (Lundquist *et al.*, 1979). In addition, insulin

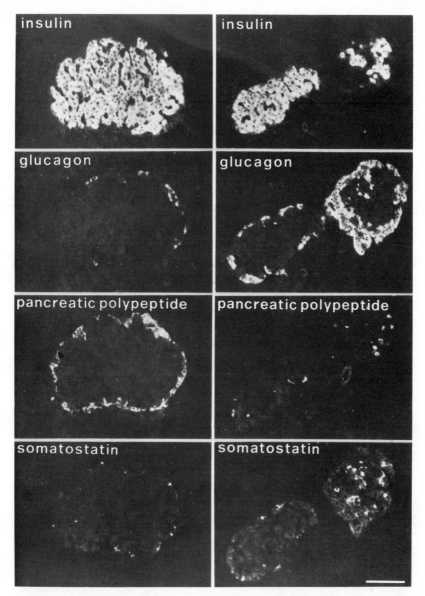

Fig. 1. Staining of four consecutive sections of control rat pancreatic islets following indirect immunofluorescence with anti-insulin, anti-glucagon, anti-pancreatic polypeptide and anti-somatostatin sera, respectively. The characteristic pattern of endocrine cells found in the body and tail of pancreas is shown in the two islets of the right panel while that found in the lower part of the pancreatic head is shown in the islet of the left panel. Note the reverse proportions of glucagon- and pancreatic polypeptide-containing cells at the periphery of the two islet types. The bar represents 100 μm.

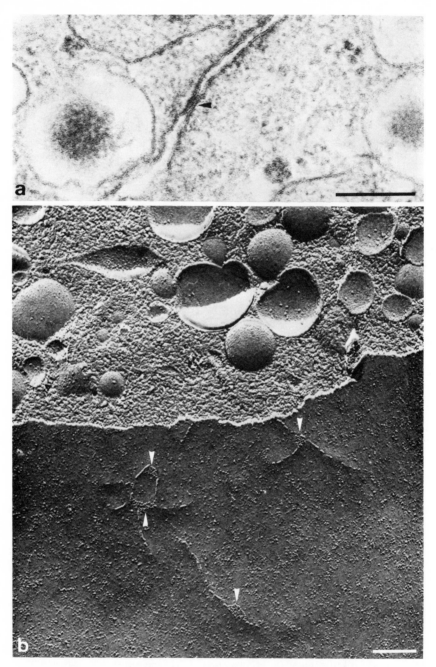

Fig. 2. (a) Thin section of the periphery of two adjacent B-cells showing the apposition
of membranes and the marked reduction of extracellular space at the site of a small gap
junction (arrow head). The bar represents 0.2 μm.

(Pace, Matchinsky, Lacy & Conant, 1977) and somatostatin (Ipp *et al.*, 1979) appear able to inhibit their own release by acting directly on B- and D-cells, respectively.

Studies with exogenous peroxidase have shown the ready access of the protein to the extracellular space of the islet, which appears as a continuous compartment surrounding all the endocrine cells (Like, 1970). It is therefore possible to hypothesize that the above summarized effects may also be elicited by endogenously secreted islet hormones which may diffuse through the extracellular space and interact with specific surface sites of islet cells. Whether such a paracrine effect (Feyrter, 1953) actually does occur within islets is still uncertain because we do not know the actual physiological concentration of hormones in the extracellular space of the islet nor whether these hormones have a free access to the receptor domains of islet cell membranes. It has been suggested that the focal tight junctions which connect islet cells and change under conditions known to modify insulin release (Orci, 1976a) may control such accessibility by opening or sealing off specific membrane compartments in relation to need (Orci, 1976a).

The identification within the islets of gap junctions (Orci, Unger & Renold, 1973), the structures believed to mediate ionic and metabolic cell-to-cell coupling (Bennet & Goodenough, 1978 and this book), has led to the additional hypothesis that islet cell interactions may also be controlled by direct exchanges of ions and low molecular weight molecules between adjacent islet cells.

JUNCTIONAL-MEDIATED COUPLING OF ISLET CELLS

The first presumptive evidence for the existence of islet cell coupling was the detection of gap junctions within the islet. Conventional (Fig. 2a) and freeze-fracture electron microscopy (Fig. 2b) have shown that gap junctions occur between both homologous (Orci, Unger & Renold, 1973) and heterologous differentiated endocrine cells (Orci *et al.*, 1975a).

A detailed quantitative analysis of gap junctions on B-cell mem-

(b) Freeze-fracture replica exposing both the plasma membrane (P face) and the cytoplasm of a B-cell from a control resting islet. Gap junctions (arrow heads) appear as small aggregates of particles associated with tight junction fibrils. Numerous secretory granules are seen in the cytoplasm. The bar represents 0.2 μm.

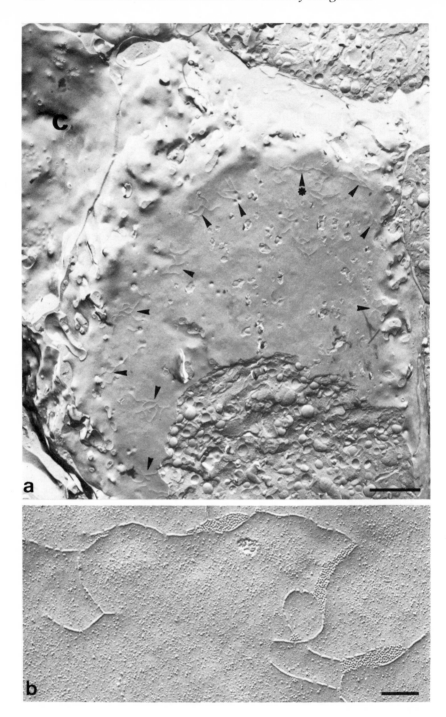

branes has been done. Under resting conditions, B-cell gap junctions (Fig. 2b) appear quite small compared to those observed in other secretory systems, their total area of $0.08-0.20 \mu m^2$ per cell representing only $0.02-0.04\%$ of the B-cell surface (Meda, Perrelet & Orci, 1979; Meda *et al.*, 1980b). Gap junctions are not randomly distributed among all B-cells of an islet nor on the membrane of individual B-cells (Meda, Denef, Perrelet & Orci, 1980a). The junctions are twice as frequent between the B-cells located at the periphery as in the centre of rat islets and they occur in small clusters often in association with tight junctions (Figs. 2b–4) on restricted areas of B-cell membranes. As seen in freeze-fracture replicas, junctional clusters form on flat portions of the 'lateral' membrane, at the limit of the B-cell pole facing the capillaries (Unger, Dobbs & Orci, 1978) (Fig. 3).

Most of the gap junctions found between B-cells are formed by particles showing a loose and irregular arrangement which varies from one junction to another and even within the same junction (Fig. 4). However, particles showing a close hexagonal arrangement can also be found (Fig. 4). It is believed that the loose irregular arrangement characterizes permeable junctions whereas the tight crystalline arrangement characterizes junctions which have been in a high-resistance state for a long time (Raviola, Goodenough & Raviola, 1980). Close packing of gap junction particles is induced by a prolonged (30–60 min), but not by a short (2–5 min), exposure of isolated islets to experimental conditions known to rapidly cause cell-to-cell uncoupling (Meda, Perrelet & Orci, 1980c).

The permeability of junctional channels connecting B-cells has been directly demonstrated by more functional approaches. Electrophysiological studies have shown that synchronized electrical activity can be

Fig. 3. (a) Freeze-fracture replica of a glucose-stimulated islet exposing the membrane and the granulated cytoplasm of a B-cell and part of an adjacent capillary (C). The arrow heads point to clusters of gap and tight junctions. Note that these clusters form a nearly continuous belt between a flat surface of the B-cell membrane and a more irregular surface of the cell which faces the capillary and shows cross-fractured microvilli. This suggests that junctional clusters may occur at precise sites on the membranes, possibly at the limit of the secretory pole of the B-cell. The arrow head with an asterisk points to a junctional cluster which is shown at higher magnification in Fig. 3b. The bar represents 2 μm.

(b) Higher magnification view of the cluster indicated by the arrow head with an asterisk in Fig. 3a, illustrating the close association of gap and tight junctions. The bar represents 0.2 μm.

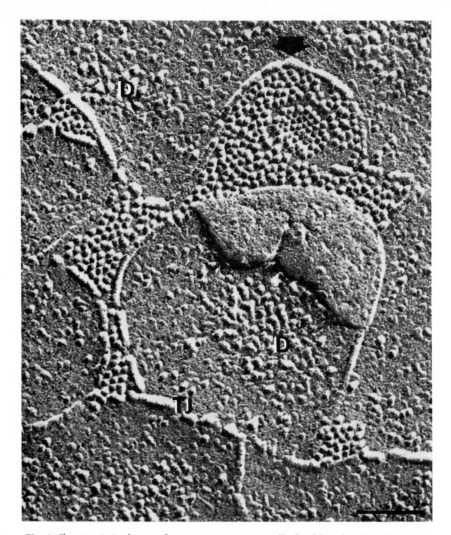

Fig. 4. Characteristic cluster of gap junctions on a B-cell of a glibenclamide-stimulated (2 days) islet. The packing of gap junction particles varies from one junction to another. The arrow points to a junction where loosely and irregularly arranged particles surround a small circular area formed by more regularly and tightly packed particles. This may reflect a change in the junction permeability. As is usual between B-cells, the gap junctions of this cluster are closely associated with tight junctions (TJ) and desmosomes (D). The bar represents 0.1 μm.

recorded in neighbouring B-cells of microdissected mouse islets (Meissner, 1976), and more recently, current injection experiments have shown that this synchronization is due to electrical coupling which probably occurs through gap junctions (Eddlestone & Rojas, 1980).

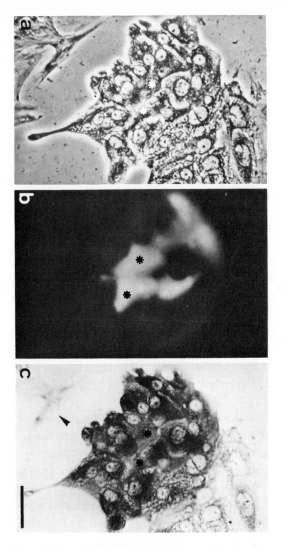

Fig. 5. (a) Two islet cells (asterisks) of this cluster in monolayer culture were successively injected with carboxyfluorescein.

(b) After both injections, the dye was rapidly transferred into some of the neighbouring islet cells and delineated two adjacent territories of communicating cells.

(c) At the end of the experiment, the cluster was fixed, incubated with an anti-
insulin serum and stained by the peroxidase–anti-peroxidase method. Dark peroxidase reaction product stains most of the cells forming the cluster, including those of
the communicating territories. Thus communication occurred between differentiated
B-cells. The arrow head points to a fibroblast which was not stained by the immuno-
peroxidase reaction. The bar represents 50 μm.

Furthermore, microinjection experiments have demonstrated that 6-carboxyfluorescein, a low molecular weight fluorescent probe to which the cell membrane is impermeable, is rapidly and directly transferred between rat islet cells in monolayer culture (Kohen *et al.*, 1979) (Figs. 5 and 6). Successive microinjections into different cells of the same cluster showed that the fluorescent probe usually did not spread into all the cells of the cluster but rather delineated communicating territories (Fig. 5) which usually comprised only two to eight cells. In several cases, the coexistence of communicating and non-communicating cells was also observed within the same cluster (Fig. 6). Additionally, islet cells *in vitro* appear able to directly exchange endogenous molecules of possible physiological significance such as glucose-6-phosphate or one of its metabolites, following microinjection (Kohen *et al.*, 1979), or ^{3}H-uridine nucleotides (Figs. 7 and 8) (Meda, Amherdt, Perrelet & Orci, 1981), following the incorporation by the cells of a ^{3}H-precursor according to the method of Pitts & Simms (1977). The observation that the impermeant molecules used were transferred only between cells in close contact (Figs. 7–9) and showed a decreasing gradient of concentration running from the loaded cells through the communicating and the non-communicating unloaded cells (Figs. 5, 7 and 8) indicated that intercellular communication did occur through permeable junctions, thus establishing that islet cells in culture are metabolically coupled.

In both dye and metabolite transfer experiments, coupling was not detected in every instance in which an islet cell loaded with an impermeant molecule contacted an unloaded islet cell (Fig. 6), suggesting that the efficiency of molecular transfer may be lower between islet cells as compared to that observed in other systems (Pitts, 1976). This lower efficiency may be due to the limited area and non-homogeneous distribution of islet cell gap junctions as well as to the occurrence of communications between similar and different islet cell types. Heterocellular communications are less readily established than homocellular ones (Pitts, 1976; Fentiman, Taylor-Papadimitriou & Stoker, 1976). Immunohistochemical (Fig. 5) and ultrastructural analyses (Figs. 7b–d) have shown that direct communications occur, in culture, between B-cells (Kohen *et al.*, 1981) and also between B- and non-B-cells (Meda & Kohen, submitted). Homologous and heterologous communications also occur in whole isolated islets microinjected with the impermeant dye Lucifer Yellow (R. Michaels and J. D. Sheridan, personal communication).

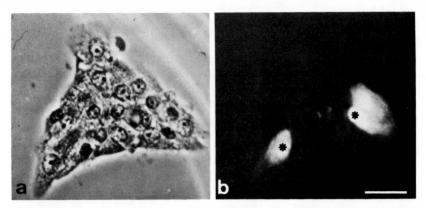

Fig. 6. (a) In this cluster, carboxyfluorescein was successively injected into two different islet cells (asterisks).

(b) After one injection (left cell), the dye failed to move into neighbouring cells while, after the second injection (right cell) it readily spread into two coupled cells. Thus, communicating and non-communicating islet cells coexisted in this cluster. The bar represents 50 μm.

Microinjections of clusters shown in Figs. 5 and 6 were performed by Drs E. and C. Kohen at the Papanicolaou Cancer Research Institute, Miami, Florida.

B-CELL COUPLING AND INSULIN SECRETION

If coupling plays a role in insulin secretion, the ability of B-cells to communicate may vary depending on their functional state. To test this idea, gap junctions were quantitatively evaluated following a short (90 min) stimulation by glucose *in vitro* and a prolonged (2 days) stimulation by glibenclamide *in vivo*. Significant changes in the frequency and size of B-cell gap junctions were observed after both stimulations (Meda, Perrelet & Orci, 1979) (compare also Fig. 2 to Figs. 3 and 4), suggesting that junctional changes generally participate in the response of B-cells to a secretagogue. If so, one would expect to observe different junctional changes depending on the efficiency and duration of each stimulation. Indeed, the relative and absolute areas occupied by gap junctions between B-cells showed a 50% increase after glucose stimulation but a larger increase after the more prolonged and intense stimulation induced by glibenclamide (Meda, Perrelet & Orci, 1979). Further experiments testing the glibenclamide challenge for various periods of time showed that the increase in gap junction area was inversely correlated with the insulin content of B-cells (Meda *et al.*, 1980b) (Fig. 9), a finding which suggests that the development of gap junctions is modulated according to the functional state of the cells. It is

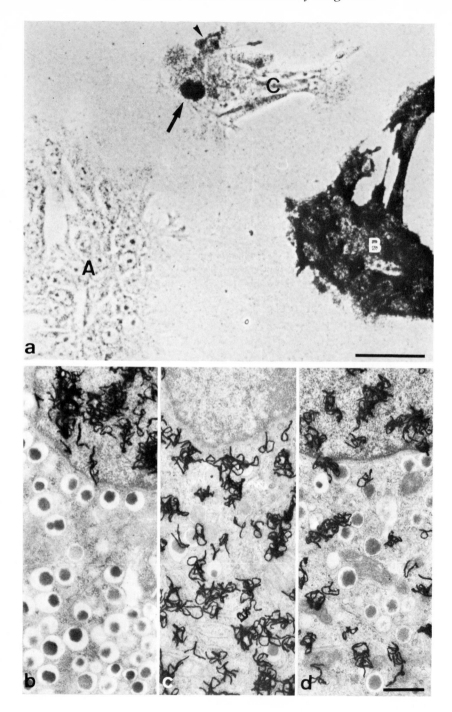

still uncertain which step(s) in the complex process of insulin secretion is associated with junctional changes. Experiments aimed at blocking the permeability of the plasma membrane to K^+, thus mimicking one of the early events in the response of B-cells to glucose (Henquin, 1979), indicated that gap junctions change with B-cell functioning even if insulin is not released (Sheppard & Meda, 1981). It is thus possible that gap junctions and coupling may participate in the sequence of metabolic, ionic and mechanical events which precede the release of insulin (Malaisse, Sener, Herchuelz & Hutton, 1979).

In view of the direct proportionality which probably exists between the area and the conductance of gap junctions (Sheridan, Hammer-Wilson, Preus & Johnson, 1978), the morphological data suggest that the ability to communicate may be enhanced between stimulated B-cells. The physiological information available is consistent with that view. The voltage deflection measured in a communicating B-cell, following the injection of a constant current pulse into a neighbouring B-cell, increases with the glucose concentration of the incubation medium and decreases upon glucose removal, indicating that junctional resistance falls between stimulated B-cells (Eddlestone & Rojas, 1980). This change is a prerequisite for the improvement of electrical coupling.

Fig. 7. (a) Light microscopic autoradiograph of a co-culture used to follow the transfer of endogenous ^{3}H-nucleotides from ^{3}H-uridine-(donor) to ^{3}H-thymidine-labelled (recipient) islet cells. Three separate clusters of endocrine cells are seen in this field observed directly through the culture dish. Cluster A comprises thymidine-unlabelled recipient cells which are separated from all donor cells and whose cytoplasmic labelling represents the cellular background. By comparison, the cytoplasm of donor cells forming cluster B is much more heavily labelled. Cluster C comprises a donor cell (arrow head) and several recipient cells, of which one (arrow) shows a thymidine-labelled nucleus. The cytoplasmic labelling of the latter cell which is intermediate between that of donors and cellular background indicates that this recipient cell has incorporated uridine nucleotides. The virtually unlabelled cytoplasm of recipients of cluster A shows that such incorporation requires a contact between a donor and a recipient cell, thus establishing that islet cells in culture are metabolically coupled. The bar represents 50 μm.

(b–d) Electron microscope autoradiographs of donor and recipient B-cells similar to those seen in Fig. 7a. In the recipient cell (b), numerous autoradiographic grains are present over the nucleus while the cytoplasm is unlabelled. By contrast, in the donor cell (c), the nucleus is unlabelled and the cytoplasm is heavily labelled. Fig. 7d shows a recipient cell which contacted donor cells in co-culture. The intermediate cytoplasmic labelling indicates that this B-cell was involved in the exchange of labelled nucleotides with these donors. Note the presence of typical beta secretory granules in the cytoplasm of the three B-cells. The bar represents 1 μm.

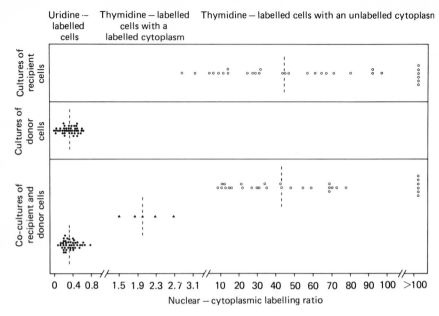

Fig. 8. Scatter diagram showing the values of nuclear–cytoplasmic labelling ratio evaluated on electron microscopic autoradiographs similar to those of Figs. 7b–d. The dotted lines represent the median value of each group. The presence of a cell population restricted to the co-cultures and characterized by a labelling ratio intermediate between that of donor and recipient cells is apparent. This population consists of recipient cells, similar to those shown in Figs. 7a (arrow) and 7d, which were involved in direct exchanges of nucleotides with adjacent donors.

Metabolic coupling may also be somewhat improved following B-cell stimulation, as judged by the increased number of cells transferring carboxyfluorescein *in vitro* in the presence of a stimulatory concentration of glucose (Kohen *et al.*, 1979).

Insulin secretagogues also increase the proportion of fractured B-cell membranes bearing gap junctions (Meda, Denef, Perrelet & Orci, 1980a) (Fig. 10) and change the topography of communicating islet cells *in vitro* (Kohen *et al.*, 1979). The two latter observations suggest that stimulation of insulin secretion is also associated with a modification of the pattern of the communicating territories.

The possible function of coupling in B-cell secretion is still unknown. It has been shown, however, that dispersed adult rat islet cells are insensitive to glucose, while still responding to other secretagogues, but they partially recover responsiveness to glucose after a 120-minute period of reaggregation (Halban, Wollheim, Blondel &

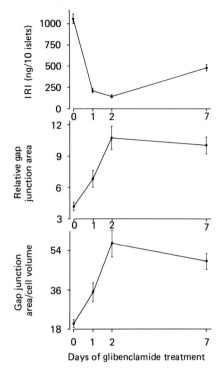

Fig. 9. Islets of Langerhans were isolated from adult rats treated with the insulin secretagogue glibenclamide (0.2 mg/100 g b.w., i.p., twice daily) for 1,2 or 7 days. A portion of the islets was used for the determination of insulin content and the rest were processed for freeze-fracture and quantitative analysis of gap junctions. Glibenclamide treatment resulted in a marked depletion of islet insulin content (upper panel) and in a significant increase in the relative gap junction area (expressed, in the middle panel, as 10^{-2} × the percentage of B-cell membrane area occupied by gap junctions) and in the ratio of gap junction area to B-cell volume (expressed, in the lower panel, as $10^{-5}/\mu$m). Using a linear regression analysis, a significant ($p < 0.05$) inverse correlation was found between the individual values of the two latter parameters characterizing gap junctions and the insulin content of the corresponding islets. This indicates that gap junction development is modulated according to the functional state of the cells they connect. All values are expressed as mean ± SEM.

Renold, 1979). One possible explanation for this is that direct intercellular communication is necessary for the normal secretory behaviour of B-cells in response to glucose. Functional asynchrony probably exists between B-cells as indicated by the different degree of granularity (Orci, 1974), the different threshold for stimulated electrical activity (Beigelman, Ribalet & Atwater, 1977) and the different response to a glucose challenge (Kolod, Meda, Perrelet & Orci, 1981)

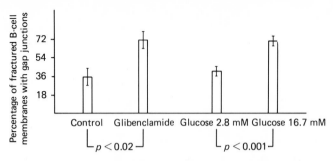

Fig. 10. In freshly isolated control islets and in islets incubated in the presence of a non-stimulatory concentration (2.8 mM) of glucose, 40% of the B-cell membrane exposed by freeze-fracture contained one or more gap junctions. This proportion was nearly doubled after stimulation of insulin secretion by glibenclamide *in vivo* or by glucose *in vitro*. Thus, previously undifferentiated areas of B-cell membrane become available for junction formation when insulin secretion is enhanced, suggesting that stimulated B-cells form permeable junctions with more neighbouring cells.

shown by different B-cells within an islet. It is thus conceivable that, at any given time, only some B-cells within an islet may respond to glucose (Meissner, 1976) and that cell coupling is required to ensure the spread of stimulation (Lawrence, Beers & Gilula, 1978; Kater & Galvin, 1978) and to coordinate the secretory response of the whole endocrine population. If this hypothesis can be substantiated, it will be of interest to assess whether a perturbation of cell coupling participates in the selective insensitivity of B-cells to glucose which has been observed in Type II (non-insulin dependent) diabetes (Cerasi, Luft & Efendic, 1971) and in an animal model which is believed to represent a rodent counterpart of this disease (Bonner-Weir, Trent, Honey & Weir, 1981).

Clearly, further work is required to understand the possible inter-relation between the control of cell coupling and secretion within the islets of Langerhans. Nevertheless, the data discussed above provide evidence in favour of the hypothesis that regulation of islet cell function occurs not only at the level of the secretory apparatus of individual endocrine cells but also through direct communications between homologous and heterologous islet cell types.

Acknowledgements

The authors thank Drs S. Bonner-Weir, D. Brown and G. Weir for critical reading of the manuscript and D. Wey, P.-A. Ruttiman and I. Bernard for technical assistance. This work was supported by grant n° 3.668.80 from the Swiss National Science Foundation and by a grant-in-aid from Hoechst AG, Frankfurt, W. Germany.

REFERENCES

ALBERTI, K., CHRISTENSEN, N. J., HANSEN, A.-A. P., LUNDBAECK, J., LUND-BAECK, K., IVERSEN, J., SEYER HAUSEN, K., ØRSKOV, H. (1973). Inhibition of insulin secretion by somatostatin. *Lancet*, 2, 1299–301.

ARIMURA, A., MEYERS, C. A., CASE, W. L., MURPHY, W. A. & SCHALLY, A. V. (1979). Suppression of somatostatin levels in the hepatic portal and systemic plasma of the rat by synthetic human pancreatic polypeptide. *Biochemical and Biophysical Research Communications*, 89, 913–18.

BAETENS, D., MALAISSE-LAGAE, F., PERRELET, A. & ORCI, L. (1979). Endocrine pancreas: three-dimensional reconstruction shows two types of islets of Langerhans. *Science*, 206, 1323–5.

BEIGELMAN, P. M., RIBALET, B. & ATWATER, (1977). Electrical activity of mouse pancreatic beta-cells. *Journal de Physiologie*, 73, 201–17.

BENNET, M. V. L. & GOODENOUGH, D. A. (1978). Gap junctions, electrotonic coupling, and intercellular communication. *Neurosciences Research Program Bulletin*, 16, 377–486.

BONNER-WEIR, S., TRENT, D. F., HONEY, R. N. & WEIR, G. C. (1981). Responses of neonatal rat islets to streptozotocin: limited B cell regeneration and hyperglycemia. *Diabetes*, 30, 64–9.

CERASI, E., LUFT, R. & EFENDIC, S. (1971). Decreased sensitivity of the pancreatic beta cells to glucose in pre-diabetic and diabetic subjects. A glucose dose-response study. *Diabetes*, 21, 224–34.

EDDLESTONE, G. T. & ROJAS, E. (1980). Evidence of electrical coupling between mouse pancreatic B-cells. *Journal of Physiology*, 303, 76P–77P.

FENTIMAN, I., TAYLOR-PAPADIMITRIOU, J. & STOKER, M. (1976). Selective-contact dependent cell communication. *Nature*, 264, 760–2.

FEYRTER, F. (1953). *Über die peripheren endokrinen (parakrinen) Drüsen des Menschen*, 2nd edn. Wien: Maudrich.

FUJITA, T., YANATORI, Y. & MURAKAMI, T. (1976). Insulo-acinar axis, its vascular basis and its functional and morphological changes caused by CCK-PZ and caerulein. In *Endocrine Gut and Pancreas*, ed. T. Fujita, pp. 347–57. Amsterdam: Elsevier.

HALBAN, P., WOLLHEIM, C. B., BLONDEL, B. & RENOLD, A. E. (1979). Evidence for the importance of contact between pancreatic islet cells in glucose-induced insulin release. *Diabetes*, 28 (suppl.), 394.

HENQUIN, J.-C. (1979). Opposite effects of intracellular Ca^{2+} and glucose on K^+ permeability of pancreatic islets. *Nature*, 280, 66–8.

HONEY, R. N. & WEIR, G. C. (1979). Insulin stimulates somatostatin and inhibits glucagon secretion from the perfused chicken pancreas-duodenum. *Life Sciences*, 4, 1747–50.

IPP, E., RIVIER, J., DOBBS, R. E., BROWN, M., VALE, W. & UNGER, R. H. (1979). Somatostatin analogs inhibit somatostatin release. *Endocrinology*, 104, 1270–3.

KATER, S. B. & GALVIN, N. J. (1978). Physiological and morphological evidence for coupling in mouse salivary gland acinar cells. *Journal of Cell Biology*, 79, 20–6.

KOERKER, D. J., RUCH, W., CHIDECKEL, E., PALMER, J., GOODNER, C. J., ENSINCK, J. & GALE, C. C. (1974). Somatostatin: hypothalamic inhibitor of the endocrine pancreas. *Science*, 184, 482–3.

KOHEN, E., KOHEN, C., THORELL, B., MINTZ, D. H. & RABINOVITCH, A. (1979). Intercellular communication in pancreatic islet monolayer cultures: a micro-fluorometric study. *Science*, **204**, 862–5.

KOHEN, E., THORELL, B., HIRSCHBERG, J. G., WOUTERS, A. W., KOHEN, C., BARTICK, P., SHELLEY, J. D., SALMON, J.-M., VIALLET, P., SCHACHTSCHABEL, D., MEDA, P., NESTOR, J. & PLOEM, J. S. (1981). Microspectrofluorometric procedures and their application in biological systems. In *Modern Fluorescence Spectroscopy*, ed. Wehry. New York: Plenum Press, (in press).

KOLOD, E., MEDA, P., PERRELET, A. & ORCI, L. (1981). Influence of intra-islet environment on B-cell function. *Experientia*, **31**, 650.

LACY, P. E. & GREIDER, M. H. (1972). Ultrastructural organization of mammalian pancreatic islets. In *Handbook of Physiology*, section 7, vol. 1, ed. D. F. Steiner & N. Freinkel, pp. 77–89. Washington: American Physiological Society.

LARSSON, L.-I., SUNDLER, F. & HÅKANSON, R. (1976). Pancreatic polypeptide – A postulated new hormone: identification of its cellular storage site by light and electron microscopic immunocytochemistry. *Diabetologia*, **12**, 211–26.

LAWRENCE, T. S., BEERS, W. H. & GILULA, N. B. (1978). Transmission of hormonal stimulation by cell-to-cell communication. *Nature*, **272**, 501–6.

LIKE, A. A. (1970). The uptake of exogenous peroxidase by the beta cells of the islets of Langerhans. *American Journal of Pathology*, **59**, 225–46.

LUNDQUIST, I., SUNDLER, F., AHREN, B., ALUMETS, J. & HÅKANSON, R. (1979). Somatostatin, pancreatic polypeptide, substance P and neurotensin: cellular distribution and effects on stimulated insulin secretion in the mouse. *Endocrinology*, **104**, 832–8.

MALAISSE, W. J., SENER, A., HERCHUELZ, A. & HUTTON, J. C. (1979). Insulin release: the fuel hypothesis. *Metabolism*, **28**, 373–86.

MEDA, P., AMHERDT, M., PERRELET, A. & ORCI, L. (1981). Metabolic coupling between cultured pancreatic B-cells. *Experimental Cell Research*, **133**, 421–30.

MEDA, P., DENEF, J.-F., PERRELET, A. & ORCI, L. (1980a). Nonrandom distribution of gap junctions between pancreatic B-cells. *American Journal of Physiology*, **238**, C114–19.

MEDA, P., HALBAN, P., PERRELET, A., RENOLD, A. E. & ORCI, L. (1980b). Gap junction development is correlated with insulin content in the pancreatic B cell. *Science*, **209**, 1026–8.

MEDA, P., PERRELET, A. & ORCI, L. (1979). Increase of gap junctions between pancreatic B-cells during stimulation of insulin secretion. *Journal of Cell Biology*, **82**, 441–8.

MEDA, P., PERRELET, A. & ORCI, L. (1980c). Gap junctions and B-cell function. In *Biochemistry and Biophysics of the Pancreatic B-cell*, ed. W. J. Malaisse & I.-B. Täljedal, pp. 157–62. Stuttgart: Georg Thieme Verlag.

MEISSNER, H. P. (1976). Electrophysiological evidence for coupling between β cells of pancreatic islets. *Nature*, **262**, 502–4.

ORCI, L. (1974). A portrait of the pancreatic B-cell. *Diabetologia*, **10**, 163–87.

ORCI, L. (1976a). The microanatomy of the islets of Langerhans. *Metabolism*, **25** (suppl. 1), 1303–13.

ORCI, L., BAETENS, D., RAVAZZOLA, M., STEFAN, Y. & MALAISSE-LAGAE, F. (1976b). Pancreatic polypeptide and glucagon: non-random distribution in pancreatic islets. *Life Sciences*, **19**, 1811–16.

ORCI, L., BAETENS, D., RUFENER, C., AMHERDT, M., RAVAZZOLA, M., STUDER, P., MALAISSE-LAGAE, F. & UNGER, R. H. (1976c). Hypertrophy and hyperplasia of somatostatin-containing D-cells in diabetes. *Proceedings of the National Academy of Sciences, USA*, 73, 1338–42.

ORCI, L., MALAISSE-LAGAE, F., RAVAZZOLA, M., ROUILLER, D., RENOLD, A. E., PERRELET, A. & UNGER, R. (1975a). A morphological basis for intercellular communication between α and β-cells in the endocrine pancreas. *Journal of Clinical Investigation*, 56, 1066–70.

ORCI, L. & UNGER, R. H. (1975b). Functional subdivision of islets of Langerhans and possible role of D-cells. *Lancet*, 2, 1243–4.

ORCI, L., UNGER, R. H. & RENOLD, A. E. (1973). Structural coupling between pancreatic islet cells. *Experientia*, 29, 1015–18.

PACE, C. S., MATCHINSKY, F. M., LACY, P. E. & CONANT, S. (1977). Electrophysiological evidence for the autoregulation of B-cell secretion by insulin. *Biochimica et Biophysica Acta*, 497, 408–14.

PATTON, G. S., IPP, E., DOBBS, R. E., ORCI, L., VALE, V. & UNGER, R. H. (1977). Pancreatic immunoreactive somatostatin release. *Proceedings of the National Academy of Sciences, USA*, 74, 2140–3.

PITTS, J. D. (1976). Junctions as channels of direct communication between cells. In *Developmental Biology of Plants and Animals*, ed. C. F. Graham & P. E. Wareing, pp. 96–110. Oxford: Blackwell Scientific Publications.

PITTS, J. D. & SIMMS, J. W. (1977). Permeability of junctions between animal cells. Intercellular transfer of nucleotides but not of macromolecules. *Experimental Cell Research*, 104, 153–63.

RAVIOLA, E., GOODENOUGH, D. A. & RAVIOLA, G. (1980). Structure of rapidly frozen gap junctions. *Journal of Cell Biology*, 87, 273–9.

SAMOLS, E. & HARRISON, J. (1976). Intraislet negative insulin-glucagon feedback. *Metabolism*, 25 (suppl. 1), 1443–7.

SAMOLS, E., MARRI, G. & MARKS, V. (1965). Promotion of insulin secretion by glucagon. *Lancet*, 2, 415–16.

SAMOLS, E., TYLER, J. M. & MARKS, V. (1972). Glucagon-insulin interrelationships. In *Glucagon: Molecular Physiology, Clinical and Therapeutic Implications*, ed. P. J. Lefebvre & R. H. Unger, pp. 151–73. Oxford: Pergamon Press.

SHEPPARD, M. S. & MEDA, P. (1981). Tetraethylammonium modifies gap junctions between pancreatic B-cells. *American Journal of Physiology*, 240, c 116–20.

SHERIDAN, J. D., HAMMER-WILSON, M., PREUS, D. & JOHNSON, R. G. (1978). Quantitative analysis of low-resistance junctions between cultured cells and correlation with gap junctional areas. *Journal of Cell Biology*, 76, 532–44.

UNGER, R. H. (1981). The milieu interieur and the islets of Langerhans. *Diabetologia*, 20, 1–11.

UNGER, R. H., DOBBS, R. E. & ORCI, L. (1978). Insulin, glucagon, and somatostatin secretion in the regulation of metabolism. *Annual Review of Physiology*, 40, 307–43.

The functional integration of cells in ovarian follicles

R. M. MOOR AND J. C. OSBORN

ARC Institute of Animal Physiology, 307 Huntingdon Road, Cambridge CB3 0JQ

INTRODUCTION

Ovarian follicles fulfil the dual physiological role of secreting steroid hormones and supporting the development of the female gamete. These functions are regulated both by pituitary hormones and by a series of signals generated from within the follicle. The importance of the intraovarian signals is evident from the early fetal stages when it can be demonstrated that interactions between the somatic and germinal elements transform the primitive gonad into the definitive ovary (Byskov, 1974; Peters, 1978). Cellular interactions in the fully differentiated ovary regulate the function of the antral follicles.

The organisation of cells within the follicle provides the mechanism by which the necessary instructional signals are transmitted between the different intrafollicular compartments. In this chapter we shall consider the cell interactions and underlying signals that specifically regulate steroid biosynthesis and oocyte function in antral follicles of mammals.

INTERACTIONS BETWEEN THE SOMATIC COMPONENTS IN FOLLICLES

The pattern of steroids secreted by antral follicles changes during development. Androstenedione and testosterone are the principal steroids produced by growing follicles, whereas progesterone is secreted in the final few hours before ovulation. However, it is only from a few fully differentiated follicles, and at selected stages in the reproductive cycle, that oestradiol-17β is secreted. This key follicular steroid is formed by the aromatisation of androgen and has traditionally been considered a product of the thecal cells (Short, 1964). This view is now disputed since oestrogen production is terminated when the theca and granulosa are physically separated (Makris & Ryan, 1975; Moor, 1977);

a finding which supports the results but not the conclusions of the pioneering study of Falck (1959). The results presented in Fig. 2 show that androgen, produced by the thecal cells, crosses the basal lamina (Fig. 1) and is aromatised by the granulosa cells (see Richards, 1980 for references). The passage of androgen between the co-operating cells is probably by diffusion since the basal lamina prevents direct inter-cellular transfer between the two cell layers. The importance of a fully differentiated granulosa layer for oestrogen synthesis is clear from the studies of Makris & Ryan (1975). They combined granulosa and theca cells from different sized follicles and showed that the aromatase system does not appear until an advanced stage of development. Granulosa cells reach this advanced stage of differentiation under the direction of the thecal cells which together with Follicle Stimulating Hormone (FSH) induce Luteinising Hormone (LH) receptor develop-ment in the granulosa (Nimrod, Tsafriri & Lindner, 1977). The result-ing increase in cellular responsiveness stimulates further oestrogen secretion within the follicle (Richards, 1980). The combined action of high levels of oestrogen and the pituitary gonadotrophins increases the growth and maturation of the follicle and ensures its rupture at ovu-lation. The cell interactions regulating oestrogen synthesis are there-fore of central importance in directing follicular function in mammals.

The precise mechanism through which thecal signals influence the granulosa cells is uncertain but probably involves the adenyl cyclase system. We and others have found that LH substantially increases the content of cyclic AMP in both the theca and granulosa cells of intact follicles (Marsh, 1976; Weiss, Seamark, McIntosh & Moor, 1976; Weiss, Armstrong, McIntosh & Seamark, 1978). However, when the two cell layers of small follicles are separated and then stimulated with LH, accumulation of cyclic AMP is restricted to the thecal layer (Fig. 3a). Since LH is without effect on cyclic AMP in granulosa cells the question arises as to how the concentration of this nucleotide increases in the granulosa compartment of the intact follicle. The direct inter-cellular transfer of cyclic AMP from theca to granulosa via permeable junctions is precluded by the interposition of the basal lamina between the two cell layers. However, a substantial proportion of the cyclic AMP synthesised by the thecal cells is secreted into the extracellular compartment (Fig. 3b) and probably reaches the granulosa by diffusion. The possibility that this nucleotide acts within the follicle as an ex-tracellular signal system is being investigated. It is worth recalling that in *Dictyostelium discoideum*, extracellular cyclic AMP binds to the

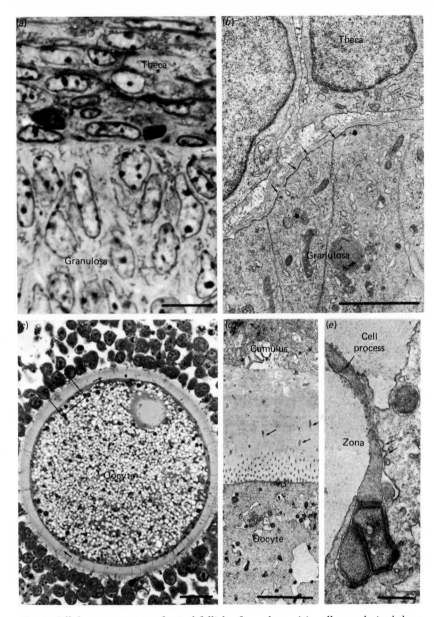

Fig. 1. Cellular components of antral follicles from sheep. (a) well vascularised theca and avascular granulosa layers (bar = 5 μm); (b) basal lamina (arrows) interposed between the theca and granulosa cells (bar = 5 μm); (c & d) oocyte and associated cumulus cells with cell processes (arrows) traversing the zona pellucida (bar = 20 & 5 μm for c & d respectively); (e) cumulus cell process with the end closely associated with oocyte membrane by a junctional complex. The arrows show coated pits (bar = 0.5 μm).

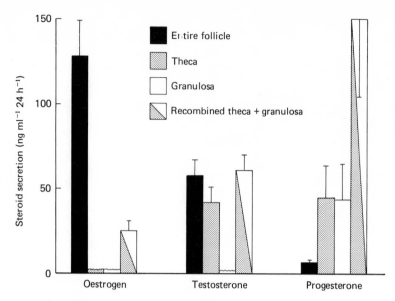

Fig. 2. Production of steroids (mean value ± SEM) by different cellular components of antral follicles cultured with FSH for 24 h. Note that testosterone is synthesised exclusively by the theca and that oestrogen production involves co-operation between theca and granulosa. (After Moor, 1977.)

membranes of target cells, activates the adenyl cyclase system and results in an increased production of intracellular cyclic AMP (see Gerisch *et al.*, 1977).

INTERACTIONS BETWEEN FOLLICLE CELLS AND OOCYTES

The oocyte exists in two distinct states while in the follicle. For most of its existence, which might extend over many years, the oocyte is arrested in the diplotene stage of meiotic prophase (the so-called 'germinal vesicle stage'). The second state, which lasts for only 12 to 48 hours, represents the period known as oocyte maturation during which the oocyte proceeds from the diplotene stage to the formation of the second metaphase plate where it is again arrested.

Cell interactions during the germinal vesicle stage

Permissive interactions in germinal vesicle oocytes. During oocyte growth, cytoplasmic processes from the cumulus cells penetrate the zona pellucida (Fig. 1c–e) and form junctional complexes, which include small gap junctions, with the oocyte membrane (Anderson &

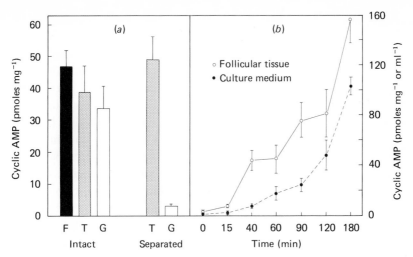

Fig. 3 (*a*) Content of cyclic AMP (mean ± SEM) in the intact follicle (F), theca (T) and granulosa (G) of small antral follicles. Luteinising hormone (LH) was added to intact follicles (left) or to the separated tissue components (right) 40 min before extraction and analysis.

(*b*) Levels of cyclic AMP in the follicular tissue and culture medium during the first 120 min after the addition of LH to antral follicles incubated *in vitro* (After Weiss *et al.*, 1976).

Albertini, 1976). Although gap junctions probably form the communicating channels (see Loewenstein, 1979) we shall use the non-specific term, permeable junctions, in our examination of the functional aspects of cell coupling. In addition, in the remainder of this chapter, we shall consider the cumulus and granulosa cells as a single coupled unit and will refer to the two cell types collectively as follicle cells. The individual components will only be identified when we have clear evidence that only one cell type is interacting with the oocyte.

The first evidence of coupling between the oocyte and its adjacent follicle cells came from the clear demonstration that the two are electrically coupled and that ionophoretically injected fluorescein dye is readily transferred between them (Gilula, Epstein & Beers, 1978). More recent findings have extended these observations in that they provide evidence for the transport of natural metabolites and for metabolic cooperation between the follicle cells and the oocyte (Moor, Smith & Dawson, 1980b). We have measured intercellular coupling between follicle cells and oocytes using intracellular markers derived from ^{3}H-labelled choline and uridine. Since these compounds are excluded by the oocyte membrane and can only enter the oocyte through junctions with follicle cells, their presence within the oocyte provides complementary evidence for the intercellular passage of compounds via permeable junctions. Measurements of intercellular coupling during maturation in sheep oocytes confirmed previous observations that communication decreases during the pre-ovulatory period (Gilula *et al.*, 1978) and showed that a decline in the entry of choline into ovine oocytes occurs between 12–15 hours after the initiation of maturation (Fig. 5*a*). Additional experiments indicate that certain steroids are also involved in the maintenance of intercellular communication between these two cell types and that FSH is responsible for the reduction in junctional transmission during maturation (Moor *et al.*, 1980b). It is of interest that the decline in coupling during maturation may not result from changes in the junctions but may instead be due to lesions within the cumulus cell processes (Moor & Cran, 1980).

Further evidence for junction-mediated co-operation between the oocyte and its immediately adjacent follicle cells has come from studies on ribonucleoside uptake by cumulus-enclosed and denuded oocytes (Heller & Schultz, 1980). Similar results from our laboratory are in complete accord with these investigations and show that the dependence of the oocyte upon metabolic co-operation with the follicle cells varies according to the nucleoside examined (Table 1). Thus uridine enters the oocyte in the presence but not in the absence of follicle cells whereas the uptake of adenosine is independent of the follicle cells.

From the foregoing discussion, it is clear that the oocyte and its associated follicle cells are linked by permeable junctions which provide a means of entry for those essential metabolic and trophic substances excluded by the oocyte membrane. These include energy substrates (Biggers, Whittingham & Donahue, 1967), precursors in the synthesis of phospholipids such as choline and inositol (Moor *et al.*, 1980b) and certain ribonucleosides (Heller & Schultz, 1980; Moor *et al.*, 1980b). The requirement for adequate nutritive support during growth probably accounts for the observation that contact between follicle cells and the oocyte is necessary both for the latter's growth and its acquisition of meiotic competence (Eppig, 1979; Bachvarova, Baran & Tejblum, 1980).

Regulation of actin synthesis. We have recently found that interactions between the somatic and germ cells are required to maintain physio-

Table 1. *The effect of cumulus cells on nucleoside uptake by single sheep oocytes after 1 h incubation* in vitro

³H-labelled compound	Cumulus cells	No. of oocytes	Mean ($\pm$ SEM) uptake (fmoles/oocyte)
Adenosine, 2.8µM	Present	59	64 $\pm$ 5
	Absent	101	64 $\pm$ 4
Guanosine, 4.5µM	Present	14	54 $\pm$ 10
	Absent	14	9.5 $\pm$ 0.8
Thymidine, 2.6µM	Present	19	2.5 $\pm$ 0.2[a]
	Absent	17	2.2 $\pm$ 0.1[a]
Uridine, 3µM	Present	35	55 $\pm$ 11
	Absent	15	0.45 $\pm$ 0.05[ab]

[a] Uptakes are effectively zero since uptakes of 2.2 fmoles/oocyte for uridine and thymidine can be accounted for by the size of the extracellular space ($= 0.72$ nl).
[b] Figures from Moor *et al.* (1980b).

logical levels of labelled actin within the oocyte (Fig. 4). Although the nature of this interaction is unclear, it is possible that actin is directly transferred from the cumulus cells to the oocyte. It is unlikely however, that permeable junctions would be the sites of transport since transmission is limited to molecules of less than 1000 daltons (Flagg-Newton, Simpson & Loewenstein, 1979). An alternative route of transport could involve absorptive endocytosis which, according to Pearse (1980), is probably the major mechanism for the uptake of macromolecules into cells. The selective uptake of yolk proteins into oocytes by endocytosis is well documented for a variety of non-mammalian species (see Giorgi, 1979) and it is known that high molecular weight proteins in the blood pass readily into the mammalian oocyte probably via the cumulus cell processes (Glass & Cons, 1968). The presence of coated vesicles in the oocyte membrane adjacent to the process (Fig. 1e) suggests that this may be a site for the selective transfer of proteins such as actin from the cumulus cells to the oocytes.

An alternative means of regulating actin synthesis could involve the direct control of synthetic activity within the oocyte by signals generated in the cumulus cells. This possibility is based on the hypothesis (detailed below) that the follicle cells play an instructional role in the suppression of meiotic activity in competent oocytes. However, the model is still highly speculative and implies a continuous need for

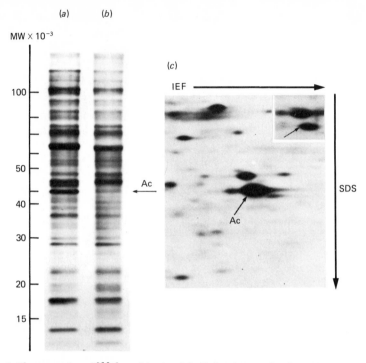

Fig. 4. Fluorographs of [^{35}S] methionine-labelled polypeptides from (*a* & *c*) oocytes labelled in the presence of cumulus cells and (*b*) oocytes labelled in the absence of cumulus cells. Oocytes were labelled for 3 h with 1 mCi/ml of [^{35}S] methionine and the polypeptides separated by one-dimensional electrophoresis on 8–15% gradient SDS polyacrylamide gels (*a*, *b*) and by two-dimensional electrophoresis (*c*) on SDS-gradient gels after isoelectric focussing. The insert in Fig. 4*c* denotes the actin region of the 2-D polypeptide pattern of denuded oocytes. In this figure, the position of actin after one- and two-dimensional electrophoresis is indicated by the letters Ac.

somatic signals since removal of the follicle cells immediately depresses actin synthesis in the oocyte. The existence of inductive stimuli would, however, provide an explanation both for the actin results and for other findings which show that the synthesis of various additional proteins in oocytes is altered by the removal of the cumulus cells (Crosby, Osborn & Moor, 1981).

Instructional interactions in germinal vesicle oocytes. It is now accepted that follicle cells are responsible for inhibiting meiosis and maintaining oocytes in the germinal vesicle stage (see Masui & Clarke, 1979). The basis for this realisation is that oocytes readily undergo meiosis when isolated from the follicle whereas oocytes retained within the follicle remain arrested at the germinal vesicle stage (Pincus & Enzmann, 1935). These authors proposed that 'the associated follicle cells serve either to maintain the egg in a nutritional state wherein nuclear maturation [meiosis] is impossible, or that they actually supply to the ovum a substance or substances which directly inhibit nuclear maturation'. Although Pincus and Enzmann recognised the possibility of alternative mechanisms of follicular control, most investigators in recent years have speculated about the involvement of a specific follicular factor which prevents oocytes from spontaneously resuming meiosis.

The first evidence for the presence of an inhibitory substance in the follicular fluid was provided by Chang (1955) and has subsequently been supported by certain workers (see Tsafriri, 1978) and disputed by others (see Leibfried & First, 1980). According to Tsafriri, Pomerantz & Channing (1976), this putative follicular fluid inhibitor is a small polypeptide (molecular weight $\leq$ 2000) whose activity is overcome by LH. The means by which this peptide could be transmitted to the oocyte and its possible action on the nucleus are not known. The size of the inhibitor molecule would prevent its passage through permeable junctions. On the other hand, the peptide is supposedly exported from the granulosa cells and its presence in the follicular fluid suggests movement by simple diffusion across the extracellular compartment. The inhibitor is apparently not able to affect the oocyte directly but instead requires the active participation of the cumulus cells (Hillensjo *et al.*, 1980). One possible action of the inhibitor could involve the suppression of selected transcriptional events within the oocyte which may be necessary for the resumption of meiosis (Rodman & Bachvarova, 1976). This suggestion is based upon the observation that follicular fluid contains a factor which specifically inhibits DNA-dependent RNA synthesis in both transformed and normal cells (Moore *et al.*, 1975; Bernard & Psychoyos, 1977).

An alternative hypothesis to that outlined above postulates that meiotic arrest in mammals is maintained by the passage of cyclic AMP from the follicle cells into the oocyte through permeable junction (Dekel & Beers, 1978; Gilula *et al.*, 1978). This idea arose from the many convincing studies showing that high intracellular levels of cyclic AMP suppress meiosis in amphibian oocytes (see Maller & Krebs, 1980) and is supported by the less convincing evidence that dibutyryl cyclic AMP and phosphodiesterase inhibitors prevent resumption of meiosis in mammalian oocytes cultured *in vitro* (Cho, Stern & Biggers, 1974;

Dekel & Beers, 1978, 1980). However, as with the putative peptide inhibitor, there is as yet no firm evidence for the involvement of cyclic AMP in the control of meiosis in mammalian oocytes, nor is there any evidence that the peptide inhibitor acts indirectly on the oocyte by stimulating the cumulus cells to produce cyclic AMP (Dekel & Beers, 1980).

Despite the confusion about the nature of the inhibitor factors, it is clear that contact between the follicle cells and oocyte is essential for the maintenance of the germinal vesicle stage in mammalian oocytes (Foote & Thibault, 1969). In addition to this instructional role, it should also be recalled that these intercellular contacts also provide the means by which important nutrients and metabolites enter the oocyte.

Cell interactions during oocyte maturation

Two processes are involved in the maturation of mammalian oocytes, namely the suppression of meiotic inhibitory factors and the generation of inductive stimuli from the follicle cells (Moor & Warnes, 1978). Considering the first of these processes, it is clear that the disagreement about the nature of the meiotic inhibitor outlined above has resulted in differing views about its suppression. On the one hand Channing and co-workers consider that meiosis is resumed when the concentration of the putative inhibitory polypeptide in the follicle decreases just before ovulation (Tsafriri & Channing, 1975). On the other hand it has been proposed that an abrupt fall in cyclic AMP within the oocyte provides the stimulus needed for the resumption of meiosis (Dekel & Beers, 1978). This intracellular fall in cyclic AMP is ascribed to a reduction in the entry of cyclic nucleotide from the follicle cells into the oocyte caused by the disruption of the intercellular communicating system. At present, we have no evidence relating to the first hypothesis. The second hypothesis, however, is not supported by our measurements of fluxes through the permeable junctions (Moor *et al.*, 1980b) nor is it supported by measurements of the content of cyclic AMP in oocytes during maturation (Moor & Heslop, 1981). Our results show that the rate of entry of molecules into oocytes via permeable junctions declines at a relatively late stage in maturation (Fig. 5*a*) and indeed can be induced without the accompanying nuclear changes (Moor, Osborn, Cran & Walters, 1981). Furthermore, there is no measurable fall in the content of cyclic AMP in the oocyte immediately after the resumption of meiosis; in fact at 1 h after stimulation the intracellular concentration is significantly elevated (Fig. 5*b*). These findings are in contrast

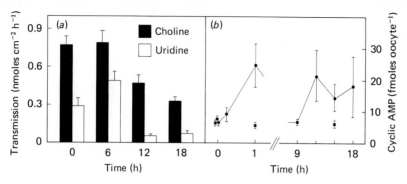

Fig. 5 (*a*) Extent of intercellular coupling between follicle cells and oocytes at selected times after the re-initiation of maturation by gonadotrophic treatment. Follicle cell–oocyte complexes were removed from antral follicles and the transmission of labelled marker into oocytes was determined after 60 min incubation in 30 μM[^{3}H] choline or 3 μM [^{3}H] uridine. (After Moor *et al.*, 1980b.)

(*b*) Content of cyclic AMP (mean value $\pm$ SEM) in oocytes obtained from follicles (●) at selected times after the re-initiation of maturation with gonadotrophins. The content of cyclic AMP in oocytes separated from their follicle cells before treatment with gonadotrophin (■) is also shown. (After Moor & Heslop, 1981.)

to those in amphibia where a rapid fall in cyclic AMP is an early and obligatory initiator of maturation (Maller & Krebs, 1980).

In addition to blocking meiosis, inhibitory signals from the follicle cells also suppress membrane transport (Moor & Smith, 1978) and protein synthesis within the oocyte (Crosby *et al.*, 1981). Separation of the oocyte from the follicle cells removes this inhibition, doubles the rate of entry of amino acids, increases by 25% the incorporation of methionine into protein and results in a number of specific changes in protein synthesis (Fig. 6*a*). It is, however, clear from this figure that the removal of follicular inhibition is not itself adequate to initiate all the changes in protein synthesis that characterise the fully matured oocyte. We suggest that inductive stimuli from the follicle cells are transmitted to the oocyte and confer upon it the capacity for subsequent development and differentiation (see Thibault, 1977; Moor & Warnes, 1978).

There is a strong possibility that steroids act as inductive signals during the critical early phase of maturation (Moor & Warnes, 1978). It has been found for example that a specific pattern of steroidal support is required for the synthesis of those cytoplasmic factors in oocytes that induce sperm decondensation and pronucleus formation during fertilisation (see Thibault, Gerard & Menezo, 1975). Other results show that oestradiol-17β, together with gonadotrophin, confers upon oocytes the capacity to cleave and differentiate into blastocysts (Moor & Trounson,

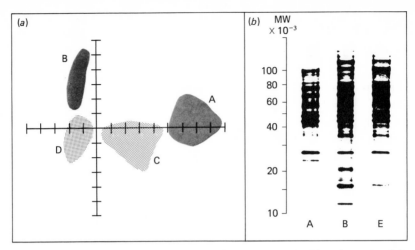

Fig. 6. (*a*) Canonical variate analysis of the effect on protein profiles in oocytes of the removal of different degrees of follicle cell support. The plot represents the first two canonical variates for oocytes cultured for 18 h as follows: (A) oocytes obtained from follicles cultured without gonadotrophins, (B) oocytes obtained from follicles cultured with LH and FSH, (C) isolated oocytes cultured outside the follicle but with an intact cumulus, and (D) isolated, LH-treated oocytes. The data used for this analysis are derived from the relative amount of labelled protein in 16 individual marker bands expressed as a percentage of the total protein synthesis, and are obtained from fluorographs similar to those shown in Fig. 6*b*. For further details of the statistical analysis, see Moor *et al.* (1981).

(*b*) Fluorograph of [^{35}S] methionine-labelled proteins synthesised by oocytes previously cultured for 18 h within follicles in the presence of the following gonadotrophins: (A) untreated controls, (B) FSH plus LH, (E) FSH, LH plus 7-chloro-3, 4-dihydro-2(3-pyridyl)-1-(2H)-naphthalenone. Polypeptide separation was by SDS linear gradient polyacrylamide gel electrophoresis. In this figure, A and B correspond to the plots shown in Fig. 6*a*.

1977). Further supporting evidence shows that the suppression of oestradiol secretion during early maturation using specific enzyme inhibitors causes gross developmental abnormalities especially at fertilisation (Moor, Polge & Willadsen, 1980a). The cause of these developmental abnormalities is probably to be found in the disrupted pattern of proteins synthesised by oocytes denied oestrogen support during maturation. This is illustrated by the electrophoretic profiles in Fig. 6*b* which show the effect on protein synthesis of blocking the 17α-hydroxylase enzyme system in follicles with 7-chloro 3,4 dihydro-2(3-pyridyl)-1-(2H)-naphthalenone. Our hypothesis is that oestrogen, generated by the interaction between the theca and granulosa cells, is transported to the oocyte where it induces specific changes in protein synthesis. Two potential methods of steroid transmission exist in the follicle. Firstly, steroids released into the extracellular compartment reach the oocyte by diffusion through the follicular fluid. Secondly, steroid molecules could readily be transmitted from the granulosa cells through the extensive system of permeable junctions directly into the oocyte cytoplasm. At present our results do not enable us to evaluate the relative importance of each of these transmission systems.

CONCLUSIONS

It has been our object to show that the expression of follicular function is regulated by a series of interactions between the cells of the theca, granulosa, cumulus and oocyte.

Follicular differentiation and steroid biosynthesis depend upon cellular co-operation across a basement membrane. These interactions involve the transfer of steroid precursors and signal molecules between the thecal and granulosa compartments. Cyclic AMP and androgen are two compounds which probably act as signals at specific stages in follicular development.

Interactions between the somatic components and oocyte involve direct contact through permeable junctions. During the germinal vesicle stage, these cellular relationships fulfil both a permissive role by providing essential substances excluded by the oocyte membrane and an instructional role in the suppression of meiosis. However, although it is accepted that follicle cells maintain meiotic arrest in oocytes, the nature of the putative inhibitory factors is uncertain. The removal of follicular inhibition results in the resumption of meiosis but is inadequate to initiate both the changes in protein synthesis and the acquisition of developmental competence that characterise the fully matured oocyte. By contrast, our evidence suggests that inductive stimuli from the follicle cells mediate full physiological maturation.

Acknowledgements

We thank Drs David Cran and Robin Harrison for generous assistance during the preparation of this manuscript. The purified gonadotrophins were donated by the National Institute of Arthritis, Metabolic and Digestive Diseases, National Institute of Health, Bethesda, Maryland. One of us (J.C.O.) is indebted to the Medical Research Council for financial support.

REFERENCES

ANDERSON, E. & ALBERTINI, D. F. (1976). Gap junctions between the oocyte and companion follicle cells in the mammalian ovary. *Journal of Cell Biology*, **71**, 680–6.

BACHVAROVA, R., BARAN, M. M. & TEJBLUM, A. (1980). Development of naked growing mouse oocytes *in vitro*. *Journal of Experimental Zoology*, **211**, 159–69.

BERNARD, J. & PSYCHOYOS, A. (1977). Inhibitory effect of follicular fluid on RNA synthesis of rat granulosa cells *in vitro*. *Journal of Reproduction and Fertility*, **49**, 355–7.

BIGGERS, J. D., WHITTINGHAM, D. G. & DONAHUE, R. P. (1967). The pattern of energy metabolism in the mouse oocyte and zygote. *Proceedings of the National Academy of Sciences, USA*, **58**, 560–7.

BYSKOV, A. G. (1974). Does the rete ovarii act as a trigger for the onset of meiosis? *Nature (London)*, **252**, 396–7.

CHANG, M. C. (1955). The maturation of rabbit oocytes in culture and their maturation, activation, fertilization and subsequent development in the fallopian tubes. *Journal of Experimental Zoology*, **128**, 379–405.

CHO, W. K., STERN, S. & BIGGERS, J. D. (1974). Inhibitory effect of dibutyryl cAMP on mouse maturation *in vitro*. *Journal of Experimental Zoology*, **187**, 383–6.

CROSBY, I. M., OSBORN, J. C. & MOOR, R. M. (1981). Follicle cell regulation of protein synthesis and developmental competence in sheep oocytes. *Journal of Reproduction and Fertility*, **62**, 575–82.

DEKEL, N. & BEERS, W. H. (1978). Rat oocyte maturation *in vitro*: relief of cyclic AMP inhibition by gonadotrophins. *Proceedings of the National Academy of Sciences, USA*, **75**, 4369–73.

DEKEL, N. & BEERS, W. H. (1980). Development of the rat oocyte *in vitro*: inhibition and induction of maturation in the presence or absence of the cumulus oophorus. *Developmental Biology*, **75**, 247–54.

EPPIG, J. J. (1979). A comparison between oocyte growth in co-culture with granulosa cells and oocytes with granulosa cell-oocyte junctional contact maintained *in vitro* (1). *Journal of Experimental Zoology*, **209**, 345–53.

FALCK, B. (1959). Site of production of oestrogen in rat ovary as studied by microtransplants. *Acta Physiologica Scandinavica*, **47** (suppl. 163), 1–101.

FLAGG-NEWTON, J., SIMPSON, I. & LOEWENSTEIN, W. R. (1979). Permeability of the cell-to-cell membrane channels in mammalian cell junctions. *Science*, **205**, 404–7.

FOOTE, W. D. & THIBAULT, C. (1969). Recherche experimentales sur la maturation *in vitro* des oocytes de truie et de veau. *Annales de Biologie Animale, Biochimie et Biophysique*, **9**, 329–49.

GERISCH, G., MAEDA, Y., MALCHOW, D., ROOS, W., WICK, W. & WURSTER, B. (1977). Cyclic AMP signals and the control of cell aggregation in *Dictyostelium discoideum*. In *Development and Differentiation in the Cellular Slime Moulds*, ed. P. Capuccinelli & J. M. Ashworth, pp. 105–24. Amsterdam: Elsevier/North-Holland Biomedical Press.

GILULA, N. B., EPSTEIN, M. L. & BEERS, W. H. (1978). Cell-to-cell communication and ovulation. A study of the cumulus-oocyte complex. *Journal of Cell Biology*, **78**, 58–75.

GIORGI, F. (1979). Coated vesicles in the oocyte. In *Coated Vesicles*, ed. C. D. Ockleford & A. Whyte, pp. 135–77. Cambridge University Press.

GLASS, L. E. & CONS, J. M. (1968). Stage dependent transfer of systemically injected foreign protein antigen and radiolabel into mouse ovarian follicles. *The Anatomical Record*, **162**, 139–56.

HELLER, D. T. & SCHULTZ, R. M. (1980). Ribonucleoside metabolism by mouse oocytes: metabolic cooperativity between the fully grown oocyte and cumulus cells. *Journal of Experimental Zoology*, **214**, 355–64.

HILLENSJO, T., POMERANTZ, S. H., SCHWARTZ-KRIPNER, A., ANDERSON, L. D. & CHANNING, C. P. (1980). Inhibition of cumulus cell progesterone secretion by low molecular weight fractions of porcine follicular fluid which also inhibit oocyte maturation. *Endocrinology*, **106**, 584–91.

LEIBFRIED, L. & FIRST, N. L. (1980). Effect of bovine and porcine follicular fluid and granulosa cells on maturation of oocytes *in vitro*. *Biology of Reproduction*, **23**, 699–704.

LOEWENSTEIN, W. R. (1979). Junctional intercellular communication and the control of growth. *Biochimica et Biophysica Acta*, **560**, 1–65.

MAKRIS, A. & RYAN, K. J. (1975). Progesterone, androstenedione, testosterone, estrone and estradiol synthesis in hamster ovarian follicle cells. *Endocrinology*, **96**, 694–701.

MALLER, J. L. & KREBS, E. G. (1980). Regulation of oocyte maturation. *Current Topics in Cellular Regulation*, **16**, 271–311.

MARSH, J. M. (1976). The role of cyclic AMP in gonadal steroidogenesis. *Biology of Reproduction*, **14**, 30–53.

MASUI, Y. & CLARKE, H. J. (1979). Oocyte maturation. *International Reviews of Cytology*, **57**, 186–282.

MOOR, R. M. (1977). Sites of steroid production in ovine Graafian follicles in culture. *Journal of Endocrinology*, **73**, 143–50.

MOOR, R. M. & CRAN, D. G. (1980). Intercellular coupling in mammalian oocytes. In *Development in Mammals*, vol. 4, ed. M. H. Johnson, pp. 3–37. Amsterdam: Elsevier/North-Holland Biomedical Press.

MOOR, R. M. & HESLOP, J. P. (1981). Cyclic AMP in mammalian follicle cells and oocytes during maturation. *Journal of Experimental Zoology*, **216**, 205–9.

MOOR, R. M., OSBORN, J. C., CRAN, D. G. & WALTERS, D. E. (1981). Selective effect of gonadotrophins on cell coupling, nuclear maturation and protein synthesis in mammalian oocytes. *Journal of Embryology and Experimental Morphology*, **61**, 347–65.

MOOR, R. M., POLGE, C. & WILLADSEN, S. M. (1980a). Effect of follicular steroids on the maturation and fertilization of mammalian oocytes. *Journal of Embryology and Experimental Morphology*, **56**, 319–35.

MOOR, R. M. & SMITH, M. W. (1978). Amino acid uptake into sheep oocytes. *Journal of Physiology*, **284**, 68–69P.

MOOR, R. M., SMITH, M. W. & DAWSON, R. M. C. (1980b). Measurement of intercellular coupling between oocyte and cumulus cells using intracellular markers. *Experimental Cell Research*, **126**, 15–29.

MOOR, R. M. & TROUNSON, A. O. (1977). Hormonal and follicular factors affecting maturation of sheep oocytes *in vitro* and their subsequent developmental capacity. *Journal of Reproduction and Fertility*, **49**, 101–9.

MOOR, R. M. & WARNES, G. M. (1978). Regulation of oocyte maturation in mammals. In *Control of Ovulation*, ed. D. B. Crighton, G. R. Foxcroft, N. B. Haynes & G. E. Lamming, pp. 159–76. London: Butterworths.

MOORE, G. P. M., LINTERN-MOORE, S., PETERS, H., BYSKOV, A. G., ANDERSEN, M. & FABER, M. (1975). The inhibition of DNA-dependent RNA synthesis in Yoshida ascites cells by bovine follicular fluid. *Journal of Cell Physiology*, **86**, 31–5.

NIMROD, A., TSAFRIRI, A. & LINDNER, H. R. (1977). *In vitro* induction of binding sites for hCG in rat granulosa cells by FSH. *Nature (London)*, **267**, 632–3.

PEARSE, B. M. F. (1980). Coated vesicles. *Trends in Biochemical Sciences*, **5**, 131–4.

PETERS, H. (1978). Folliculogenesis in mammals. In *The Vertebrate Ovary*, ed. R. E. Jones, pp. 121–44. New York: Plenum Press.

PINCUS, G. & ENZMANN, E. V. (1935). The comparative behaviour of mammalian eggs *in vivo* and *in vitro*. I. The activation of ovarian eggs. *Journal of Experimental Medicine*, **62**, 665–75.

RICHARDS, J. S. (1980). Maturation of ovarian follicles: actions and interactions of pituitary and ovarian hormones on follicular cell differentiation. *Physiological Reviews*, **60**, 51–89.

RODMAN, T. C. & BACHVAROVA, R. (1976). RNA synthesis in preovulatory mouse oocytes. *Journal of Cell Biology*, **70**, 251–7.

SHORT, R. V. (1964). Ovarian steroid synthesis and secretion *in vivo*. *Recent Progress in Hormone Research*, **20**, 303–33.

THIBAULT, C. (1977). Are follicular maturation and oocyte maturation independent events? *Journal of Reproduction and Fertility*, **51**, 1–15.

THIBAULT, C., GERARD, M. & MENEZO, Y. (1975). Acquisition par l'ovocyte de lapine et de veau du facteur de decondensation du noyau du spermatozoide fecondant (MPGF). *Annales de Biologie Animale, Biochimie et Biophysique*, **15**, 705–14.

TSAFRIRI, A. (1978). Oocyte maturation in mammals. In *The Vertebrate Ovary. Comparative Biology and Evolution*, ed. R. E. Jones, pp. 409–42. New York & London: Plenum Press.

TSAFRIRI, A. & CHANNING, C. P. (1975). An inhibitory influence of granulosa cells and follicular fluid upon porcine oocyte meiosis *in vitro*. *Endocrinology*, **96**, 922–7.

TSAFRIRI, A., POMERANTZ, S. H. & CHANNING, C. P. (1976). Inhibition of oocyte maturation by porcine follicular fluid: partial characterisation of the inhibitor. *Biology of Reproduction*, **14**, 511–16.

WEISS, T. J., ARMSTRONG, D. T., McINTOSH, J. E. A. & SEAMARK, R. F. (1978). Maturational changes in sheep ovarian follicles: gonadotrophic stimulation of cyclic AMP production by isolated theca and granulosa cells. *Acta Endocrinologica*, **89**, 166–72.

WEISS, T. J., SEAMARK, R. F., McINTOSH, J. E. A. & MOOR, R. M. (1976). Cyclic AMP in sheep ovarian follicles: sites of production and response to gonadotrophins. *Journal of Reproduction and Fertility*, **46**, 347–353.

The origins of cell diversity in the early mouse embryo

C. A. ZIOMEK, H. P. M. PRATT AND M. H. JOHNSON

Department of Anatomy, Downing Street, Cambridge CB2 3DY, UK

INTRODUCTION

The generation of cell diversity during development appears to depend upon the interaction between a temporal programme within a cell and positional recognition between cells. The ability of cells to recognise their different positions presumably involves some sort of interaction between them that probably occurs primarily at their cell surfaces. In this chapter we analyse in detail three sorts of cell interaction which occur between the 2- and 8-cell stage of development of the mouse embryo. Each sort of interaction has claims to be considered important for the development at the 16- and 32-cell stages of the two tissues of the blastocyst, the inner cell mass (ICM) and trophectoderm (Fig. 1). These claims will be examined closely.

The embryo undergoes cleavage to 8 cells over a $2\frac{1}{2}$ day period. Compaction then occurs at the 8-cell stage to form a morula, and involves polarisation of individual cells, flattening of cells against each other to maximise cell contact and formation of specialised intercellular junctions. At division to 16 cells, some cells become surrounded totally by others and the exterior (or outside) population are thought to found the trophectodermal lineage whereas the enclosed (or inside) population are thought to found the ICM lineage (see Johnson, 1981a,b for review of evidence). After division to 32 cells, zonular tight junctions form between cells of the outside population, which secretes fluid centrally to create a fluid-filled cavity, the blastocoel, within which an eccentrically placed cluster of inside cells forms the ICM.

In this chapter, we examine the allocation of cells to ICM and trophectodermal lineages in respect of (a) the effects of varying the numbers of contacts between cells, (b) the induction and maintenance

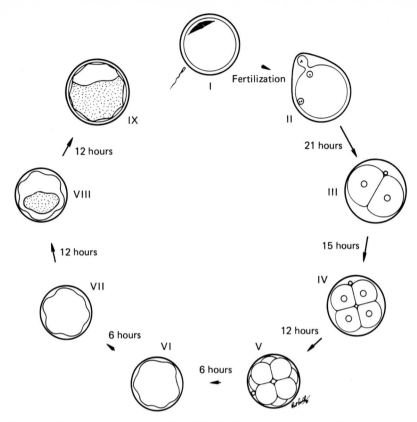

Fig. 1. Outline of blastocyst formation. (i) Unfertilised, ovulated egg with chromosomes on second metaphase spindle; surrounded by zona pellucida which persists until the blastocyst stage. (ii) At fertilisation, 2nd metaphase completed leading to formation of polar body and female pronucleus; male pronucleus forms. (iii) Pronuclei come together and chromosomes line up on spindle for the first cleavage division; 2-cell embryo results. (iv) Cleavage to 4-cells, and (v) 8-cells. At 8-cell stage embryo undergoes compaction (vi) in which individual cells polarise to become epithelial-like, flatten on each other thereby maximising cell contacts and form specialised focal tight and gap junctions. (vii) The cells in the morula so formed divide to give a compact mass of 16 cells, which in turn divide (viii) to yield a 32-cell morula in which fluid is secreted internally to form a small blastocoel. (ix) By division to 64-cells, the blastocoel is enlarged, surrounded by trophectoderm cells and contains an eccentrically placed inner cell mass (ICM). The trophectoderm gives rise to the trophoblast and chorion whereas the ICM yields all endoderm, mesoderm and embryonic ectoderm derivatives.
(Authors' copyright on figure and legend).

of polarity in individual 1/8 cells and (c) the effects of varying the degree of intercellular flattening between cells.

(a) The effect of cell interactions on the allocation of cells to different developmental lineages

Graham and his colleagues have studied the contacts between cells in intact or fragmented early mouse embryos, and have proposed that the number of contacts made by a cell affects the relative disposition of cells and cytoplasm at the subsequent division (Graham & Deussen, 1978; Kelly, Mulnard & Graham, 1978; Graham & Lehtonen, 1979; Lehtonen, 1980). They demonstrate that division of one 1/2 blastomere at the 2-cell stage occurs earlier than the other by up to several hours. Subsequently, this temporal advantage is retained by the progeny of the early dividing cell up to the 8- to 16-cell stage and beyond (Kelly *et al.*, 1978). The 2/4 couplet formed from the early dividing 1/2 cell can pack with the second 1/2 blastomere in a variety of ways (the variety being greater in zona-free embryos) and the contacts which are made with the second 1/2 blastomere influence the subsequent division plane of this cell and/or the packing of its progeny. In general, the more contacts the first 2/4 makes with the second 1/2, the more contacts will exist between the first 2/4 and the last 2/4 (Graham & Lehtonen, 1979). In addition, the progeny of the first cell to divide have more total contacts than the progeny of the second cell to divide. What these observations suggest is that cell contacts (or some 'memory' of them) are retained and recognised during division. Such a conclusion is reinforced by examination of subsequent divisions. In the transition from 4 to 8 cells, one of the 2/8 cells formed by division of each 1/4 cell was observed to retain in many cases all the intercellular contacts of the 1/4 cell, whereas the other 1/8 cell usually had fewer cell contacts (Graham & Deussen, 1978). As a result, one 1/8 cell of any 2/8 pair will tend to occupy a deeper position within the embryo compared to the other, and since the 2/4 from the early dividing 1/2 cell has more total contacts than the 2/4 from the late dividing 1/2, the early 2/4 will tend to generate more 1/8 cells with more contacts and a 'deeper' position.

Groups of 1/8 cells isolated from the embryo and flattened into a monolayer under oil, show a similar pattern – namely one of their progeny tends to maintain all the contacts of the parent (Graham & Lehtonen, 1979). Again therefore, the 1/8 cells with more contacts will tend to contribute to the deeper cell population. Since it is at the 8- to 16-cell transition that totally surrounded cells are first detected (Mulnard, 1967; Barlow, Owen & Graham, 1972; Graham & Lehtonen, 1979; Handyside, 1981; Johnson & Ziomek, 1981a), this series of observations leads Graham *et al.* to conclude that continuing cell interactions guide cells to a final inside or outside position and therefore into the ICM or trophectodermal lineages.

Now since the earlier dividing 1/2 cell forms 2/4 cells that tend to retain more of their contacts on division to 4/8 than do the later dividing cells thereby ending up 'deeper', Graham and his colleagues argue that the early dividing 1/2 cell should make a disproportionately larger contribution to the ICM than the late dividing 1/2 cell. This idea was tested in two ways. First, individual 4-cell embryos were disaggregated to single cells, and either the first or last to divide to 2/8 was labelled with thymidine and the blastomeres reaggregated. It was found that the early dividing 1/4 (presumptive progeny of the early dividing 1/2) contributed disproportionately to the ICM (Kelly *et al.*, 1978). However, a contribution from late dividing 1/4 cells was often also detected. Second, oil droplets were injected either peripherally or centrally into 1/8 cells with different numbers of cell contacts. All peripherally injected oil droplets were subsequently found in the trophectoderm. This result suggests that all 1/8 cells contribute to the trophectoderm (Graham & Deussen, 1978). A similar conclusion was reached by Wilson, Bolton & Cuttler (1972). In contrast, cells injected with a centrally placed oil droplet contributed at least 1 cell to the ICM in 91.6% of cases when they had 5 or more contacts (3.4 1/8 cells per embryo on average), and in only 25% of cases when they had 4 or less contacts (6.7 1/8 cells per embryo) (Graham & Deussen, 1978). These results support the notion that the ICM is derived mainly from subsets of cells which are either early dividing or which have more contacts, and since the progeny of the early dividing 1/2 cell tend to be both early dividing and to have more contacts, they may indeed contribute preferentially to the ICM. However, the results do not preclude the possibility that all cells can and do contribute to the ICM, since the oil marker used is not distributed to all progeny but *can* only be distributed to one of the progeny. Thus, these results only give us a clue about the relative amount of cytoplasm distributed to the ICM.

Two types of experiment are required to test the hypothesis of Graham *et al.* further. First, we need to know how many cells are present in the ICM and trophectodermal lineages at their foundation; these are usually assumed to be the inner and outer cell populations at the 16-cell stage. Second, we need to know where the cells in these two populations come from and to confirm that they do indeed give rise to ICM and trophectoderm, i.e. the lineage of cells from the 2-cell to the 16-cell and blastocyst stages must be traced directly.

The numerology poses problems, because estimates of inside and outside cell numbers at the 16-cell stage differ widely. Graham and his colleagues (Barlow *et al.*, 1972; Graham & Lehtonen, 1979) used fixed and sectioned material from various strains of mice to arrive at estimates of rarely more than 2 inside cells. In contrast, a variety of approaches in which cells isolated from live 16-cell embryos were sorted for presence of inside or outside markers, had led to estimates of 6 to 8 inside cells in 93% of embryos (Handyside, 1981; Johnson & Ziomek, 1981). It seems possible that fixation may cause shrinkage leading to a 'popping-out' of inner cells (such as occurs after brief exposure to low Ca^{2+} – see Fig. 7 Reeve & Ziomek, 1981) thereby causing underestimates of inside cell numbers. If this is so, then the estimates of 6 to 8 inside cells may be more reliable, since they do not rest solely on maintenance of relative inside : outside position throughout processing and analysis but are arrived at independently of relative position and by more than one technique.

When patterns of division from 8- to 16-cell stages were examined, it was found that on average 1 out of $8\frac{1}{8}$ cells contributed 2 outside cells and 7 out of $8\frac{1}{8}$ cells contributed 1 inside and 1 outside cell. However, in some embryos all 1/8 cells contributed 1 inside and 1 outside cell, and in most of the others, $6\frac{1}{8}$ cells did so, with the remaining 2 each contributing 2 outside cells. In no case did a 1/8 cell generate 2 inside cells (Johnson & Ziomek, 1981a). These results are in agreement with the observations of Kelly *et al.* (1978) and Wilson *et al.* (1972), in that every 1/8 cell contributes to the outside, trophectodermal lineage. For the inside, ICM lineage, the two sets of results are not readily compatible. For example, in embryos in which every 1/8 cell contributes 1 inside and 1 outside cell (29% of embryos; total 8 inside cells) there can be no preferential contribution to the inside cell lineage at the 16-cell stage by progeny of the early dividing 1/2 cell. In the bulk of the remaining embryos in which there are only 7 or 6 inside cells, it would be necessary to assume that the 1/8 cells derived from the early dividing 1/2 blastomere each contributed an inside cell and that only 2 or 3 of the progeny of the later dividing 1/2 blastomeres contributed inside cells. There should then be a relationship between division order and division plane at the 8- to 16-cell transition, the earlier dividing 1/8 cells giving inside and outside cells and the later cells dividing to give two outside cells. There are two ways of analysing this relationship. Firstly, embryos with cell numbers intermediate between 8 and 16 can be examined for the frequency of divisions yielding two outside cells

Table 1. *Numbers of divisions of 1/8 blastomeres to yield one 2/16 blastomeres, both of which are outside, in embryos of different total cell number*

No. of cells per embryo	No. of embryos examined	No. of 1/16 cells per embryo	Total no. of divisions (1/8 to 2/16)	No. of divisions yielding 2 outside cells	No. of divisions yielding 2 outside cells ÷ Total no. of divisions
9	3	2	3	0	0/3 (0)
10	4	4	8	1	1/8 (0.12)
11	3	6	9	1	1/9 (0.11)
12	4	8	16	4	4/16 (0.25)
14	12	12	72	12	12/72 (0.17)
15	6	14	42	7	7/42 (0.17)
16	14	16	112	12	12/112 (0.11)

in relation to the total number of divisions. Data shown in Table 1 (taken from Johnson & Ziomek, 1981a; Handyside, 1981) does not provide evidence for a major difference in this frequency ratio when embryos having 9 to 12 cells are compared with embryos having 13 to 16 cells. Thus, the generation of outer cells is not significantly higher in the first four 1/8 to 2/16 cell divisions than in the last four. However, there is some overlapping of cell cycles i.e. the first 1/8 derived from the late-dividing 1/2 cell may divide before the last 1/8 cell from the early dividing 1/2 cell. A second, more direct, approach to this question is to mark the early dividing 1/2 blastomere at the 2-cell stage and follow the division order and planes in the progeny of both it and its unmarked companion through the 8- to 16-cell transition. This experiment is underway.

There is thus at present no data to support the notion that early dividing 1/8 cells contribute preferentially to the inside cell population at the 16-cell stage and thus to the presumptive ICM lineage. Could the presumption of a direct lineage relationship between inside cells and ICM be wrong? The discrepancy between these data and those of Graham *et al.* could be resolved if the inside population at the 16-cell stage did *not* constitute the foundation of a strict ICM lineage but rather that some of its cells (those derived from the later dividing 1/2 blastomere) regularly transferred to the trophectodermal lineage at some later stage of development. What evidence bears on this question?

First, inside cells at the 16-cell stage have several features resembling ICM cells and thus appear to anticipate an ICM fate (Ziomek & Johnson, 1981). Moreover, the ratios of the average numbers of inside:outside cells at the 16- and 32-cell stages are 7.0:9.0 and 13.8:18.2 (Johnson & Ziomek, 1981a; Handyside, 1978), which is compatible with the notion that inside and outside populations 'breed true'. However, it is known that inside cells at both 16- and 32-cell stages are developmentally labile and can be induced *in vitro* to develop as trophectoderm (Johnson *et al.*, 1977; Handyside, 1978; Johnson, 1979; Hogan & Tilly, 1978; Spindle, 1978). Moreover, the ratio of inside:outside cells at the 64-cell blastocyst stage is nearer 19.2:44.8. This shift in ratio at the 32- to 64-cell transition has led Handyside (1978) to suggest that some inside cells may indeed regularly feed out to form trophectoderm at this stage. Were this proved to be the case, and were those cells shown to be progeny of late dividing 1/8 cells, then the present conflict of data would be reconciled. However, it is equally possible that differential rates of cell division or death could yield the same result, and the data for discriminating between these alternatives are not available.

The work of Graham and his colleagues attempting to relate continuing cell interactions to the foundation of distinct tissue lineages can be summarised as follows. Cell interactions up to the 16-cell stage do seem to influence the relative disposition of cells and cytoplasm within the 16-cell morula. However, there is at present no direct evidence to support the view that these interactions influence the allocation of cells to two different cell lineages in the 16-cell morula. Perhaps the weakness of the analysis by Graham *et al.* lies in the assumption that the surfaces of cells are uniform and equivalent. Although this may be the case at the 2-, 4- and early 8-cell stages it is not true subsequently. From the mid-8-cell stage onwards there is heterogeneity of surface properties within the membranes of individual cells, and between the membranes of different cells, and these surface differences seem to have a decisive influence on cell packing, interaction and allocation to different lineages as we will demonstrate in the next section. It is thus not possible to conclude on present evidence that the numbers of contacts a cell makes, and the interactions which it engages in, are in themselves adequate to explain the generation of two cell lineages.

(b) The effects of cell interactions on the induction and maintenance of polarity in 8-cell blastomeres

Examination of blastomeres from early non-compacted 8-cell mouse embryos has revealed that they have a uniform distribution of microvilli and ligand-binding sites on their cell surfaces (Handyside, 1980; Reeve & Ziomek, 1981) and develop a random, dispersed distribution of endocytotic vesicles in their cytoplasm upon exposure to horseradish peroxidase (HRP) (Reeve, 1981). In contrast, cells from the late compacted 8-cell embryo are polarised functionally and morphologically along a radial axis through the embryo. Microvilli (Ducibella & Anderson, 1975; Reeve & Ziomek, 1981) and ligand binding sites (Handyside, 1980; Ziomek & Johnson, 1980; Reeve & Ziomek, 1981) are concentrated at the apical, exposed surface of each blastomere, microtubules are aligned along the lateral membranes of each cell (Ducibella *et al.*, 1977) and cytoplasmic vesicles containing endocytosed HRP concentrate beneath the apical microvillous pole (Reeve, 1981). The pole of intense fluorescent ligand-binding observed on late 8-cell blastomeres is coincident with the apical pole of microvilli (Fig. 2) and may therefore be explained by a uniform distribution of

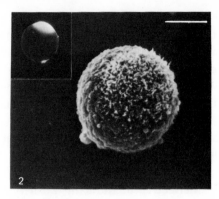

Fig. 2. An SEM micrograph of a single blastomere taken from a late, compacted 8-cell embryo. Notice the pole of microvilli (scale bar = 10 μm). Inset: a similar cell stained with FITC-concanavalin A to show the pole of ligand binding (X 580).

ligand-binding sites in the plasma membrane but a non-uniform distribution of membrane, i.e. microvilli, over the cell surface (Reeve & Ziomek, 1981). Aside from the integral membrane protein, alkaline phosphatase, the activity of which is reported to be restricted to the lateral and basal plasma membranes of 4- and 8-cell blastomeres (Mulnard & Huygens, 1978; Izquierdo et al., 1980), there is no evidence for a restriction of specific membrane proteins solely to the microvillous or non-microvillous membranes of the polarised 8-cell blastomeres (discussed Johnson, 1981a, b).

The observation that blastomeres polarise at the 8-cell stage has led Johnson, Pratt & Handyside (1981) to propose that polarisation is a prerequisite for the divergent differentiation of the cell lineages leading to blastocyst formation. They propose that division of the radially polarised 8-cell blastomeres could generate two distinct inside and outside cell populations which might constitute the foundation of the ICM and trophectodermal lineages. This speculation has led us to examine in some detail the polarisation process and its developmental consequences.

We have shown that polarisation of early non-polar 8-cell blastomeres is a process that is mediated by asymmetric cell contacts. Non-polar 1/8 cells do not polarise when cultured as single cells for up to 9 h but do polarise when cultured together with a companion 1/8-cell for as little as 5 h. The axis of polarity generated in each cell of such pairs is perpendicular to the point of contact between the cells in the pairs (Fig. 3) and independent of the previous plane of cleavage (Ziomek & Johnson, 1980). In aggregates of 3 or more 1/8 cells, all cell contacts are respected in establishing the axis of polarity (Fig. 4), the pole forming on each cell at the point geometrically most distant from all cell contacts (Johnson & Ziomek, 1981b). The notion that an asymmetry of cell contacts is required for the expression of polarity is supported by the observations that polarisation of a 1/8-cell can be suppressed by surrounding it on all surfaces with companion 1/8 cells (Johnson & Ziomek, 1981b).

Aggregation of 1/8 cells to carrier blastomeres of various embryonic stages has revealed that there is cell specificity in the interactions leading to polarisation (Johnson & Ziomek, 1981b). Unfertilised and fertilised eggs are ineffective at inducing polarity, whereas 2-, 4-, 8- and 16-cell blastomeres induce the polarisation of their companion 1/8 cells. This inducing capacity is increasingly effective in carrier cells of later embryonic age up to a maximum at the 8-cell stage (i.e. $2 < 4 < 8 = 16$). It is tempting to speculate that the appearance of inducing capacity at the 2-cell stage may correlate with the switching on of the embryonic genome seen at this time (Flach, Johnson, Bolton, Braude & Taylor, in preparation). The 2- and 4-cell carrier blastomeres, although possessing the capacity to induce the polarisation of their companion 1/8 cells, do not themselves polarise and it thus appears that the ability to respond fully to the polarity-inducing signal does not arise until the 8-cell stage. That 4-cell membranes should have inducing capacity makes biological sense since it would allow embryos displaying any asynchrony of cell division at the 4- to 8-cell transition to begin polarising even before the last 1/4 blastomere had divided. Since *all* inducing contacts are recognised in establishing the axis of polarity, it would ensure that a pole formed only on the outer free surface of each 1/8 blastomere, imparting a radial symmetry to the embryo, as a whole.

We have also been able to show that once polarity is established in a 1/8 cell it is stable for the lifetime of that cell. For example, if pairs of polarised 1/8 cells are disaggregated, the axis of polarity persists whether the cells are cultured for several hours as single cells or reaggregated in pairs such that the point of intercellular contact is radically shifted (Fig. 5; Johnson & Ziomek, 1981b). Even at division from the 8- to the 16-cell stage, elements of polarity are conserved during cytokinesis (Johnson & Ziomek, 1981). Each polar 1/8 cell gives rise to either 2 polar daughter cells (a division through the apical pole) or 1 larger, polar and 1 smaller, apolar daughter cell (a division plane at some angle with respect to the apical pole). In the latter case, the polar

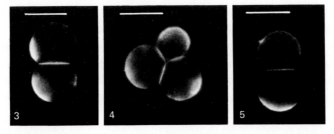

Fig. 3. Two 1/8 cells which were isolated within an hour of their formation and aggregated together (note remains of midbody on lower of the two cells). The pair were cultured for 7 h. Note both are polarising opposite to the point of contact, the top one is more advanced than the bottom one in the development of polarity. Scale bar = 20 μm.

Fig. 4. Three 1/8 cells, isolated within an hour of their formation, aggregated together and cultured for $8\frac{1}{2}$ h. Note that the pole forms at a point most distant from points of contact with other cells. Scale bar = 20 μm.

Fig. 5. Two 1/8 cells, isolated within an hour of their formation, aggregated together for 5 hours, disaggregated, reaggregated together with a new point of contact, and then cultured for a further 4 h. Note that the axes of polarity were determined by the *first* point of contact and remain stable despite altered contacts subsequently. Scale bar = 20 μm.

daughter forms from the apex of the polar parent and the apolar cell from its base. These two cell types, polar and apolar, are morphologically and behaviourally distinct (Ziomek & Johnson, 1981b) and as a result, come to occupy different positions within the 16-cell morula (Handyside, 1981; Reeve & Ziomek, 1981). Polar cells (presumed trophectoderm precursors) reside on the outside of the embryo, whereas the apolar cells (presumed contributors to the ICM) are located internally within the 16-cell morula. Thus, the different surface features of the two types of cell appear to determine their internal or external position, and these features are derived by differential inheritance from a polarised 1/8 cell. This conclusion need not conflict with the data of Graham and his colleagues (section *a*) but is much more specific. Thus, apolar (internal) cells, which are more adhesive (Ziomek & Johnson, unpublished) will tend to make more cell contacts, whereas polar (outside) cells, which are less adhesive will tend to have fewer cell contacts.

The results, taken together, offer support for the suggestion that polarisation of the blastomeres of the 8-cell embryo is a prerequisite for the subsequent divergent differentiation of the cell lineages leading to the development of a blastocyst. Polarisation of 8-cell blastomeres is induced by asymmetric, cell-specific interactions and the axis of its expression is defined by the number and location of its cell contacts. Once induced, both the fact of polarity and the axis of its expression are stable during the 8-cell stage and beyond, through division to the 16-cell stage, thereby generating two morphologically and behaviourally distinct cell types, that are believed to found the ICM and trophectodermal cell lineages of the blastocyst.

(c) The significance of cell flattening at compaction of the 8-cell embryo

Compaction involves three types of process: the cell-specific, contact-mediated induction of polarity described in section (*b*), extensive cell flattening to maximise cell contacts and minimise intercellular space (Ducibella *et al.*, 1975) and the formation of specific intercellular junctions (Lo, this volume). In this section, we analyse the process of cell flattening and its relationship to the induction of polarity and the morphogenesis of the blastocyst.

The pattern of intercellular adhesion of the compacting mouse 8-cell embryo can be modified in a variety of ways. Reduction of cell flattening may be achieved by modifying the cholesterol composition of embryos (Pratt, Keith & Chakraborty, 1980), exposing them to low Ca^{2+} (Ducibella & Anderson, 1976, 1979), concanavalin A (Reeve, in preparation), cytochalasin D (Pratt, Chakraborty & Surani, unpublished), colcemid (Pratt, Ziomek, Reeve & Johnson, in preparation) or tunicamycin (Surani, 1979). In all of these cases although intercellular flattening is inhibited, polarisation nonetheless occurs (Pratt, Ziomek, Reeve & Johnson, in preparation). This result suggests that flattening, or increased intercellular contact on a large scale, is not an important component of the polarity-inducing signal, and therefore is not important in the initial generation of two cell lineages. What role then does flattening play in the formation of a blastocyst? Use of another inhibitor of cell flattening, an antiserum directed against antigens expressed on the surface of nullipotent teratocarcinoma cells (reviewed Babinet & Condamine, 1980), suggests that flattening may be involved primarily in morphogenetic interactions.

Monovalent Fab fragments of rabbit anti-F-9 IgG (Kemler *et al.*, 1977; Ducibella, personal communication) as well as whole antibody raised against F-9 (Ducibella, 1980) or LS5770 (Johnson *et al.*, 1979) can inhibit cell flattening in a compacting embryo or reverse the process once it is established. Cleavage and cell division continue normally in these uncompacted aggregates to produce 'grape-like' structures which consist of 30 to 60 morphologically similar cells instead of the conven-

tional blastocyst with its two component tissues ICM and trophectoderm (Kemler et al., 1977; Johnson et al., 1979). This effect on
intercellular adhesion is reversible up to the early blastocyst stage
(approx. 96 h post-hCG) and removal from the antibody prior to this
time leads to restoration of normal intercellular associations and the
generation of normal blastocysts which outgrow in vitro (Johnson et
al., 1979) and give rise to normal offspring when reimplanted into
foster mothers (Kemler et al., 1977). If antibody treatment is prolonged
beyond approx. 96 h post-hCG, the cells start to accumulate fluid
intracellularly but do not compact or assume a blastocyst morphology
when removed from the antibody (Kemler et al., 1977; Johnson et al.,
1979; Ducibella, 1980; Johnson, 1981a).

It is unlikely that widespread non-specific cross-linking of surface
determinants is the molecular basis of intercellular adhesion in the
normal compacting morula, since divalent antibodies do not induce
intercellular adhesion within morulae (in common with a number of
other multivalent ligands)(Ducibella, 1980). Furthermore the inhibition
of compaction observed with anti-F-9 or anti-LS5770 is unlikely to be
due to prevention of cell–cell recognition by non-specific coating of
the cell surface, since Fab fragments or whole antibody from a variety
of sera, including some directed against morula surface antigens other
than F-9, do not inhibit compaction. It seems possible that a specific
contact molecule(s) is involved in the adhesion of compacting cells
since the inhibitory activity in whole sera and in IgG or Fab fragments
can be absorbed out by whole cells or the plasma membrane fractions
of embryonal carcinoma cells and their early differentiated derivatives
but not fully differentiated embryonal carcinoma cell lines which do
not express the F-9 antigen (Kemler et al., 1977; Ducibella, 1980;
Johnson et al., 1979). A glycoprotein (gp84, MW 84 000) has been
isolated from the plasma membranes of embryonal carcinoma cells and
is thought to represent the anti-F-9 Fab target since it co-purifies with
the inhibitory activity, binds to rabbit anti-embryonal carcinoma IgG
and has the same tissue distribution (defined by absorption) as the Fab
target (Hyafil et al., 1980). Culture in gp84 alone has no effect upon
compaction and development but in association with anti-F-9 it can
overcome the decompacting activity, suggesting that gp84 (or a cross-
reactive molecule) is involved in the intercellular recognition and
adhesion occurring during compaction of the morula (Hyafil et al.,
1980).

Although culture in these inhibitory antibodies prevents large-scale intercellular flattening, some regions of intercellular apposition do develop (Johnson *et al.*, 1979) and polarisation of individual blasto-meres (as assessed by the concentration of microvilli and FITC-Con A binding sites at the apices of the cells: Ziomek & Johnson, 1980; Reeve & Ziomek, 1981; Pratt, Ziomek, Reeve & Johnson, in preparation) also occurs normally. Furthermore embryos which are prevented from compacting also exhibit the normal programme of protein synthesis, including the synthesis of polypeptides which are specific to ICM and trophectoderm (Handyside & Johnson, 1978; Johnson, unpublished). Present evidence therefore suggests that the anti- F-9 target adhesion molecules are not required for the induction or polarisation of 8-cell blastomeres, nor for initiation or continuation of molecular maturation of the blastocyst.

The effect of the antibody on gap junctional communication be-tween blastomeres has not yet been established. However, when em-bryonal carcinoma cells are treated with anti-F-9 Fab they round up and disengage their intercellular contacts, their gap and tight junctions are internalised within 6 h and prolonged treatment (30 h) inhibits irreversibly the reassembly of these junctions (Dunia *et al.*, 1979). This inhibition of junctional assembly is associated with a diminished ability to cooperate metabolically with other embryonal carcinoma cells (Nicolas *et al.*, 1981). Further experiments are necessary to as-certain whether anti-F-9 Fab has similar effects on gap junctional assembly and intercellular communication in embryos to those ob-served in embryonal carcinoma cells. However, if the results from both systems are analogous then polarisation and molecular maturation of the embryo would be occurring in the absence of gap junctional communication as well as specific intercellular adhesion. Indeed, there is independent evidence to suggest that polarisation of some 8-cell blastomeres precedes both cell flattening and the establishment of gap junctional communication (Ziomek & Johnson, 1980; Goodall & John-son, in preparation). This would suggest that these latter intercellular features perform primarily an integrating function later in develop-ment by maintaining the radial symmetry of the embryo, organising the morphology of the blastocyst and orientating the two developing tissue types within it.

CONCLUSIONS

Interactions between the cells of the mouse embryo are clearly important to the development of the blastocyst. The present evidence suggests that at least two types of interaction, with distinct developmental consequences, are operating. First, a specific cell interaction leads to the induction of polarity in blastomeres of the 8-cell mouse embryo. The induction of polarity relies on focal contact and is not dependent upon cell flattening or junctional communication. The polarity is stable, even through division, and constitutes the mechanism by which the two cell lineages of the blastocyst are founded. Second, intercellular flattening and formation of specialised junctions follow polarisation, involve at least one specific and identified surface determinant and appear to be important for the continued orderly differentiation of the two cell lineages and their eventual organisation into a blastocyst.

Acknowledgements

We wish to acknowledge valuable discussion with Chris Graham and John Reeve. The original work reported in this chapter was supported by grants to M.H.J. from the Medical Research Council, Cancer Research Campaign and Ford Foundation.

REFERENCES

BABINET, C. & CONDAMINE, H. (1980). Antibodies as tools to interfere with developmental processes. In *Development in Mammals*, vol. 4, ed. M. H. Johnson, pp. 267–304. Amsterdam, London: Elsevier/North-Holland.

BARLOW, P., OWEN, D. A. & GRAHAM, C. F. (1972). DNA synthesis in the preimplantation mouse embryo. *Journal of Embryology and Experimental Morphology*, 27, 431–45.

DUCIBELLA, T. (1980). Divalent antibodies to mouse embryonal carcinoma cells inhibit compaction in the mouse embryo. *Developmental Biology*, 79, 356–66.

DUCIBELLA, T., ALBERTINI, D. F., ANDERSON, E. & BIGGERS, J. D. (1975). The preimplantation mammalian embryo: characterization of intercellular junctions and their appearance during development. *Developmental Biology*, 45, 231–50.

DUCIBELLA, T. & ANDERSON, E. (1975). Cell shape and membrane changes in the eight-cell mouse embryo: prerequisites for morphogenesis of the blastocyst. *Developmental Biology*, 47, 45–58.

DUCIBELLA, T. & ANDERSON, E. (1976). The effect of calcium on tight junction formation and fluid transport in the developing mouse embryo. *Journal of Cell Biology*, 70, 95a.

DUCIBELLA, T. & ANDERSON, E. (1979). The effects of calcium deficiency on the formation of the zonula occludens and blastocoel in the mouse embryo. *Developmental Biology*, 73, 46–58.

DUCIBELLA, T., UKENA, T., KARNOVSKY, M. J. & ANDERSON, E. (1977). Changes in cell surface and cortical cytoplasmic organization during early embryogenesis in the preimplantation mouse embryo. *Journal of Cell Biology,* **74**, 153–67.

DUNIA, I., NICOLAS, J. F., JAKOB, H., BENEDETTE, E. L. & JACOB, F. (1979). Functional modulation in mouse embryonal carcinoma cells by Fab fragments of rabbit anti-embryonal carcinoma cell serum. *Proceedings of the National Academy of Sciences, USA,* **76**, 3387–91.

GRAHAM, C. F. & DEUSSEN, Z. A. (1978). Features of cell lineage in preimplantation mouse development. *Journal of Embryology and Experimental Morphology,* **48**, 53–72.

GRAHAM, C. F. & LEHTONEN, E. (1979). Formation and consequences of cell patterns in preimplantation mouse development. *Journal of Embryology and Experimental Morphology,* **49**, 277–94.

HANDYSIDE, A. H. (1978). Time of commitment of inside cells isolated from preimplantation mouse embryos. *Journal of Embryology and Experimental Morphology,* **45**, 37–53.

HANDYSIDE, A. H. (1980). Distribution of antibody and lectin binding sites on dissociated blastomeres from mouse morulae: evidence for polarization at compaction. *Journal of Embryology and Experimental Morphology,* **60**, 99–116.

HANDYSIDE, A. H. (1981). Immunofluoresce techniques for determining the numbers of inner and outer blastomeres in mouse morulae. *Journal of Reproductive Immunology,* **2**, 339–50.

HANDYSIDE, A. H. & JOHNSON, M. H. (1978). Temporal and spatial patterns of the synthesis of tissue-specific polypeptides in the preimplantation mouse embryo. *Journal of Embryology and Experimental Morphology,* **44**, 191–9.

HOGAN, B. & TILLY, R. (1978). *In vitro* development of inner cell masses isolated immunosurgically from mouse blastocysts. I. Inner cell masses from 3.5 day pc. blastocysts incubated for 24 hr before immunosurgery. *Journal of Embryology and Experimental Morphology,* **45**, 93–105.

HYAFIL, F., MORELLO, D., BABINET, C. & JACOB, F. (1980). A cell surface glycoprotein involved in the compaction of embryonal carcinoma cells and cleavage stage embryos. *Cell,* **21**, 927–34.

IZQUIERDO, L., LOPEZ, T. & MARTICORENA, P. (1980). Cell membrane regions in preimplantation mouse embryos. *Journal of Embryology and Experimental Morphology,* **59**, 80–102.

JOHNSON, M. H. (1979). Molecular differentiation of inside cells and inner cell masses isolated from the preimplantation mouse embryo. *Journal of Embryology and Experimental Morphology,* **53**, 335–44.

JOHNSON, M. H. (1981a). Membrane events associated with the generation of a blastocyst. *International Review of Cytology,* (suppl. 12), 1–37.

JOHNSON, M. H. (1981b). The molecular and cellular basis of preimplantation mouse development. *Biological Reviews,* (in press).

JOHNSON, M. H., CHAKRABORTY, J., HANDYSIDE, A. H., WILLISON, K. & STERN, P. (1979). The effect of prolonged decompaction on the development of the preimplantation mouse embryo. *Journal of Embryology and Experimental Morphology,* **54**, 241–61.

JOHNSON, M. H., HANDYSIDE, A. H. & BRAUDE, P. R. (1977). Control mechanisms in early mammalian development. In *Development in Mammals,* vol. 2, ed. M. H.

Johnson, pp. 67–98. Amsterdam, London: Elsevier/North-Holland.

JOHNSON, M. H., PRATT, H. P. M. & HANDYSIDE, A. H. (1981). The generation and recognition of positional information in the preimplantation mouse embryo. In *Cellular and Molecular Aspects of Implantation*, ed. S. R. Glasser & D. W. Bullock, pp. 55–74. Plenum Press.

JOHNSON, M. H. & ZIOMEK, C. A. (1981a). The foundation of two distinct cell lineages within the mouse morula. *Cell*, **24**, 71–80.

JOHNSON, M. H. & ZIOMEK, C. A. (1981b). Properties of polar and apolar cells from the 16-cell mouse morula. *Wilhelm Rouxs' Archiv of Developmental Biology*, (in press).

KELLY, S. J., MULNARD, J. G. & GRAHAM, C. F. (1978). Cell division and cell allocation in early mouse development. *Journal of Embryology and Experimental Morphology*, **48**, 37–51.

KEMLER, R., BABINET, C., EISEN, H. & JACOB, F. (1977). Surface antigen in early differentiation. *Proceedings of the National Academy of Sciences, USA*, **74**, 4449–52.

LEHTONEN, E. (1980). Changes in cell dimensions and intercellular contacts during cleavage-stage cell cycles in mouse embryonic cells. *Journal of Embryology and Experimental Morphology*, **58**, 231–49.

MULNARD, J. G. (1967). Analyse microcinematographique de developpement de l'oeuf de sours du stade II au blastocyste. *Archives Biologie (Liege)*, **78**, 107–38.

MULNARD, J. & HUYGENS, R. (1978). Ultrastructural localization of nonspecific alkaline phosphatase during cleavage and blastocyst formation in the mouse. *Journal of Embryology and Experimental Morphology*, **44**, 121–31.

NICOLAS, J. F., KEMLER, R. & JACOB, F. (1981). Effects of anti-embryonal carcinoma serum on aggregation and metabolic cooperation between teratocarcinoma cells. *Developmental Biology*, **81**, 127–32.

PRATT, H. P. M., KEITH, J. & CHAKRABORTY, J. (1980). Membrane sterols and the development of the preimplantation mouse embryo. *Journal of Embryology and Experimental Morphology*, **60**, 303–19.

REEVE, W. J. D. (1981). Cytoplasmic polarity develops at compaction in rat and mouse embryos. *Journal of Embryology and Experimental Morphology*, **62**, 351–67.

REEVE, W. J. D. & ZIOMEK, C. A. (1981). Distribution of microvilli on dissociated blastomeres from mouse embryos: evidence for surface polarization at compaction. *Journal of Embryology and Experimental Morphology*, **62**, 339–50.

SPINDLE, A. I. (1978). Trophoblast regeneration by inner cell masses isolated from cultured mouse embryos. *Journal of Experimental Zoology*, **203**, 482–9.

SURANI, M. A. H. (1979). Glycoprotein synthesis and inhibition of glycosylation by tunicamycin in preimplantation mouse embryos: compaction and trophoblast adhesion. *Cell*, **18**, 217–27.

WILSON, I. B., BOLTON, E. & CUTTLER, R. H. (1972). Preimplantation differentiation in the mouse egg as revealed by microinjection of vital markers. *Journal of Embryology and Experimental Morphology*, **27**, 467–79.

ZIOMEK, C. A. & JOHNSON, M. H. (1980). Cell surface interaction induces polarization of mouse 8-cell blastomeres at compaction. *Cell*, **21**, 935–42.

ZIOMEK, C. A. & JOHNSON, M. H. (1981). Induction of polarity in mouse 8-cell blastomeres: specificity, geometry and stability. *Journal of Cell Biology*, (in press).

Gap junctional communication compartments and development

C. W. LO

University of Pennsylvania, Biology Department, Philadelphia, Pa. 19104

INTRODUCTION

Gap junctions have been found in all animal cell types which have been examined and yet despite their widespread abundance throughout the animal kingdom, their physiological function remains unknown. Since gap junctional channels permit the direct cytoplasmic exchange of molecules between cells, it has been suggested that perhaps they may be involved in regulating differentiation (Loewenstein, 1968; Sheridan, 1976). It would be especially interesting to consider the possibility that if chemical gradients were involved in the formation and/or maintenance of developmental compartments that gap junctions may provide an intracellular pathway for mediating their formation (Lo & Gilula, 1980a, b; Tickle, 1980; Wolpert, 1978).

The presence of gap junctions between cells can be directly detected by the use of microelectrodes to inject ions or fluorescent tracer molecules. Since gap junctions behave as molecular sieves with exclusion limits of approximately 1500 in MW (Simpson, Rose & Loewenstein, 1977), ions or small molecules should readily pass between cells linked via gap junctional channels. If gap junctional channels are present, the passage of ions can be detected by positioning another electrode in a second cell to monitor the simultaneous voltage deflection expected from the current injected in the first cell. This type of exchange is referred to as ionic coupling or electrical coupling. Similarly, the spread of injected fluorescent molecules will allow the direct visualization of the gap junction mediated cell–cell exchange. As a result of the sieve-like nature of the junctional channels, in most cases, whenever ionic coupling has been detected, the spread of injected fluorescent dyes or dye coupling has similarly been observed.

GAP JUNCTION COMMUNICATION
IN THE EARLY MOUSE EMBRYO

The gap junctional communication properties of the early mouse embryos are consistent with and suggestive of a possible role of gap junctions in the regulation of developmental processes (Lo & Gilula, 1980a, b). In the early preimplantation embryo, gap junctional communication is expressed for the first time at the late eight-cell stage (Fig. 1) and starting at this stage, all the cells of the embryo are completely linked to one another by gap junctional channels. Thus dye injected into one cell resulted in spread to all the cells of the embryo. It is also at about this time that the first determination/differentiation event takes place – that of trophectoderm formation (see Gardner & Rossant, 1976, for review). It has been determined that the early determination/differentiation events in the mouse embryo are position dependent such that the outside cells differentiate to form trophoblast early on (Tarkowski & Wroblewska, 1967) and later the outside cells of the inner cell mass (ICM) cells differentiate to form extraembryonic endoderm (Rossant, 1978). This has been referred to as the 'inside–outside' hypothesis since the geographical position of each blastomere determines its ultimate developmental fate. It has been proposed that perhaps gap junctions which interlink all the cells of the late eight-cell embryo might allow the passive exchange of molecules between the cells of the late embryo such that a gradient can be established to distinguish the inside position from the outside position (Lo & Gilula, 1980a). In addition, tight junctions are also formed for the first time at this stage (Magnuson, Jacobson & Stackpole, 1978) and they eventually form a complete permeability barrier around the entire apical surfaces of the outermost cell layer. This permeability seal can then further enhance differences in the microenvironment of the outside cells from the inside cells and gap junctions may play a passive role in mediating the formation of an intracellular inside–outside gradient. The creation of a special microenvironment by the tight junctions cannot by itself explain the inside–outside signalling involved in trigerring specific differentiation since experiments by Pedersen *et al.* (Pedersen & Spindle, 1980) have shown that the normal development of a mouse embryo to the blastocyst stage can take place inside the cavity of a giant blastocyst. In normal mouse embryogenesis, the embryo forms the blastocyst at around day four. If several mouse embryos are aggregated together early on (day one or two), all the blastomeres will undergo

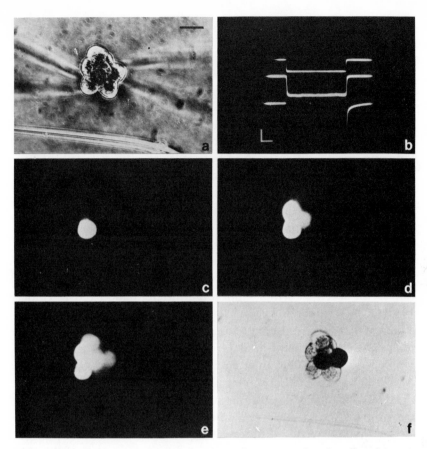

Fig. 1. Intercellular communication in the early compacted eight-cell embryo via junctional channels. (a) Two cells of an early compaction eight-cell embryo were each impaled with a microelectrode. (b) As current (top trace) was injected into the blastomere on the left, a voltage deflection was detected in that cell (bottom trace) and also in the impaled cell on the right (middle trace). The calibration bars are the same as for Fig. 3. (c–e) Fluorescein was injected into the blastomere on the left and fluorescence images were recorded at various times after the start of injection. (c) 4 min, (d) 16 min and (e) 25 min after the start of injection. (f) HRP was injected into the blastomere on the right. The dark reaction product indicated the presence of HRP in two blastomeres, the blastomere into which HRP was injected and also an adjacent blastomere. The scale bar in (a) = 46 μm in (a) and (c–f). From Lo & Gilula (1980a). © MIT.

regulation and give rise to one giant blastocyst. It was observed that young embryos (day two) inserted into such giant blastocysts will develop normally as independent embryos and form normal blastocysts as long as they remained afloat in the cavity. However, if cell–cell contact was established with the giant blastocyst, the embryos underwent development in conjunction with the giant blastocyst and formed only a secondary ICM, not a complete blastocyst. Thus it is clear that the 'inside' microenvironment alone is not sufficient for inducing 'inside' differentiation but rather specific cell–cell interactions are necessary for establishing the inside–outside positional signalling. All of these observations are suggestive of a role for gap junction mediated cell–cell exchange in regulating the early development of the mouse embryo.

Furthermore, from these studies, it was found that 'cell–cell communication compartments' are formed during later stages of mouse embryo development and this correlated temporally and geographically with specific further embryonic differentiation (Lo & Gilula, 1980b). Thus although in the early mouse embryo, from the eight-cell stage to the early postimplantation stages (see Fig. 2), dye injected into one cell resulted in spread to all the cells of the embryo, as development proceeded, the spread of injected dye became more and more limited. At the later postimplantation stages, dye injected into cells of the inner cell mass did not pass into the surrounding trophoblast cells even though they were in direct cell–cell contact (Fig. 3). In addition, as the cells of the ICM further differentiated, they underwent further segregation in dye spread so that dye injected into one region of the ICM resulted in spread to some of the adjacent cells while other regions of the ICM were impervious to the dye spread (Fig. 4). This segregation of dye spread to specific groups of cells is referred to as the formation of communication compartments. These compartments are characterized by extremely sharp and distinct boundaries beyond which there is no dye spread. Interestingly, although dye spread was confined to specific regions of the ICM, electrical coupling was still detected across the boundary beyond which no dye spread was observed. Thus the trophoblast cells are electrically coupled to the ICM cells although no dye spread occurs between them. Therefore these communication compartments are said to be semi-independent since the maintenance of ionic coupling indicates the persistence of some level of communication. It is thought that communication compartments are probably formed as a result of a smaller number of junctional channels linking the cells at the boundary regions v. the cells within the compartment itself. This could

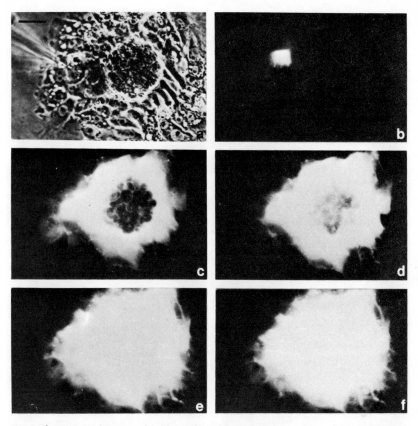

Fig. 2. Fluorescein dye spread in the early implanted mouse embryo. (a) A trophoblast cell of an early attached mouse embryo was impaled with a microelectrode. (b–f) Fluorescein was injected into the trophoblast cell and the fluorescence images were recorded at various times after the start of injection: (b) 4 min, (c) 8 min, (d) 16 min, (e) 20 min and (f) 31 min. Scale bar = 46 μm in (a–f). From Lo & Gilula (1980b). © MIT.

result in the drastic reduction in flux of molecules exchanged across the cells at the boundary regions and account for the apparent lack of dye transfer. Presumably dye may still be transferred across the cells at the boundary region but at such a low level that with continuous quenching, no dye spread is visible. On the other hand, ionic coupling, being a much more sensitive assay method, may still permit the detection of ionic exchange. The formation of these communication compartments can be roughly correlated in time with the differentiation of the ICM cells to embryonic ectoderm and extraembryonic endoderm. From these observations, it was hypothesized that perhaps the communication compartments formed in the mouse embryo are

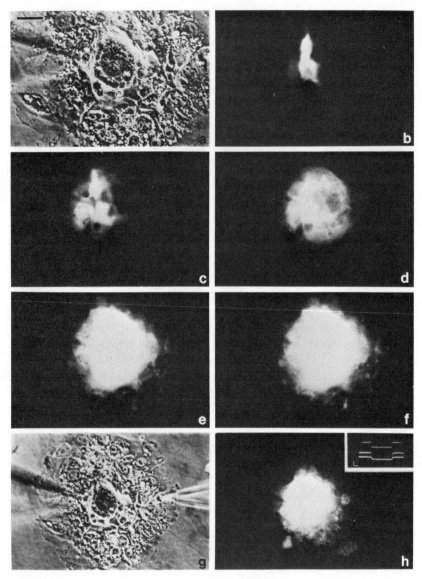

Fig. 3. Ionic coupling across the boundary of dye spread in a mouse embryo at a later stage of development. (a)The embryo was impaled with microelectrodes in a tropho-blast cell and in a cell bordering on the ICM region. (b–f) Fluorescein was injected into the cell in the ICM region and the fluorescence images were recorded at (b) 4 min, (c) 9 min, (d) 12 min, (e) 19 min and (f) 31 min after the start of injection. (g–h) Lower magnification of (a) and (f). Note that the impaled trophoblast cell is beyond the region filled by the dye. (Inset) As a current pulse (top trace) was passed during the fluorescein injection from the microelectrode on the left, a voltage deflection was detected in that cell (middle trace) and also in the trophoblast cell (bottom trace) on the right. The horizontal calibration bar = 100 ms and the vertical calibration bar = 5 nA and 10 mV. The scale bar in (a) = 46 μm in (a–f).

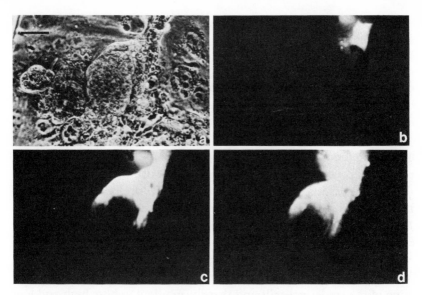

Fig. 4. Segregation of fluorescein dye transfer in the ICM of a well developed late implanted embryo. (a) A late attached mouse embryo was impaled at the ICM region with a microelectrode. (b–d) Fluorescein was injected into the impaled cell and fluorescence images were recorded at various times after the start of dye injection: (b) 4 min, (c) 13 min and (d) 31 min. Scale bar = 46 μm. From Lo & Gilula (1980b). © MIT.

involved in mediating the formation of gradients via the passive cytoplasmic exchange of molecules through gap junctional channels. Thus within each communication compartment separate gradients could be formed, each leading to the establishment of a developmental compartment. The determination and differentiation of each cell within such compartments can then be specified by its position relative to the boundaries established. The continued presence of a low level of communication between compartments may be important for allowing the coordination and regulation of determination/differentiation events between different compartments.

Similar to the mouse embryos, apparent changes in cell–cell communication as detected by dye coupling are also observed to accompany the development of the molluscan embryo (de Laat et al., 1980; also see the chapter by van den Biggelaar in this volume) and can also be correlated with specific cell differentiation processes, namely the determination of the mesentoblast and the establishment of dorsal–ventral polarity. Thus in light of the observations in the mouse and the molluscan embryo, it would seem highly plausible that communication compartments may play an important role in development.

GAP JUNCTIONAL COMMUNICATION AND DEVELOPMENTAL COMPARTMENTS

In the mouse embryo, it was not possible to directly test this hypothesis of gap junctional involvement in the regulation of development since there are no experimental means at the present for the detection and mapping of developmental compartments in the mouse.

Another system in which it would be interesting and important to determine the pattern of gap junctional communication is that of the epidermal cells of insect cuticle. The pattern of insect cuticle is specified by the underlying epidermal cells which are responsible for its biogenesis. A comparison of the normal cuticular patterns of the insect *Rhodnius* with those in which the integument has been transplanted or rotated (Locke, 1959, 1960, 1967; Lawrence, Crick & Munro, 1972) revealed that within each segment there is a gradient of positional information which determines the polarity and the specific differentiation of the epidermal cells and that the gradient, whose high and low points are defined by the segmental boundary, is repeated within each segment. In addition, cell lineage studies revealed that the segmental boundary represents a compartment border so that clones generated after the blastoderm stage never cross the segmental boundary (Lawrence, 1973). Thus each segment is derived from a small group of primordial cells which once having been set aside for one segment (compartment) will never cross over to an adjacent segment (compartment). If gap junctions were involved in mediating the transmission of morphogenetic gradients, which presumably would direct the differentiation within each compartment, then to a first approximation, cell–cell communication ought not to be present between the epidermal cells of two adjacent segments at the segmental boundary. Ultrastructural analysis demonstrated the presence of gap junctions between cells at the boundary (1975) and, moreover, studies by Warner & Lawrence (1973) demonstrated the existence of ionic coupling across the segmental boundary. Such findings are extremely provocative and suggest that either gap junctions are not involved in mediating the formation of compartments or that cell–cell communication at the segmental boundary must somehow be different from the cell–cell communication present within each segment (and thus providing for a discontinuity at the boundary regions). Studies of cell–cell communication in the developing mouse embryo have suggested that in fact the latter alternative may be correct. As discussed above, it was found that quantitative differences in communication at presumptive compartment boundaries can lead to the formation of semi-independent communication compartments. This was suggested by observations that boundary lines were detected by dye injection experiments which were not otherwise visible by ionic coupling studies. From these studies, it became clear that in evaluating the role of gap junctional communication in development, what may be important in many cases is not necessarily the qualitative presence or absence of cell–cell communication but rather the quantitative level or the degree of coupling present. Thus if gap junctional communication is important in mediating the formation of segmental compartments in *Rhodnius*, one would predict that the use of ionic coupling measurements, which is a more sensitive assay method, would allow the detection of coupling across the segmental boundary but that the spread of injected dye would not be observed or would demonstrate a discontinuity at the boundary.

The wing disc of *Drosophila* is also well suited for testing the hypothesis of gap junctional involvement in the regulation of development since studies on fate mapping of the imaginal discs (Bryant, 1975) in conjunction with the use of X-ray induced mitotic recombination for tagging cells have allowed the clear demarcation of developmental compartment boundaries (Garcia-Bellido, Repoll & Morata, 1973; Steiner, 1976). If cell–cell communication is involved in mediating the formation of developmental compartments, a map of the communication compartments, present in the wing disc at any stage of development should be directly superimposable on a map of the disc's developmental compartments. In other words, if this hypothesis is on the right track, the boundaries of both compartments should be identical in the simplest case or in more complex cases, one should at least be a subset of the other. It should be emphasized here that it is entirely possible that dye injection studies may reveal the presence of compartments which are not detected by cell lineage analysis since cell lineage studies are limited by the necessity of having large enough clone patches for allowing the visualization of compartment borders. From ultrastructural study of discs, it is known that gap junctions are present between cells of the imaginal discs (Poodry & Schniederman, 1970). It seems that the distribution of gap junctions in the wing disc is not only polarized but in fact there are fewer junctions present in the folded regions (Ryerse, 1980). This would be consistent with a role of gap junction in the regulation of pattern formation since the available fate maps of imaginal discs demonstrate that folded regions of discs often serve as dividing lines separating different cuticular structures which correspond to different developmental compartments (see Poodry, 1980 for review). Preliminary studies of dye coupling in imaginal wing disc of *Drosophila melanogaster* have demonstrated the presence of communication compartments reminiscent of those observed in the mouse embryo (see Fig. 5). Second and third larval instar wing discs were impaled with microelectrodes at different regions and fluorescein was injected by iontophoresis. Fig. 5 illustrates a mid third larval instar wing disc impaled at the wing pouch area. The injected dye initially filled a small group of cells very brightly and more dimly a surrounding group of cells circumscribing a trapezoid-like shape. With continued dye injection over time, the dye brightly filled all the cells within the originally dimly outlined trapezoid shaped region. Dye however did not readily pass beyond the confines of this group of cells. Observations of other impalements seem to further indicate that in fact there may be boundaries formed at some regions of folds present in the disc; that is some folds seem to impede the movement of dye so that communication compartments are observed which have boundaries that coincide with specific folds in the disc. All these observations are consistent with the presence of communication compartments in the developing wing disc. These communication compartments are probably also semi-independent in nature such that as in the case of the communication compartments in the mouse embryo, the boundaries revealed by the dye spread probably represent regions of reduced flux. Thus dye molecules probably can cross the boundaries but at a much reduced rate. In fact, using a more sensitive illumination and optical system, limited dye spread is observed beyond the boundaries in the wing disc but at a much lower flux. Hopefully, with further careful and detailed analysis, it will be possible to determine whether these communication compartments are involved in mediating the formation and/or the maintenance of developmental compartments in the developing wing disc. The study of discs of homoeotic mutants will be especially useful in these analyses (Morata & Garcia-Bellido, 1976).

From these studies, it is clear that gap junctional communication compartments are formed and maintained during development. Perhaps the understanding of how they are formed and what their importance is may lead us ultimately to understanding what role cell–cell communication plays in developmental processes.

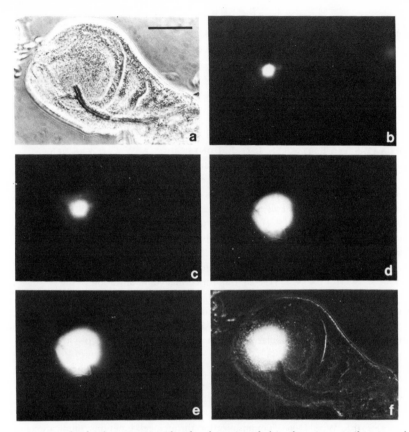

Fig. 5. An early third instar wing disc has been impaled in the wing pouch area and fluorescein iontophoresis was carried out for an extended time period. (a) Phase image of the wing disc at the start of injection. (b–f) Fluorescence images at different times after the start of injection: (b) 6 min, (c) 9 min, (d) 86 min, (e) 110 min (f) 113 min after the start of injection. (f) Is a combined phase and darkfield fluorescence image. Note that the shape and limits defined by the pattern of dye spread is already visible in (b) and (c) as a faint fluorescence delineating a trapezoid-shaped region. By 86 min, this pattern has become fully established and further injection did not increase the extent of dye spread. Scale bar = 100 μm.

REFERENCES

BRYANT, P. (1975). Pattern formation in the imaginal wing disc of *Drosophila melanogaster*. Fate map regeneration and duplication. *Journal of Experimental Zoology*, **193**, 49–77.

GARCIA-BELLIDO, A., REPOLL, P. & MORATA, G. (1973). Developmental compartmentalization of the wing disc of *Drosophila*. *Nature*, **245**, 251–3.

GARDNER, R. L. & ROSSANT, J. (1976). Determination during embryogenesis. In *Embryogenesis in Mammal: Ciba Foundation Symposium 40*, pp. 5–25. Amsterdam: Elsevier.

DE LAAT, S. W., TERTOOLEN, L. G. J., DONESTEIGN, A. W. C. & VAN DEN BIGGELAAR, J. A. M. (1980). Intercellular communication patterns are involved in cell determination in early molluscan development. *Nature*, **287**, 546–8.

LAWRENCE, P. A. (1973). A clonal analysis of segment development in *Oncopeltus* (Hemiptera). *Journal of Embryology and Experimental Morphology*, **30**, 681–99.

LAWRENCE, P. A., CRICK, F. H. C. & MUNRO, M. (1972). A gradient of positional information in an insect, *Rhodnius*. *Journal of Cell Science*, **11**, 815–53.

LAWRENCE, P. A. & GREEN, S. M. (1975). The anatomy of a compartment border. The intersegmental boundary in *Oncopeltus*. *Journal of Cell Biology*, **65**, 373–82.

LO, C. W. & GILULA, N. B. (1980a). Gap junctional communication in the preimplantation mouse embryo. *Cell*, **18**, 399–409.

LO, C. W. & GILULA, N. B. (1980b). Gap junctional communication in the postimplantation mouse embryo. *Cell*, **18**, 411–22.

LOCKE, M. (1959). The cuticular pattern in an insect, *Rhodnius prolixus*. *Journal of Experimental Biology*, **36**, 459–77.

LOCKE, M. (1960). The cuticular pattern of an insect. The intersegmental membrane. *Journal of Experimental Biology*, **27**, 398–406.

LOCKE, M. (1967). The development patterns in the integument of insects. *Advances in Morphogenesis*, **6**, 33–8.

LOWENSTEIN, W. R. (1968). Communication through cell junctions: implications in growth and differentiation. *Developmental Biology*, (suppl. 2), 151–83.

MAGNUSON, T., JACOBSON, J. B. & STACKPOLE, C. W. (1978). Relationship between intercellular permeability and junction organization in the preimplantation mouse embryo. *Developmental Biology*, **67**, 214–24.

MORATA, G. & GARCIA-BELLIDO, A. (1976). Developmental analysis of some mutants of the bithorax system of *Drosophila*. *Wilhelm Roux's Archives*, **179**, 125–43.

PEDERSEN, R. A. & SPINDLE, A. I. (1980). The role of the blastocoel microenvironment in early mouse embryo differentiation. *Nature*, **284**, 550.

POODRY, C. A. (1980). Imaginal discs: morphology & development. In *The Genetics & Biology of Drosophila*, ed. M. Ashburner & E. Novitski. London, New York: Academic Press.

POODRY, C. A. & SCHNIEDERMAN, H. A. (1970). The ultrastructure of the developing leg of *Drosophila melanogaster*. *Wilhelm Roux's Archives*, **166**, 1–44.

ROSSANT, J. S. (1978). Cell commitment in early rodent development. In *Development in Mammals*, vol. 4, ed. Martin Johnson. Amsterdam: Elsevier.

RYERSE, J. S. (1980). Gap junction distribution in *Drosophila melanogaster* wing discs. *American Zoologist*, **20**, 914.

SHERIDAN, J. D. (1976). Cell coupling and cell communication during embryogenesis. In *The Cell Surface in Animal Embryogenesis*, ed. Poste & G. L. Nicolson. Amsterdam: Elsevier.

SIMPSON, I., ROSE, B. & LOWENSTEIN, W. R. (1977). Size limit of molecules permeating the junctional membrane channels. *Science*, **195**, 294–6.

STEINER, E. (1976). Establishment of compartments in developing leg discs of *Drosophila melanogaster*. *Wilhelm Roux's Archives*, **180**, 9–30.

TARKOWSKI, A. K. & WROBLEWSKA, J. (1967). Development of blastomeres of mouse eggs isolated at the four and eight cell stage. *Journal of Embryology and Experimental Morphology*, **18**, 155–80.

TICKLE, C. (1980). The polarizing region and limb development. In *Development in Mammals*, vol. 4, ed. Martin H. Johnson. Amsterdam: Elsevier.

WARNER, A. E. & LAWRENCE, P. A. (1973). Electrical coupling across developmental boundaries in insect epidermis. *Nature*, **245**, 47–8.

WOLPERT, L. (1978). Gap junctions: channels for communication in development. In *Intercellular Junctions and Synapses*, ed. N. B. Gilula & J. Pitts.

Cellular organisation in the early molluscan embryo

J. A. M. VAN DEN BIGGELAAR AND A. W. C.
DORRESTEIJN

Zoological Laboratory, University of Utrecht, Padualaan 8, 3508 TB Utrecht, The Netherlands

We start at a macroscopic level, and all the results of our measurements, even those of the microscopic world, at some time refer back to the macroscopic world. (Ilya Prigogine, 1980, p. 15)

INTRODUCTION

Any reversed film or retrospection of the embryonic development of whatever multicellular organism will precisely show the cell lineage of the different parts of that particular embryo. For species with a mosaic development, it is characteristic that once the cell lineage of one individual embryo has been retraced, one can predict how each individual cell will contribute to the development of the different parts of any other embryo of the species. This property depends on the constancy of the geometrical relations between the successive generations of cells. Without any variability all embryos of the same species are compelled to repeat exactly the same cell configurations at successive cleavage stages.

The possibility to predict the developmental fate of a cell according to its relative position in the cell pattern, may depend on two mechanisms of cell determination. First, it may indicate that the specific contribution of a particular blastomere is the result of its special position in the embryo. Second, the contribution of a particular cell to a well-defined part of the embryo may be the result of the inheritance of a special part of the ooplasm. The latter possibility presupposes the formation of qualitatively different regions which are subsequently compartmentalised according to a precisely determined cleavage pattern. A great number of observations on molluscan development actually demonstrate remarkable examples of ooplasmic segregation leading to a high degree of cytoplasmic localisations (Wilson, 1904; Ries & Gersch, 1936; Raven, 1945, 1958; Dohmen & Lok, 1975; Dohmen & Verdonk, 1974). This ooplasmic segregation reflects a longitudinal organisation of the egg. Each level along the egg axis gradually obtains different developmental capacities. This is especially clear for the vegetal part, as its removal leads to specific developmental deficiences (Wilson, 1904; Clement, 1952; van Dongen, 1975, 1976; van Dongen & Geilenkirchen, 1974, 1975). Experimental evidence for a definite organisation of the undivided egg in transverse or dorsoventral direction is lacking. Dorsoventral organisation requires cellular interactions (van den Biggelaar, 1976, 1977; van den Biggelaar & Guerrier, 1979).

CLEAVAGE PATTERN

A characteristic aspect of the cleavage pattern in molluscs is that only the first two cleavages are meridional. For the sake of convenience the minor deviation that may be observed at second cleavage, is neglected. In the following we will only consider the species with an equal four-cell stage. An embryo in which the first two cleavages are equal has four qualitatively equal blastomeres, and lacks any differentiation in transverse direction.

In a number of animal species the third cleavage is perpendicular to the egg axis. During further division an alternation of vertical and horizontal cleavages can be observed. This regular alternation produces radially symmetrical embryos that can be divided into two symmetrical halves by an infinite number of meridional planes. In the molluscan embryo there is no such alternation of cleavages parallel or perpendicular to the egg axis. From the four-cell stage the subsequent generations of mitotic spindles have an oblique position with respect to the egg axis and each blastomere is divided into a slightly more animal and a slightly more vegetal daughter cell. The regularity in this so-called spiral cleavage is an alternation of clockwise (dextral) and anti-clockwise (sinistral) divisions. As a consequence of this alternation of sinistral and dextral cleavages, the descendants of each of the blastomeres of the four-cell stage show the same zigzag superposition in animal-vegetal direction (Fig. 1). Because of the constancy of the geometrical relations between the successive generations of cells, the original quadrants can be distinguished continuously.

Because of the longitudinal organisation of the egg, the zigzag cleavage pattern produces qualitatively unequal daughter cells. The previously mentioned possibility to predict the presumptive capacities and the presumptive fate of the individual blastomeres may thus, at least partly, be related to the inheritance of a particular part of the original egg structure.

(a)   (b)   (c)

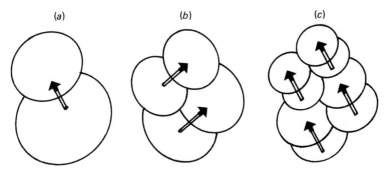

Fig. 1. Cleavage pattern of one quadrant showing the alternation of dextral and sinistral cleavages. (a) Third cleavage, 8-cell stage; (b) fourth cleavage, 16-cell stage; (c) fifth cleavage, 32-cell stage.

EPIGENETIC DETERMINATION OF THE CELL LINES

From cell lineage studies it is known that the major parts of the foot and the shell gland are mainly formed by descendants of the second quartet cell of only one quadrant. The stem cell of the mesoderm appears to be produced by the same quadrant (Raven, 1958). Experimental morphological experiments have shown that the quadrants originally have identical developmental capacities (van den Biggelaar & Guerrier, 1979). This means that up to a certain moment each quadrant has the developmental potency to develop any quadrant of the embryo. For each special function each quadrant has an equally qualified candidate.

The following is a phenomenological description of the way in which the embryo succeeds in making the inevitable choice without making use of gene products. The diversification of the quadrants is realised in two steps. The first step is carried out at the four-cell stage. Theoretically, the cells may then form two configurations. One in which one pair of opposite quadrants makes a cross-furrow at the animal pole, and the other pair at the vegetal pole. In the second configuration two quadrants make both cross-furrows, forcing the others into a more lateral position. Usually the first configuration is realised with the two cross-furrows in a criss-cross position. This is not illogical as only in the criss-cross position can the embryo approach the spherical form. This pattern appears to have a morphogenetic significance as the two cells at the vegetal cross-furrow will develop the median quadrants of the future embryo. Thus the first step in the diversification of the quadrants is based upon the position at the vegetal cross-furrow. This is a purely topographical determination as the four cells still have the same developmental capacities.

In a sense, the embryo remains radially symmetrical at the following 8-cell and 16-cell stage. All blastomeres are localised at the periphery, there are no inside or insular cells. Depending on the species, the 16-cell stage may lead to a 24-cell or a 32-cell embryo. In *Lymnaea* and *Physa* the eight cells of the first quartet of micromeres skip one division round, and the embryo passes from a 16-cell into a 24-cell stage. In *Gibbula* (*Trochus*), *Haliotis* and *Patella* all cells divide at fifth cleavage, and 32-cell embryos are formed. Up to this stage the cleavages have been almost synchronous. Whether the embryo is built up of 24 or 32 cells, after fifth cleavage the cleavage cavity disappears and a more or less solid mass of cells is formed with one of the median macromeres in a nearly insular position (Fig. 2). In an attempt to simulate the successive cleavage stages with soap bubbles, Robert (1902) did not succeed in obtaining configurations in which more than 16 bubbles all occupy a peripheral position, none of them being insular. Herbert & Graham (1974) observed that during the early development of the mouse embryo the 16-cell stage was the most advanced cell stage without inside cells. It is quite plausible that in any solid aggregation of cells in which each member tries to combine a maximal volume with a minimal surface, it becomes more and more difficult to obtain an equilibrium configuration without inside cells if the number of cells passes the approximately critical value of 16. Embryos can only form inside cells if one or more cells can disrupt the superficial contacts with the neighbouring cells. This condition is not fulfilled in the molluscan embryo and an intermediate configuration is realised, in which one of the two cross-furrow macromeres is in a nearly inside position and makes contacts with a large number of micromeres. If the pattern of the closely packed cells after fifth cleavage would merely be the result of surface tension or an external pressure by the vitelline membrane, it would be difficult to explain how all embryos consistently reproduce one and the same configuration. The invariability with which only one out of a great number of theoretically possible cell patterns is realised, necessarily implies the presence of additional guiding forces that depend on the properties of the individual cells. A real candidate for this second pattern-generating force is a differential adhesivity between different cells.

In *Lymnaea* and *Patella* the equatorial blastomeres become less adhesive after fifth cleavage (van den Biggelaar, 1976, 1977). Simultaneously, the adhesivity in the animal and vegetal tiers remains relatively strong or increases. As a consequence, the animal and vegetal cells become

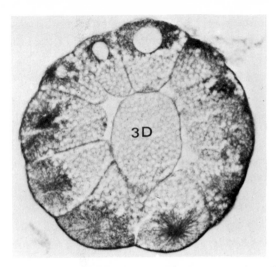

Fig. 2. Section perpendicular to the egg axis of a 32-cell embryo of *Patella* showing the central position of the macromere 3D.

columnar and protrude far inside the embryo. The less adhesive equatorial cells are pulled apart into a more peripheral position. The animal micromeres and vegetal macromeres approach each other. Finally, the only possible equilibrium position appears to be reached when one of the two cross-furrow macromeres is in a nearly inside position. As the two cross-furrow macromeres are located more centrally from the outset, it is not surprising that one of them becomes central and not one of the two more peripheral non-cross-furrow macromeres. Thus the second step in the diversification of the quadrants is the topographical differentiation of the two cross-furrow macromeres into a central and non-central macromere during the interval between fifth and sixth cleavage.

Summarising, it can be concluded that the embryo's problem of how to make equipotential cells unequal without changing their material composition is resolved by the purely mechanical impossibility for equipotential cells to occupy equivalent topographical positions. Essentially, it is the geometrical differences within the cell patterns that initiate inequality between initially equal counterparts (van den Biggelaar *et al.*, 1981). As in most analogous situations, a privileged position must be exploited by making use of the relational possibilities for communication with members which are not in contact with the counterparts. It appears that by the central position of one macromere, the others are excluded from making contact with the overlying micromeres. The contacts with the animal micromeres are essential for the induction of the central macromere to develop the stem cell of the mesoderm, whereas the other macromeres will only contribute to the formation of the endoderm. Once the macromere that will produce the stem cell of the mesoderm is induced, the corresponding quadrant is determined to develop the dorsal part of the embryo, the opposite quadrant becomes ventral and the other two the right and left part. This shows that the diversification of the four quadrants, and henceforth the determination of the different cell lines is the epigenetic result of merely geometrical differences.

CELLULAR INTERACTIONS IN RELATION TO THE DETERMINATION OF THE QUADRANTS

With the assumption of an inductive influence from the animal micromeres upon the central macromere, we enter the domain of cellular interactions. The central problem now becomes, which kind of message is transmitted between the micromeres and the central 3D macromere, and which kind of communication channel is used. We will restrict ourselves to an analysis of the following transmission mechanisms that are possibly involved: (1) the formation of intercellular junctions, especially gap junctions, (2) the resorption of microvilli protruding from one cell into another one. Diffusion of signal substances, and the action of surface components of the micromeres upon the cell surface of the 3D macromere, has not been investigated.

Intercellular junctions

The possibility of cell communication by means of specialised contacts motivated an analysis of the distribution pattern of cellular junctions. Elbers (1959) and Bluemink (1967) have described the presence of intermediate junctions in the *Lymnaea* embryo. Berendsen (1971) reported the formation of septate junctions at the two-cell stage of *Lymnaea*. The presence of both types of junctions has been confirmed in a detailed analysis of the successive cleavage stages (Dorresteijn, personal observations). The septate junctions probably seal the peripheral boundaries. They may also keep the blastomeres in a coherent peripheral layer and prevent the formation of complete island cells as discussed in the preceding paragraph. In *Patella* we have found intermediate junctions but no septate junctions. Although septate or tight junctions may be involved in the transmission of messages from one cell to another (see the chapter by Sheridan, this volume), the generally accepted candidate for this function is the gap junction. In *Lymnaea* as well as in *Patella* we have found gap junctions from the four-cell stage onwards.

In *Lymnaea* the distribution of gap junctions between the different pairs of blastomeres has been investigated up to the sixth cleavage. For each cell stage two embryos have been examined. We will distinguish two types of gap junctions: homotypic junctions between cells from one tier interconnecting different quadrants, and heterotypic gap junctions coupling cells of different tiers. It should be realised that the necessarily limited number of embryos that could be studied, implies that the observed distribution pattern of gap junctions has no more than a relative significance. As the obtained pattern may very well represent no more than a snapshot of constantly appearing and disappearing junctions, the decision to investigate only two embryos renders the result still more relative.

At the eight-cell stage homotypic junctions have been found between the macromeres but not between the micromeres. Heterotypic junctions are present between the first quartet of micromeres and the macromeres. As the macromeres are involved in both categories of gap junctions, it may be concluded that gap junctions are relatively more frequently formed in the vegetal hemisphere than in the animal hemisphere.

The 16-cell stage is composed of four tiers of cells, the animal micromeres $1a^1-1d^1$, the lower tier of sister cells $1a^2-1d^2$, the second quartet cells 2a–2d, and the vegetal macromeres 2A–2D. Homotypic junctions are only possible at the two poles of the egg axis, because only there are corresponding cells in touch with each other. In both of these tiers, $1a^1-1d^1$ and 2A–2D, homotypic gap junctions have been found. Heterotypic junctions have been observed between all successive tiers, except with the animal tier. At this moment we do not have much information on whether gap junctions formed during preceding cell stages are preserved during later stages and parcelled out between the daughter cells. During fourth cleavage of the *Lymnaea* embryo we have found gap junctions between dividing cells.

The 24-cell stage in *Lymnaea* is reached by division of the second quartet cells 2a–2d and the macromeres 2A–2D; the cells $1a^1-1d^1$ and $1a^2-1d^2$ do not divide. The homotypic junctions found between the members of the tier $1a^1-1d^1$ as well as heterotypic junctions between members of these two tiers may already have been formed in the preceding 16-cell stage. All types of heterotypic junctions have been found, except between members of the animal quartet $1a^1-1d^1$ and the quartet $2a^1-2d^1$. We have repeatedly postulated that the quadrants are initially equal. Therefore we have made a preliminary comparison of the distribution of gap junctions between members of the same quadrant and between members of different quadrants. The results did not point to preferred combinations with one particular quadrant. After the determination of the central 3D macromere relatively few gap junctions appeared to be formed at the sites of contact between the central macromere and the overlying micromeres.

An analysis of the distribution pattern of gap junctions in *Patella* is not yet available. The results obtained so far have not shown gap junctions between the central macromere and the micromeres of the animal hemisphere.

Summarising, it may be concluded that the presence of gap junctions during each of the successive cleavage stages after second cleavage provides the embryo with a possible mechanism for intercellular communication of morphogenetic signals. At least in *Lymnaea* the distribution of gap junctions between the different pairs of blastomeres is rather uniform. The occurrence *per se* gives little information on their possible function in determinative processes. It is probable that by a selective regulation of their permeability they may function as a mechanism for a communication pattern between cells. For *Patella* the permeability of the gap junctions can be inferred from dye-coupling experiments that have been performed during the early cleavage stages (de Laat *et al.*, 1980). In this study the interest was focussed on the existence of communication channels between the macromeres and the animal micromeres. Therefore, only the transport of the fluorescent marker (Lucifer Yellow) from the macromeres to the neighbouring cells was investigated. It was shown that after labelling one of the macromeres at the 8-cell or 16-cell stage, the dye was not transferred to any of the adjacent cells. Apparently, the already present gap junctions do not transport the dye. This situation changes completely after fifth cleavage. About 40–50 minutes after the beginning of the 32-cell stage dye transfer has been observed from each of the macromeres to adjacent cells. Every macromere borders on the corresponding lower second quartet cell, on its third quartet cell, and on two or three other macromeres. The pattern of dye-coupling of the central macromere 3D differs from the transport pattern of the macromeres 3A, 3B and 3C. The latter three macromeres appeared to be coupled with each of the surrounding cells, whereas the transfer of the marker molecules from 3D to the adjacent second quartet cell was blocked (Fig. 3). As the macromeres were impaled shortly after fifth cleavage, thus before the determination of one macromere to the central 3D, the difference in the transport pattern cannot be attributed to previous differences between the macromeres. This implies that the permeability of the homotypic gap junctions between the macromeres as well as the heterotypic gap junctions between the macromeres and the micromeres is controlled by the interaction between the animal micromeres and the central 3D macromere. It also indicates that the topographical unique position of the central macromere is amplified by differences in the communication pattern with the neighbours. It is still a problem how to explain the absence of dye transfer from 3D to its derivative of the second quartet, $2d^2$, whereas the macromeres 3A, 3B, and 3C are coupled with the corresponding sister cells $2a^2$, $2b^2$ and $2c^2$, respectively. It also remains to be explained how the opening of the sluices between 3D and other cells is triggered. Once the gap junctions interconnecting 3D with its vegetal neighbours are opened, the gap junctions between the other macromeres and micromeres not bordering on 3D become permeable.

From earlier cell deletion experiments we know that the diversification of the four macromeres and the differentiation of the quadrants fail to appear if at the eight-cell stage the micromeres of the first quartet are removed (van den Biggelaar & Guerrier, 1979). If similarly, the animal micromeres control the permeability of the gap junctions of the central 3D and secondarily of the other macromeres, then removal of the first quartet cells should also result in the absence of dye-coupling between the four macromeres, and between the latter and neighbouring micromeres at the vegetal pole. We have been able to demonstrate that when all four micromeres are removed at the eight-cell stage, and the macromeres are injected with Lucifer Yellow after fifth cleavage, the dye is not transferred from the impaled macromere to any of the surrounding cells. One might object that the gap junctions of the macromeres remain closed because of the artificial opening between the intraembryonic and extraembryonic medium. Therefore, in a series of control experiments a second quartet cell was deleted after fourth cleavage to exclude the possibility that the contact between the internal and external medium could have closed the gap junctions. In these embryos the normal transport pattern was observed from the macromeres to surrounding cells.

Re-examining the coupling between the animal micromeres and the

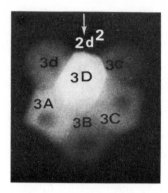

Fig. 3. Micrograph of a living 32-cell embryo of *Patella* in which the macromere 3D has been labelled with Lucifer Yellow after the beginning of fifth cleavage. Note the absence of label in the micromere 2d^2.

central 3D, we could not confirm our previous results together with de Laat *et al.* (1980). The fluorescence observed at the animal pole of living embryos with an impaled 3D macromere has been interpreted erroneously. After repeating these experiments we again obtained a luminescent spot at the animal pole. After sectioning these embryos we found the animal micromeres to be negative (Fig. 4). The luminescence at the animal pole must be explained by transmission of light by the central 3D. The impermeability of the gap junctions between 3D and the inducing animal micromeres shows that the transmission of the inducing signal is not mediated by gap junctions. This conclusion is only valid if the permeability of the gap junctions is not unidirectional. This possibility is excluded as injection of Lucifer Yellow into one of the animal micromeres after fifth cleavage did not result in a fluorescent 3D cell. The latter may not be conclusive if the dilution of the dye from a relatively small micromere into an approximatively seven-times larger 3D cell is too strong.

Transport of membrane vesicles

A second mechanism for the transmission of signal molecules from the micromeres to the central 3D might be the resorption of microvilli from the micromeres by 3D. This mechanism has not been investigated systematically. Dohmen & van der Mast (1978) have observed the constriction of annular nexuses from the macromere 3D into the overlying micromeres in the embryo of *Lymnaea*. In *Patella* we have seen the protrusion of microvilli from a great number of blastomeres into

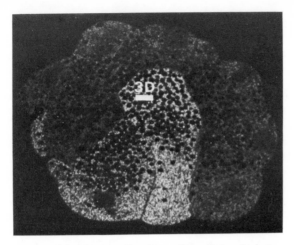

Fig. 4. Median section through a 32-cell embryo of *Patella* in which the macromere 3D has been labelled with Lucifer Yellow after the beginning of fifth cleavage. Note the absence of label in the animal micromeres contacting 3D.

adjacent cells. Further investigations will be necessary to obtain more information about the function of these microvilli.

SUMMARY

The development of the molluscan embryo is a classic example of the progressive determination of originally equipotential cells. Ooplasmic segregation is the driving force in the diversification of the blastomeres along the longitudinal egg axis. The diversification in transverse direction is the result of differences in the geometrical relations of equipotential cells. Intercellular communication as revealed by the transfer of Lucifer Yellow from the macromeres to surrounding cells is controlled by a preceding interaction of the central 3D macromere with the cells of the first quartet of micromeres. This latter contact is not accompanied by dye-coupling.

Acknowledgement

We thank Professor Dr N. H. Verdonk and Dr M. R. Dohmen for critically reading the manuscript.

REFERENCES

BERENDSEN, W. (1971). Morphologische analyse van de eerste celdeling van het ei van *Lymnaea stagnalis* L. Een electronenmicroscopisch onderzoek. Ph. D. thesis. University of Utrecht.

BIGGELAAR, J. A. M. VAN DEN (1976). Development of dorsoventral polarity preceding the formation of the mesentoblast in *Lymnaea stagnalis*. *Proceedings of the Koninklijke Nederlandse Akademie van Wetenschappen, Amsterdam,* **C79,** 112–26.

BIGGELAAR, J. A. M. VAN DEN (1977). Development of dorsoventral polarity and mesentoblast determination in *Patella vulgata*. *Journal of Morphology,* **154,** 157–86.

BIGGELAAR, J. A. M. VAN DEN, DORRESTEIJN, A. W. C., LAAT, S. W. DE & BLUEMINK, J. G. (1981). The role of topographical factors in cell interaction and determination of cell lines in molluscan development. In *International Cell Biology 1980–1981,* ed. H. G. Schweiger, pp. 526–38. Berlin: Springer-Verlag.

BIGGELAAR, J. A. M. VAN DEN & GUERRIER, P. (1979). Dorsoventral polarity and mesentoblast determination as concomitant results of cellular interactions in the mollusk *Patella vulgata*. *Developmental Biology,* **68,** 462–71.

BLUEMINK, J. G. (1967). The subcellular structure of the blastula of *Lymnaea stagnalis* L. (Mollusca) and the mobilization of the nutrient reserve. Thesis, University of Utrecht.

CLEMENT, A. C. (1952). Experimental studies on germinal localizations in *Ilyanassa.* I. The role of the polar lobe in determination of the cleavage pattern and its influence in later development. *Journal of Experimental Zoology,* **121,** 593–625.

DOHMEN, M. R. & LOK, D. (1975). The ultrastructure of the polar lobe of *Crepidula fornicata* (Gastropoda, Prosobranchia). *Journal of Embryology and Experimental Morphology,* **34,** 419–28.

DOHMEN, M. R. & MAST, J. M. A. VAN DE (1978). Electron microscopical study of RNA-containing cytoplasmic localizations and intercellular contacts in the early cleavage stages of eggs of *Lymnaea stagnalis* (Gastropoda, Pulmonata). *Proceedings of the Koninklijke Nederlandse Akademie van Wetenschappen, Amsterdam,* **C81,** 403–14.

DOHMEN, M. R. & VERDONK, N. H. (1974). The structure of a morphogenetic cytoplasm, present in the polar lobe of *Bithynia tentaculata* (Gastropoda, Prosobranchia). *Journal of Embryology and Experimental Morphology,* **31,** 423–33.

DONGEN, C. A. M. VAN (1975). The development of *Dentalium* with special reference to the significance of the polar lobe. VI. Differentiation of the cell pattern in lobeless embryos of *Dentalium vulgare* (da Costa) during late larval development. *Proceedings of the Koninklijke Nederlandse Akademie van Wetenschappen, Amsterdam,* **C79,** 245–66.

DONGEN, C. A. M. VAN (1976). The development of *Dentalium* with special reference to the significance of the polar lobe. VII. Organogenesis and histogenesis in lobeless embryos of *Dentalium vulgare* (da Costa) as compared to normal development. *Proceedings of the Koninklijke Nederlandse Akademie van Wetenschappen, Amsterdam,* **C79,** 454–65.

DONGEN, C. A. M. VAN & GEILENKIRCHEN, W. L. M. (1974). The development of

Dentalium with special reference to the significance of the polar lobe. I, II, III. Division chronology and development of the cell pattern in *Dentalium dentale* (Scaphopoda). *Proceedings of the Koninklijke Nederlandse Akademie van Wetenschappen, Amsterdam*, **C77**, 57–100.

DONGEN, C. A. M. VAN & GEILENKIRCHEN, W. L. M. (1975). The development of *Dentalium* with special reference to the significance of the polar lobe. IV. Division chronology and development of the cell pattern in *Dentalium dentale* after removal of the polar lobe at first cleavage. *Proceedings of the Koninklijke Nederlandse Akademie van Wetenschappen, Amsterdam*, **C78**, 358–75.

ELBERS, P. F. (1959). Over de beginoorzaak van het Li-effect in de morfogenese; een electrononmicroscopisch onderzoek aan eieren van *Lymnaea stagnalis* en *Paracentrotus lividus*. Thesis, University of Utrecht.

GUERRIER, P., BIGGELAAR, J. A. M. VAN DEN, DONGEN, C. A. M. VAN & VERDONK, N. H. (1978). Significance of the polar lobe for the determination of dorsoventral polarity in *Dentalium vulgare* (da Costa). *Developmental Biology*, **63**, 233–42.

HERBERT, M. C. & GRAHAM, C. F. (1974). Cell determination and biochemical differentiation of the early mammalian embryo. In *Current Topics in Developmental Biology*, vol. 8, ed. A. A. Moscona & A. Monroy, pp. 151–78. New York: Academic Press.

LAAT, S. W. DE, TERTOOLEN, L. G. J., DORRESTEIJN, A. W. C. & BIGGELAAR, J. A. M. VAN DEN (1980). Intercellular communication patterns are involved in cell determination in early molluscan development. *Nature*, **287**, 546–8.

PRIGOGINE, I. (1980). *From Being to Becoming. Time and Complexity in Physical Sciences*. Oxford: Freeman.

RAVEN, CHR. P. (1945). The development of the egg of *Lymnaea stagnalis* L. from oviposition till first cleavage. *Archives Néerlandaises de Zoologie*, **7**, 353–434.

RAVEN, CHR. P. (1958). *Morphogenesis: The Analysis of Molluscan Development*. London: Pergamon Press.

RIES, E. & GERSCH, M. (1936). Die Zelldifferenzierung und Zellspezialisierung während der Embryonalentwicklung von *Aplysia limacina* L. Zugleich ein Beitrag zu Problemen der vitalen Färbung. *Publicazione della Stazione Zoologica di Napoli*, **15**, 223–73.

ROBERT, A. (1902). Recherches sur le développement des troques. *Archives de Zoologie Expérimentale et Générale*, 3ᵉ Série, **10**, 269–359.

WILSON, E. B. (1904). Experimental studies on germinal localization. *Journal of Experimental Zoology*, **1**, 197–268.

Interactions of teratocarcinoma-derived cells in culture – a useful model?

M. L. HOOPER

Department of Pathology, University Medical School, Teviot Place, Edinburgh EH8 9AG

Study of the mechanisms which underlie the development of the mammalian embryo is complicated by its small size and limited accessibility to experimental intervention. For this reason the use of model systems imvolving mouse teratocarcinomas has become popular. Teratocarcinomas are tumours whose malignant stem cell, the embryonal carcinoma cell, shares with cells of the early embryo the capacity to differentiate into a variety of cell types, which may include derivatives of all three germ layers. Indeed, experimental evidence indicates that embryonal carcinoma cells may be regarded as intrinsically normal embryonic cells endowed with malignant properties by virtue of an abnormal microenvironment (reviewed by Solter & Damjanov, 1979; Martin, 1980). In particular, when placed in the environment of the normal embryo by injection into the blastocyst, embryonal carcinoma cells undergo normal differentiation and contribute to the tissues of the resulting chimeric embryo. Their study is thus highly relevant to an understanding of processes occurring in the normal embryo, and has the advantage that tumour lines and tissue culture lines may be established, providing a means of obtaining large quantities of experimental material in a controlled environment. Embryonal carcinoma lines isolated during early work were incapable of *in vitro* differentiation, but could be propagated as an essentially homogeneous population of stem cells (e.g. Fig. 1*a*) which on subcutaneous injection into a syngeneic mouse would produce differentiated tumours (Fig. 1*b*). Such lines have subsequently been found to respond *in vitro* to treatment with retinoic acid by differentiation into endoderm-like cells (Strickland & Mahdavi, 1978; Fig. 1*c*). More recently, embryonal carcinoma lines have been isolated which can be maintained as an undifferentiated stem cell population by growth on feeder layers of mitomycin C-treated fibroblasts (Fig. 1*d*), but which on removal from the feeder undergo extensive *in vitro* differentiation into a wide range of cell types (Fig. 1*e, f*). Cell interactions appear to play an essential role not only in the maintenance of the undifferentiated state by growth on the feeder layer, but also in the differentiation process, the first stage of which involves the formation of three-dimensional cell aggregates, so-called 'embryoid bodies', whose surface layer differentiates into endodermal cells (Martin & Evans, 1975a). This provides an obvious parallel with the development of the normal embryo, where the role of cell interactions is well established (see chapters by Ziomek *et al.*, Lo and van den Biggelaar, this volume).

One possible mechanism by which cell interactions might influence developmental processes would be the passage of developmental signals between cells by metabolic co-operation through the gap junction (reviewed by Hooper & Subak-Sharpe, 1981; see also chapter by Finbow, this volume). The presence of gap junctions between cells is the rule rather than the exception in developing systems, and although some features of their distribution in the normal embryo make them attractive candidates for the transmission of developmental signals (see chapters by Lo and van den Biggelaar, this volume) the evidence remains somewhat circumstantial due to the difficulty of specifically inhibiting metabolic co-operation in a developing system. The range of methods available for studying the role of gap-junction-mediated communication is, however, broadened by use of the teratocarcinoma model. First, embryonal carcinoma cell lines have been used to raise antisera directed against cell surface antigens: one such antiserum inhibits metabolic co-operation, compaction and differentiation in embryonal carcinoma cells (reviewed by Babinet & Condamine, 1980). Second, the possibility of isolating cell lines in various states of differentiation enables one to investigate changes in histiotypic specificity of metabolic co-operation accompanying differentiation (reviewed by Hooper & Subak-Sharpe, 1981). This article will focus on a third approach, namely the use of somatic cell genetic techniques to isolate communication-defective variants from embryonal carcinoma cell lines whose developmental capacity may then be studied.

Somatic cell genetic analysis of communication-deficiency is made possible by the existence of selective systems for the isolation of communication-defective variants and communication-competent revertants (reviewed by Hooper & Subak-Sharpe, 1981). Fig. 2 illustrates how the experimental design employed in the detection of metabolic

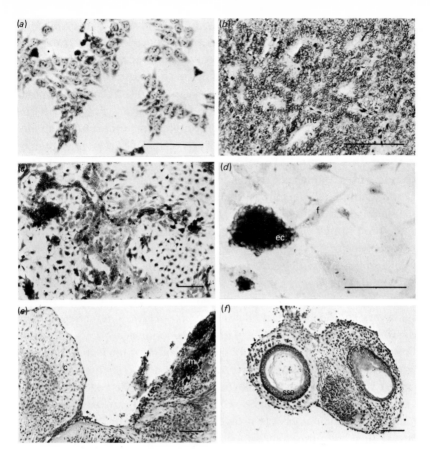

Fig. 1. (*a*) Undifferentiated culture of PC13TG8 embryonal carcinoma cells. (*b*) Section of tumour formed by subcutaneous inoculation of PC13TG8 cells into a syngeneic mouse. ne, neuroepithelium. (*c*) Culture of PC13TG8 cells after 1 week of growth in the presence of 3 µg/ml retinoic acid. (*d*) Undifferentiated culture of PSA4TG12 cells growing on feeder layer of mitomycin C-treated STO cells. ec, embryonal carcinoma cells; f, feeder cells. (*e*, *f*) Sections parallel to the substratum through differentiated outgrowths formed *in vitro* by PSA4TG12. c, cartilage; m, melanocytes; sse, stratified squamous epithelium. In (*a*–*f*), bar = 100 µm.

co-operation may be modified to produce conditions in which co-operation-defective cells are at either a selective growth advantage or a selective disadvantage. To detect metabolic co-operation, cocultures of wild-type and HGPRT$^-$ cells are exposed to ^{3}H-hypoxanthine, fixed, extracted with trichloracetic acid (TCA) and autoradiographed. HGPRT$^-$ cells remain unlabelled when in isolation, but can become labelled as a result of the passage through gap junctions of labelled purine nucleotides formed in wild-type cells (Fig. 2a; see chapter by Finbow, this volume). To select in favour of co-operation-defective cells one uses, in place of radioactive hypoxanthine, its toxic analogue 6-thioguanine, which kills wild-type cells as a result of incorporation via the nucleotide into nucleic acid. HGPRT$^-$ cells in isolation survive since they cannot metabolise thioguanine to the nucleotide, but they become sensitive as a result of gap junction formation with wild-type cells and consequent passage of the nucleotide through the junction ('kiss of death', Fig. 2b). Thus if HGPRT$^-$ cells are cocultured with an excess of wild-type cells at high cell density in 6-thioguanine, the survivors should be enriched for any co-operation-defective cells in the HGPRT$^-$ population. Alternatively, selection against co-operation-defective cells can be achieved by the use of HAT medium (Fig. 2c) in which HGPRT$^-$ cells depend entirely upon purine nucleotides received through gap junctions to fulfil their purine requirement for growth ('kiss of life'). If the wild-type cell is pretreated with mitomycin C, which prevents growth but not gap junction formation, surviving communication-competent HGPRT$^-$ cells will grow to form colonies on a background of non-dividing wild-type cells. Selective systems based on the intercellular transfer of molecules other than nucleotides have also been devised: Fig. 3 shows how, by the use of the compound ouabain which inhibits the sodium pump in the cell membrane, it is possible to select in favour of cells capable of transferring sodium and potassium ions through the junction. In principle, combinations of these selective systems are also possible to allow the isolation of temperature-sensitive variants and, conceivably, variants whose junctions are selectively permeable.

The parent cell line used for the isolation of the variants produced in our laboratory was the HGPRT$^-$ embryonal carcinoma clone PC13TG8 (Hooper & Slack, 1977; Fig. 1a). This feeder-independent cell line grows in standard tissue-culture media without overt differentiation, but on injection into syngeneic mice gives differentiated tumours, which in addition to embryonal carcinoma contain principally nervous

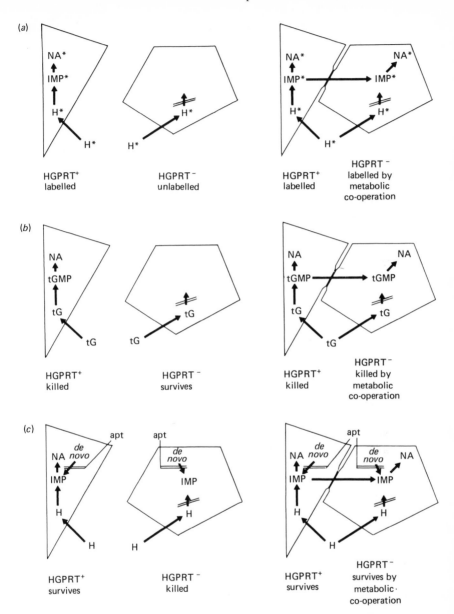

Fig. 2. (*a*) Postulated mechanism of metabolic co-operation for nucleotides derived from (³H) hypoxanthine. (*b*) 'Kiss of death' between HGPRT⁺ and HGPRT⁻ cells in 6-thioguanine. (*c*) 'Kiss of life' between HGPRT⁺ and HGPRT⁻ cells in HAT medium. H, hypoxanthine; apt, aminopterin; tG, 6-thioguanine; NA, nucleic acid. Asterisks denote ³H-labelled compounds. From Hooper & Subak-Sharpe (1981) with permission.

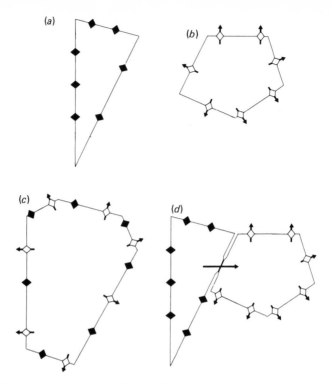

Fig. 3. Postulated mechanism of rescue from ouabain toxicity of sensitive cell by coculture with resistant cell. (*a*) Ouabain-sensitive (oua^S) cell. In the presence of ouabain the Na$^+$, K$^+$-ATPase ($\blacklozenge$) is inhibited and the cell dies due to an inability to pump Na$^+$ out of the cell. (*b*) Ouabain-resistant (oua^R) cell. In ouabain, the Na$^+$, K$^+$-ATPase of this cell ($\lozenge$) remains active. Arrows indicate movement of Na$^+$, which is accompanied by an equal and opposite flow of K$^+$. (*c*) Fusion hybrid between oua^S and oua^R cell. Membrane contains ATPase of both parents so that in ouabain only a fraction of the ATPase is active, resulting in an intermediate level of resistance. (*d*) oua^R and oua^S cells connected by a junction permeable to Na$^+$. Since Na$^+$ can pass freely from one cell to the other, the situation is formally analogous to (*c*). From Hooper & Subak-Sharpe (1981) with permission.

tissue and neuroepithelium (Fig. 1*b*), with small quantities of other epithelia and cartilage. After addition of retinoic acid to PC13TG8 cells *in vitro*, cultures initially appear heterogeneous (Fig. 1*c*) but eventually are dominated by flat, triangular epithelial cells. These cells have been characterised as endoderm, although the question of whether they represent visceral or parietal endoderm has not yet been fully resolved (Strickland & Mahdavi, 1978; Adamson, Gaunt & Graham, 1979; Rees, Adamson & Graham, 1979; Adamson & Graham, 1980).

PC13TG8 cells participate extensively in metabolic co-operation both at homotypic cell contacts and at contacts with cells of a number of different lines (Hooper & Slack, 1977; Hooper & Morgan, 1979a). Using the selective procedure of Fig. 2a, a co-operation-defective variant R5/3 has been isolated from PC13TG8 (Slack, Morgan & Hooper, 1978). When R5/3 is compared with PC13TG8 in capacity to act as a recipient of labelled nucleotides from wild-type cells after incubation in ^{3}H-hypoxanthine, a reduced grain count index is seen (Fig. 4) although some recipient cells are heavily labelled even in the case of R5/3. As the assay does not quantify the molecules transferred directly but rather the result of their incorporation into nucleic acid, it is important to exclude the possibility that a reduction in grain count index is due to an increased purine nucleotide pool size in the variant cell as a result of *de novo* synthesis rather than due to a change in the amount of labelled nucleotide transferred. The possibility has been excluded by two approaches; first, transfer of the same metabolites, viz. purine nucleotides, has been investigated using a colony formation assay based on the kiss of life (Fig. 2c; Slack *et al.*, 1978); second, transfer of a range of different small molecules has been investigated (Hooper & Morgan, 1979a). Thus it may be concluded that R5/3 carries a lesion affecting its ability to form gap junctions, although the grain count distribution observed suggests the hypothesis that the deficiency results not in a total inability to make junctions but in a reduced probability of junction formation (Slack *et al.*, 1978). This hypothesis would predict that the rate of junction formation between R5/3 cells should be reduced compared with that between PC13TG8 cells, and preliminary kinetic analysis indicates that this is indeed so (T. A. Smith, personal communication). The situation is, however, complicated by the fact that in addition to its communication defect, R5/3 shows increased thioguanine-resistance and increased ploidy compared with PC13TG8.

In order to investigate whether these altered properties of R5/3 are the result of a common genetic lesion or of multiple lesions, a communication-competent revertant H2T12 has been isolated from R5/3 using the selective system of Fig. 2c (Hooper & Morgan, 1979b). In H2T12 the ability to participate in intercellular transfer of both nucleotides and alkali metal ions is restored to a level similar to that of PC13TG8 (Tables 1, 2), indicating that in R5/3 the deficiencies in communication observed with different categories of small molecule result from a common genetic lesion. This supports the concept of the

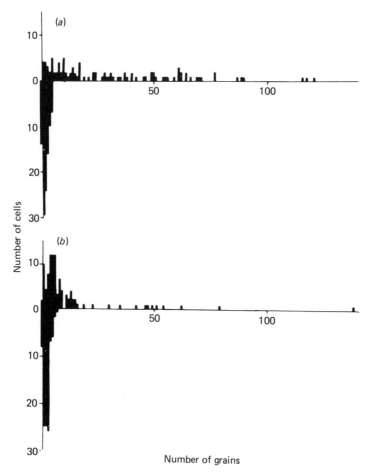

Fig. 4. Distribution in a single experiment of grain numbers over recipient cell cocultured with Don donor cells in the presence of ^{3}H-hypoxanthine. (*a*) PC13TG8 recipients; (*b*) R5/3 recipients. In each panel the upper histogram gives the distribution of grain numbers over recipients in direct contact with donors, whereas the lower, inverted histogram gives the distribution over isolated recipients. Median grain counts: (*a*) upper, 22; lower, 2; (*b*) upper, 6; lower, 2. Thus in this experiment the grain count index for R5/3 = $(6 - 2) \div (22 - 2) \times 100\% = 20\%$. (From Slack *et al.*, 1978, with permission.)

gap junction as an array of hydrophilic pores without substrate speci-ficity except with regard to size. In contrast, the increased thioguanine-resistance and increased ploidy seen in R5/3 remain in H2T12, indicating that these are not a consequence of the lesion resulting in communication-deficiency (Hooper & Morgan, 1979b).

Table 1. *Transfer of uridine nucleotides between PC13TG8,*
R5/3 and H2T12 cells

		Grain count index .		
Donor	Recipient	PC13TG8	R5/3	H2T12
PC13TG8		100	83 ± 13	155 ± 22
R5/3		78*	20 ± 9	104 ± 4
H2T12		105 ± 15	67 ± 2	144 ± 7

Grain count indices are obtained by subtracting the median grain count for isolated recipients from the median grain count for recipients in contact with donors, and expressing the resulting difference as a percentage of the corresponding value for the coculture where both donor and recipient are PC13TG8. Values are quoted as mean $\pm$ half-range of duplicate determinations with the exception of the entry marked with an asterisk, which is the result of a single determination. From Hooper & Morgan (1979b), with permission.

Thus by comparing the properties of PC13TG8, R5/3 and H2T12 the consequences of communication-deficiency can be dissociated from the consequences of secondary lesions.

The gap junction deficiency present in R5/3 can be observed ultra-structurally as a reduction in observable gap junction area per unit cell volume estimated from quantitative morphometry of thin-section electron-microscope preparations (Hooper & Parry, 1980; Table 3). This analysis also reveals an increased incidence of surface microvilli in R5/3. Both changes are reversed in H2T12, indicating that they are causally related to the co-operation-deficiency of R5/3. However, while the reduced gap junction incidence is not unexpected and merely provides additional evidence for the role of the gap junction in metabolic co-operation, the significance of the increased incidence of microvilli is unclear. One possibility is that both it and the communication-deficiency result from an underlying cytoskeletal defect (Hooper & Parry, 1980).

A clue to the nature of the R5/3 defect may be provided from the analysis of polyacrylamide gel profiles of the total cell protein from PC13TG8, R5/3 and H2T12. A major cell protein present in PC13TG8 with an estimated molecular weight of 44 000 is markedly reduced in amount, if not absent, in R5/3 but reappears in H2T12 with a slightly altered mobility. However, a role for this protein in metabolic co-

Table 2. *'Kiss of life' rescue of PC13TG8 and derivatives*
from ouabain toxicity

	Plating efficiency (%)			
	Without mitomycin C- treated STO cells		With mitomycin C- treated STO cells	
Cell line	− ouabain	+ ouabain	− ouabain	+ ouabain
PC13TG8	20.6	0.0	40.5	34.7
R5/3	49.6	0.0	65.0	7.9
H2T12	38.5	0.0	54.3	34.9

From Hooper & Morgan (1979b), with permission.

Table 3. *Stereological analysis of thin sections of*
PC13TG8,R5/3 and H2T12

Cell line	Gap junction area per unit cell volume (μm^{-1})	Microvillar surface area per unit cell volume (μm^{-1})
PC13TG8	$(5.7 \pm 2.0) \times 10^{-4}$ (4)	0.071 ± 0.017 (4)
R5/3	$(1.2 \pm 0.6) \times 10^{-4}$ (4)	0.371 ± 0.156 (4)
H2T12	$(3.9 \pm 0.8) \times 10^{-4}$ (2)	0.072 ± 0.008 (2)

Results are presented as mean $\pm$ SEM, with number of experiments (each on a different preparation of cells) in parentheses. Data from Hooper & Parry (1980), with permission.

operation seems unlikely as it is absent from some communication-competent clones which are sibs of PC13TG8. An alternative possibility is that it represents a protein whose synthesis requires metabolic co-operation, and indeed there is evidence that cell contact is required for its synthesis (T. E. J. Buultjens, C. M. MacDonald & M. L. Hooper, in preparation).

We are currently investigating the developmental consequences of the communication-deficiency of R5/3. As a starting-point we have examined the developmental capacity of the three lines by the two direct assays discussed above, viz. subcutaneous injection into mice and treatment with retinoic acid in culture. As mentioned above, PC13TG8 forms differentiated tumours on subcutaneous injection. No differentiated elements have been seen in three tumours formed by R5/3 (unpublished results). However, in order to interpret this result it is crucial to examine the properties of revertant tumours as embryonal carcinoma lines frequently lose developmental capacity after repeated subculture, probably because the culture environment selects in favour of fast-growing variant cells. Unfortunately we have been unable to obtain tumours by injection of H2T12 and it has therefore been necessary to isolate further revertant clones from the selected population H2T (see Hooper & Morgan, 1979b). From these clones we have to date obtained two tumours which are both undifferentiated. It would appear, therefore, that R5/3 has lost the ability to form differentiated tumours for reasons unrelated to its co-operation-deficiency. In the second direct assay of developmental capacity, PC13TG8, R5/3 and H2T12 respond to retinoic acid in identical fashion (unpublished results). It therefore appears that co-operation deficiency does not interfere with this, limited, part of the developmental programme: this is perhaps not surprising, as the response to retinoic acid occurs in monolayer without obvious requirement for cell interaction of any kind.

However, the differentiation of feeder-dependent embryonal carcinoma lines such as PSA4 (Martin & Evans, 1975b; cf. Figs, $1d-f$) does appear to depend on cell interaction. In order to investigate the role of metabolic co-operation in this process we have examined the behaviour of mixed cultures of PSA4 with PC13TG8, R5/3 and H2T12. Rosenstraus & Levine (1979) found that mixed cultures of the pluripotent line PSA1 and the nullipotent line F9 failed to differentiate in culture, and presented evidence which they interpreted as sorting-out of the cells from the two lines at the embryoid body stage, resulting in the channelling of the PSA1 cells into the surface endoderm layer and leaving no pluripotent cells in the residual core of embryonal carcinoma cells from which subsequent differentiation proceeds. Our preliminary results indicate that PC13TG8 and H2T12 cells both inhibit the differentiation of PSA4, while R5/3 cells do so less efficiently. If confirmed these results suggest that the R5/3 lesion interferes with the signals which control sorting-out. This conclusion is also suggested by observations on aggregation chimeras formed between cells of these lines and normal morulae. When PC13TG8 cells are aggregated with normal morulae and the resulting composite embryos cultured to the blastocyst stage, the PC13TG8 cells are completely surrounded by cells from the normal embryo in the majority of cases. A similar result is observed with H2T12, but R5/3 cells are almost invariably excluded from the embryo and form a separate aggregate outside the blastocyst

(C. L. Stewart & S. Kimber, in preparation). Whether these effects on cell sorting are due to the gap junction deficiency resulting from the R5/3 lesion, to the increased incidence of microvilli or to some other effect of the lesion cannot at present be evaluated. This will probably require the isolation of a series of independent co-operation-defective variants resulting from different underlying lesions. Such variants would be most profitably isolated in feeder-dependent lines such as PSA4TG12: this would enable direct tests of the effect of the lesion on *in vitro* differentiation in contrast to the indirect test described above, and would enable studies of chimeras to be extended to the embryo proper of the post-implantation conceptus, to which PC13TG8 cells do not contribute appreciably (Papaioannou, 1979). From such work, elucidation of the role of gap junction-mediated signalling in the development of the embryo should soon be forthcoming.

REFERENCES

ADAMSON, E. D., GAUNT, S. J. & GRAHAM, C. F. (1979). The differentiation of teratocarcinoma stem cells is marked by the types of collagen which are synthesised. *Cell*, **17**, 469–76.

ADAMSON, E. D. & GRAHAM, C. F. (1980). Loss of tumorigenicity and gain of differentiated function by embryonal carcinoma cells. In *Differentiation and Neoplasia*, ed. R. G. McKinnel, M. A. DiBerardino, M. Blumenfeld & R. D. Bergad, pp. 290–7. Berlin: Springer Verlag.

BABINET, C. & CONDAMINE, H. (1980). Antibodies as tools to interfere with developmental processes. In *Development in Mammals*, vol. 4, ed. M. H. Johnson, pp. 267–304. Amsterdam: Elsevier/North-Holland Biomedical Press.

HOOPER, M. L. & MORGAN, R. H. M. (1979a). The lesion in a metabolic co-operation-defective embryonal carcinoma variant is of broad specificity with regard to metabolite and to contiguous cell type. *Experimental Cell Research*, **119**, 410–14.

HOOPER, M. L. & MORGAN, R. H. M. (1979b). Isolation of a revertant clone from a variant embryonal carcinoma cell line deficient in metabolic co-operation. *Experimental Cell Research*, **123**, 392–6.

HOOPER, M. L. & PARRY, J. E. (1980). Incidence of gap junctions and microvilli in variant cell lines with altered capacity for metabolic co-operation. *Experimental Cell Research*, **128**, 461–6.

HOOPER, M. L. & SLACK, C. (1977). Metabolic co-operation in HGPRT$^+$ and HGPRT$^-$ embryonal carcinoma cells. *Developmental Biology*, **55**, 271–84.

HOOPER, M. L. & SUBAK-SHARPE, J. H. (1981). Metabolic co-operation between cells. *International Review of Cytology*, **69**, 45–104.

MARTIN, G. R. (1980). Teratocarcinomas and mammalian embryogenesis. *Science*, **209**, 768–75.

MARTIN, G. R. & EVANS, M. J. (1975a). Differentiation of clonal lines of teratocarcinoma cells: formation of embryoid bodies *in vitro. Proceedings of the National Academy of Sciences of the USA*, **72**, 1441–5.

MARTIN, G. R. & EVANS, M. J. (1975b). In *Teratomas and Differentiation*, ed. M. I. Sherman & D. Solter, p. 169. New York: Academic Press.

PAPAIOANNOU, V. E. (1979). Interactions between mouse embryos and teratocarcinomas. In *Cell Lineage, Stem Cells and Cell Determination*, INSERM Symposium no. 10, ed. N. Le Douarin, pp. 141–55. Amsterdam: Elsevier/North-Holland Biomedical Press.

REES, A. R., ADAMSON, E. D. & GRAHAM, C. F. (1979). Epidermal growth factor receptors increase during the differentiation of embryonal carcinoma cells. *Nature*, **281**, 309–11.

ROSENSTRAUS, M. J. & LEVINE, A. J. (1979). Alterations in developmental potential of embryonal carcinoma cells in mixed aggregates of nullipotent and pluripotent cells. *Cell*, **17**, 337–46.

SLACK, C., MORGAN, R. H. M. & HOOPER, M. L. (1978). Isolation of metabolic cooperation-defective variants from mouse embryonal carcinoma cells. *Experimental Cell Research*, **117**, 195–205.

SOLTER, D. & DAMJANOV, I. (1979). Teratocarcinoma and the expression of oncodevelopmental genes. In *Methods in Cancer Research*, vol. 18, ed. W. H. Fishman & H. Busch, pp. 277–332. New York: Academic Press.

STRICKLAND, S. & MAHDAVI, V. (1978). The induction of differentiation in teratocarcinoma stem cells by retinoic acid. *Cell*, **15**, 393–403.

Interactions at basement membranes

MARJORIE A. ENGLAND

Department of Anatomy, Medical Sciences, University of Leicester, Leicester LE1 7RH

One of the most puzzling problems in gastrulation has been how cells reach their appointed destinations in the embryo since similar morphogenetic patterns are duplicated by each embryo during gastrulation. For example the mesoderm cells migrate between the epiblast and hypoblast layers in a direction away from the primitive streak area and towards the area pellucida/area opaca border. This sheet of cells when viewed in whole mount preparations always assumes a similar pattern and distribution in the embryo. (Hamilton, 1952; Rosenquist, 1966).

Transmission electron microscopy (TEM) reveals that the only material visible beside the three cell types is extracellular. The possible role of extracellular material (ECM) in gastrulating embryos began to emerge a little over a decade ago. Trelstad et al. (1967) published an account of how mesoderm cells in the early chick embryos might use the basal lamina lining the epiblast as a directional substrate. Hay (1968, 1973) described the first recognisable collagen in the chick embryo as an incomplete basal lamina under the epiblast and hypoblast. She suggested that this collagen guides the migrating primitive streak mesoderm cells which aggregate into chordamesoderm and induce the overlying epiblast to form neural folds. Furthermore she thought that the collagenous basal lamina served as a foothold for elongating neural tube cells, guiding migrating scleratoma cells and possibly stabilising the chondrogenic bias already present in the cells.

This early work (Grobstein, 1955; Trelstad, 1973; Hay, 1973; Bernfield et al., 1973) directed further attention to the importance of the extracellular matrices in both developing and adult systems. Previous workers had commented on the presence and possible importance of ECM but with more advanced techniques it was now possible to examine the materials in greater detail. Matrix components have since been shown to influence cell morphogenesis and differentiation in many systems, i.e. sclerotome cells (Ebendal, 1977; Trelstad, 1977); corneal epithelium (Meier & Hay, 1975; Bard & Hay, 1975); limb-bud chondrogenesis (Dessau *et al.*, 1980); neural crest cells (Weston *et al.*, 1977; Lofberg & Ahlfors, 1978); somite chondrogenesis (Lash & Vasan, 1978; Belsky *et al.*, 1980).

The extracellular matrix is composed of water, ions, collagenous and noncollagenous elements, i.e. glycoproteins (including fibronectin and laminin) and glycosaminoglycans (GAGs). These elements vary from tissue to tissue and alter with the sequence of development.

COLLAGENOUS ELEMENTS

The major molecular component of extracellular matrix is collagen which is present as five distinct genetically determined types. The collagen molecule is a triple-helix composed of three polypeptide chains. Type I collagen is found in skin, tendon, bone and dentin. Type II is the major collagen in cartilage though some type II is also present in neural retinal tissues and the vitreous body. Type III collagen is present in lung, liver, arteries and muscle.

The major components of the basement membrane are types IV and V collagen. Type IV, which has been identified in a variety of tissues, is usually laid down as a procollagen in a non-fibrillar, non-striated, amorphous form (Wartiovaara *et al.*, 1980). Type V has been located in skin, the corneal stroma and placenta.

NONCOLLAGENOUS ELEMENTS

Glycoproteins

Fibronectin is a glycoprotein of 240–250 000 molecular weight which is found in connective tissue matrices, body fluids, and basement membranes. It interacts with collagen (Hahn & Yamada, 1979), glycosaminoglycans (Ruoslahti & Engvall, 1980), fibrin, heparin in the cold, hyaluronic acid and heparan sulphate (but not chondroitin sulphate) (Yamada *et al.*, 1980). It is similar to but not identical with plasma cold insoluble globulin (Yamada & Kennedy, 1979). Fibronectin has been found to influence the shape, attachment and motility of cells *in vitro* (Ali & Hynes, 1978; Yamada & Olden, 1978).

Laminin, a large glycoprotein with two chains of 220 000 and 440 000 molecular weight linked by disulphide bonds, is present in all basement membranes studied and is produced by several cell types *in vitro*. It also has been thought to increase cell adhesiveness in embryonic kidney formation during the early aggregation of nephrogenic mesenchyme (Ekblom *et al.*, 1980) and to act as a structural element of the basement membrane since it occurs in similar quantities to type IV collagen (Timpl *et al.*, 1979). Madri *et al.* (1980) have suggested that laminin serves as an adhesive factor between epithelial cells and the basement membrane or as an electrostatic barrier for negatively charged protein since it has a high sialic acid content.

Elastic fibres contain elastin and microfibrillar protein.

Glycosaminoglycans (GAGs)

GAGs such as chondroitin sulphate are large linear polymers of carbohydrates covalently linked to polypeptides. If this complex is then attached to a protein core it is known as a proteoglycan. The possible exception to this is hyaluronate which can exist alone (Manasek, 1975).

GAGs (proteoglycans) are closely associated with the collagen fibrils which are present as a three-dimensional lattice. While water and small molecules in solution can move through this matrix, large molecules may be unable to enter or are trapped. Functionally the matrix may serve as an exclusion filter or as a sponge, retaining and concentrating large molecules.

Water may also be trapped to some extent producing tissue turgor (Scott, 1975).

EXTRACELLULAR MATERIALS IN THE CHICK GASTRULA

Transmission electron microscopy

Contacts between migrating mesoderm cells and the basement membranes of the epiblast and hypoblast were noted in several reports (Revel, 1974; Bancroft & Bellairs, 1975; England & Cowper, 1976; Ebendal, 1976). However, by TEM there was no apparent pattern to the basement membrane materials. The basal lamina lining the epiblast was a thin layer which was interrupted in the primitive streak. Fibrils 4–4.5 nm in diameter made up the very irregular network of the lamina which was also overlaid by coarser, sparser fibrils in the gastrulating embryo (Low, 1967, 1968). Solursh (1976) demonstrated that large amounts of GAGs, especially hyaluronic acid were produced between stages 4 and 8 (Hamburger & Hamilton, 1951). He suggested that the GAGs might open up spaces between cells leaving the primitive streak and thereby facilitating mesodermal migration. Sanders (1979) re-examined the developing basal lamina by TEM and its associated extracellular materials and found that the basal lamina consisted of a lucent lamina interna (approximately 20 nm wide) and a lamina densa (approximately 20 nm wide). In addition, fibrils 20 nm thick were located in the region of the primitive streak at stage 5. His studies with hyaluronidase confirmed Solursh's results showing that hyaluronic acid was a major component of the basal lamina. He suggested that an enzyme resistant framework of collagen was present with dense period-icities of GAGs built onto this substrate.

THE PATTERN OF EXTRACELLULAR FIBRES

Recent results using scanning electron microscopy have revealed that the extracellular fibres lying on the epiblast basement membrane are organised into patterns (Wakely & England, 1979). There are three major patterns of fibres present in the stage 4 embryo (Fig. 1): area a, a grid pattern in the area pellucida; area b, a fibrous band on the anterior area pellucida/area opaca border; and area c, vertically oriented fibres in the proamnion region. These fibres guide the primitive streak mesoderm cells in gastrulation and the movement of the primordial germ cells (PGCs).

At stage 4 the first pattern of fibres radiates from the primitive streak regions (area a) in a grid pattern across the area pellucida on the ventral epiblast basement membrane. These fibres are long with several side-branching fibres interconnecting them (Fig. 2). The origin of this fibrous pattern is not clear, but it matches to some degree, the pattern of the migrating primary hypoblast. The primary hypoblast migrates over the area pellucida as the secondary hypoblast moves away from the primitive streak in a secondary wave. As the second wave moves outward the primary hypoblast assumes a position along the area pellucida/area opaca border. Migrating mesoderm cells are in contact with the grid pattern of extracellular materials on the area pellucida basement membrane layer. The cells move as a sheet over the area pellucida epiblast grid of extracellular materials. With further incuba-tion the stage 4 pattern of extracellular fibres alters at early stage 5 and directs the prenotochordal cells (chordamesoderm) into position ante-rior to Hensen's node (England, 1981) (see below).

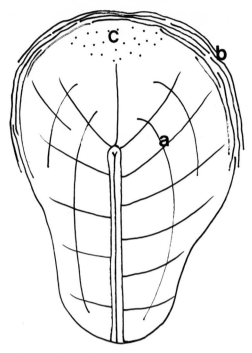

Fig. 1. Three areas of extracellular fibres present in the stage 4 chick embryo. (a) Long fibres radiate from the primitive streak and form a grid pattern in the area pellucida; (b) a fibrous band present on the area pellucida/area opaca border; (c) a small area consisting of short rods in the proamnion region of the area pellucida.

The second pattern of fibres (area b) is a prominent band on the area pellucida/area opaca border (Fig. 3) which curves in an arc along that border as far posterior as Hensen's node. With further incubation this band extends more posteriorly into the area pellucida toward the primitive streak. The band is composed of many fibres (circa 30–40) which run approximately parallel to one another. Each fibre varies in thickness throughout its length and often side-branches connect a fibre to its neighbours (Fig. 4). It is difficult to follow a single fibre throughout its length. In association with the fibres are numerous granular structures which cover their surfaces. Adjacent to the fibrous band in the area pellucida toward Hensen's node are numerous granular structures similar in appearance to those on the fibrous band. These granules often are lined up to form incomplete fibres. In stage 3 embryos the fibrous band is very incomplete. Some fibres are present, but large spaces occur between the fibres, and there are several areas

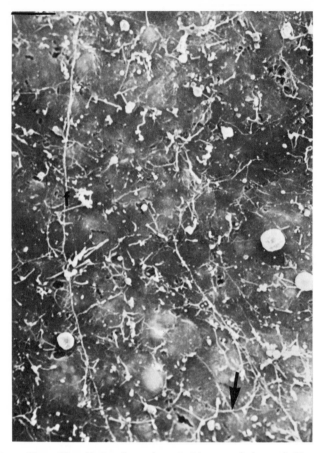

Fig. 2. Long fibres (f) radiating from the primitive streak (area a). Note the side-branching fibres (arrow). Scale bar = 10 μm.

where the fibres have not yet formed (Fig. 5). The basal lamina lying on the epiblast is a very thin sheet through which can be seen the epiblast cell perimeters. In several areas the granular structures are lined up in the general shape of a fibre.

In the stage 3 embryo primordial germ cells are present on the ventral epiblast migrating toward the fibrous band (Fig. 6). Once in contact with the band these cells assume a rounded shape with small basal projections. As the fibrous band has a high concentration of fibronectin present (Fig. 7) and the area pellucida a low concentration (Critchley *et al.*, 1979) it can be argued that the primordial germ cells are attracted by the fibronectin in the band (England, 1980). Also

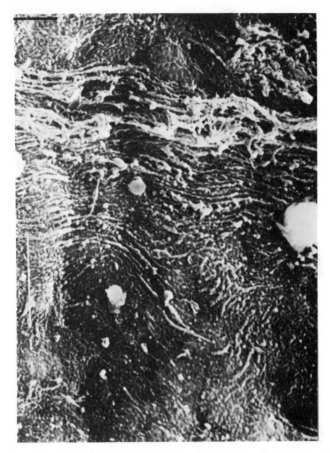

Fig. 3. Fibrous band (area b) on the area pellucida/area opaca border. The fibres are approximately parallel to one another. Scale bar = 5 μm.

present in this area is a high concentration of type I collagen whose staining pattern is similar to that of fibronectin. This combination of fibronectin–collagen would promote close association of the primordial germ cells and their substrate. The fibrous band also stains strongly for sulphated glycosaminoglycans with alcian blue at varying concentrations of $MgCl_2$.

When examined by TEM the fibrous band appears as large loops of basal lamina projecting into the space between epiblast and hypoblast (Fig. 8). Granular structures of condensed tufts of basal lamina material are present on the outer border of the loops. With further incubation the loops become more complex and folded back on themselves and one another. This appearance is similar to that described by Mayer &

Fig. 4. Higher magnification of fibres in the fibrous band (area b). The fibres have numerous granules on their surfaces. Scale bar = 2 μm.

Packard (1978) in the region of the terminal sinus and edge cells in the 48-h chick. They argue that this appearance in the older embryo was associated with a vascular expansion in which ECM is secreted to maintain a suitable substrate for cell migration. A high concentration of glycosaminoglycans was present in this area.

A third region of extracellular materials, area c, is present in the stage 4 embryo. Anterior to the primitive streak in the area pellucida is a region characterised by short rods of extracellular material. With further incubation the fibrous band on the anterior area pellucida/area opaca increases in width into this region of short rods. Additionally, as the fibrous band increases in length with age so too does the fibronectin content of the new area of band offering an excellent opportunity to study fibronectin deposition *in vivo*. This anterior area is small in

Fig. 5. The stage 3 fibrous band (area b) is incompletely formed, but the pattern of fibres is present. Scale bar = 2 μm.

comparison with the other two areas described. In addition to being encroached upon by the expanding width of the fibrous band, it also is maintained in its own right in the proamnion region where the future mouth will form. Mesoderm cells do not enter this area and the vertical orientation of the stubs of extracellular material present suggests this region is a functional barrier to migrating mesoderm cells.

PRIMARY NEURAL INDUCTION

In the stage 4 area pellucida the extracellular fibres form a grid pattern and by early stage 5, the time of primary neural induction, the notochord forms as a midline structure from mesoderm cells anterior to Hensen's node. In the interval between these stages pre-notochordal cells in the epiblast invaginate and move anteriorly to Hensen's node. As the epiblast cells move toward the node to invaginate their extracellular fibres are carried with them as shown by the pattern of fibres changing from a grid to a midline fan-shape anterior to the node (Fig. 9). Several cells are contacting this fan-shape of fibres (Fig. 10). In addition, fibres are present as a halo around the base of the node. The fan-shaped fibres stain strongly for the presence of sulphated GAGs

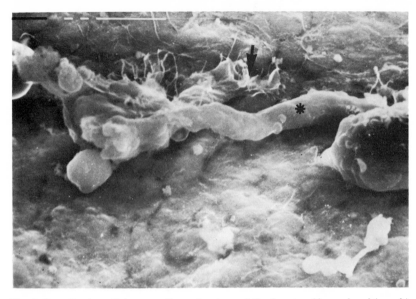

Fig. 6. Stage 3 primordial germ cell moving toward the forming fibrous band (area b). Note the long cellular extension (*) and the numerous leading edges (arrow). Scale bar = 6 μm.

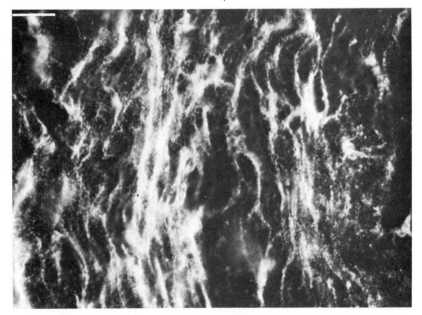

Fig. 7. The fibrous band (area b) stained for fibronectin. The embryo was incubated with rabbit anti-fibronectin antisera diluted 1 : 25 in phosphate buffered saline, and subsequently with fluorescein isothiocyanate-labelled goat anti-rabbit diluted 1 : 10. Preparation by D. R. Critchley and M. A. England. Scale bar = 5 μm.

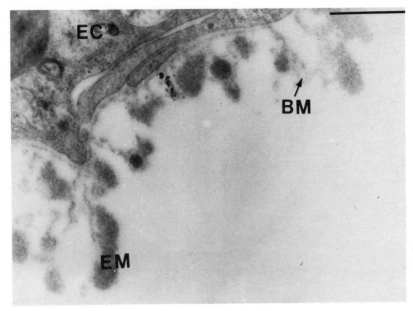

Fig. 8. Transmission electron micrograph of the fibrous band (area b). The basement membrane is projected into loops with dense areas of extracellular material on its outer surface. EC, epiblast cells; BM, basement membrane; EM, extracellular material. (Preparation by L. Kordylewski and M. A. England.) Scale bar = 1 μm.

and fibronectin. In the stage 9 embryo remnants of this fan-shape remain as a band of fibres on the ventral surface of the neural tube (Fig. 11).

FIBRE FORMATION

As previously mentioned, the origin of the fibrous patterns is unclear. The grid pattern in the area pellucida is somewhat similar to the movement of the primary hypoblast layer which migrates away from the primitive streak toward the area opaca border. Evidence to support the belief that the primary hypoblast may be involved in fibre formation was obtained in the stage 4 embryo since some of the primary hypoblast at this stage is overlying the anterior area pellucida/area opaca border. When hypoblast overlying the fibrous band is transplanted into a stage 4 embryo whose area opaca hypoblast has been dissected away and is allowed to further incubate, a new fibrous band is formed between the transplanted hypoblast and host epiblast. The new fibrous band is oriented according to the graft hypoblast and this

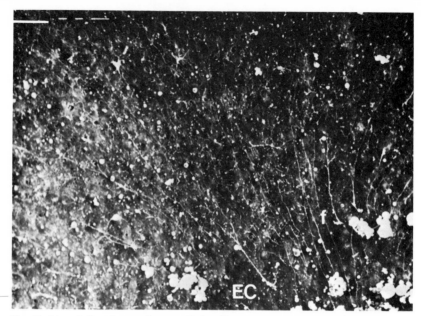

Fig. 9. Stage 5 fan-shape of extracellular fibres on the ventral epiblast basement membrane anterior to Hensen's node. The fibres are long. f, fibres; EC, epiblast. Scale bar = 20 μm.

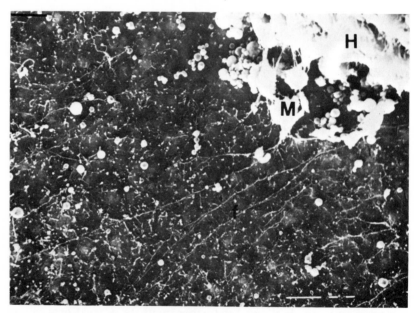

Fig. 10. Mesoderm cell contacting the fan-shape of extracellular fibres immediately anterior to the stage 5 Hensen's node. f, fibres; H, Hensen's node; M, mesoderm cell. Scale bar = 10 μm.

Fig. 11. Fibrous band (arrow) on the ventral surface of the stage 9 neural tube. NT, neural tube. Scale bar = 3 μm.

would suggest the fibrous band forms as a result of a specific interaction between the epiblast and the hypoblast of the area pellucida/ area opaca border. The band must be epithelial in origin since, at this time, no mesoderm cells are present.

FIBRE MALLEABILITY

The fibres may be physically altered by the cells which are in close proximity to them. The stage 4 grid pattern of extracellular fibres is changed to a fan-shape by the migration of the underlying epiblast cells. The fibres are also altered in the band by the primordial germ cells. The parallel array of fibres is disturbed by the primordial germ cells moving through them. Hay (1978) has suggested that basement membranes are quickly organised and disorganised during development. The present studies would confirm this. Each pattern alteration is dependent upon cell movements and changing micro-environments so that the normal patterns are the result of several co-ordinated events. If any one element fails the resulting disturbance is momentarily abnormal. Once the pattern has changed in sequence, it is not fully duplicated again even in extreme instances such as wounding.

Overton (1976), for example, demonstrated that following wounding the extracellular fibres formed in a normal distribution though they were disorganised and never exactly duplicated the original pattern.

FUNCTION OF FIBRES

To ascertain whether the extracellular fibres are necessary for cell migration, unincubated embryos were treated with *cis*-hydroxyproline which is known to prevent the extracellular deposition of pro-collagen (Lauscher & Carlson, 1975). Following incubation the embryos were compared with normal control embryos. The extracellular fibres were not present nor was the basement membrane on the ventral epiblast layer. The mesoderm cells were clumped in the region of the primitive streak and did not migrate across the area pellucida. Primordial germ cells were not in evidence on the ventral epiblast surface. This would suggest the fibres are collagen precursors (Wakely & England, 1979).

CELL–SUBSTRATE INTERACTIONS

The mesoderm cells in the area pellucida and primordial germ cells both utilise the extracellular fibres as a contact guidance system in the gastrulating chick. Both cell types contact the fibres, and where they are experimentally denied this substrate by the use of *cis*-hydroxyproline, cell migration does not occur in the area pellucida. Trelstad (1973) and Hay (1973) demonstrated that the mesoderm cells contact the extracellular matrix during gastrulation possibly forming a contact guidance system. This observation has been confirmed and the morphogenetic patterns recognised in early development are seen to be the result of cells using the patterns of extracellular fibres present on the ventral basement lamina of the epiblast layer. High concentrations of fibronectin and collagen type I are present in the fibrous band. Both are visible as parallel arrays of fibres arcing along the area pellucida/area opaca border. Further, a high concentration of sulphated GAGs is also present. Pratt *et al.* (1979), Sanders & Anderson (1979) and Vanroelen *et al.* (1980) report similar findings in the early chick embryo. The PGCs are known to move in a direction contrary to the main cell movements in the area pellucida. Initially, these cells move toward the fibrous band from an area of low fibronectin in the area pellucida to high fibronectin in the fibrous band. This suggests the band attracts the PGCs initially and subsequently they use it as a contact guidance system. (A more detailed account of cell guidance is given in the following chapter by Wylie *et al.*) The combination of the components of the extracellular matrix also suggests there is a fibronectin–collagen complex which mediates cell adhesion (Kleinman *et al.*, 1979). Kleinman *et al.* have suggested the cell surface receptors for fibronectin are probably sialic acid containing glycoconjugates. The PGCs would attach to the fibronectin–collagen complex and their subsequent movement along this fibrous tract would explain their contralateral movements. Fibronectin also stimulates cell motility *in vitro* (Ali & Hynes, 1978) and *in vivo* (Critchley *et al.*, 1979; Wakely & England, 1979; Wylie *et al.*, 1979). High concentrations of GAGs are known to influence differentiation (Hay & Meier, 1974) and large amounts of sulphated GAGs are also present in the fibrous band.

MICROENVIRONMENTAL VARIATIONS

Studies of guidance systems in the neural crest have demonstrated that the changing microenvironment of ECM is an important influence on neural crest morphogenesis (Weston & Butler, 1966; Derby, 1978; Pintar, 1978; Erickson *et al.*, 1980). The ECM will not support a second population of neural crest cells migrating through this environment. Meier & Hay (1974) have shown that the presence of GAGs (chondroitin and heparin sulphates) stimulates further GAG synthesis. This was demonstrated in neural crest migration by Weston *et al.* (1978) who showed changes in the pattern of GAG distribution as the neural crest cells moved through the ECM.

As the PGCs move through the fibrous band, large whorls of fibres are left in their trail disorganising the parallel array of fibres. This would suggest that similar mechanisms to those in the neural crest might operate in the region of the fibrous band.

PRIMARY NEURAL INDUCTION

The possible role of ECM in induction systems has been extensively studied (Grobstein, 1955, 1967; Ekblom *et al.*, 1980). Saxen and co-workers (1976) introduced both Millipore and Nucleopore filters between the inducer and the epithelium and concluded that cells do not normally contact one another across Millipore filters, but that spinal cord and metanephric mesenchyme can contact one another by long

processes through Nucleopore filters during induction. They concluded that cell contact is normally present in induction systems. Hay
(1977) points out that Saxen *et al.* did not consider the possibility that
ECM had crossed the filter. It is also known that a small number of cells
introduced into this type of *in vitro* preparation are naturally damaged
by the inherent physical trauma. These damaged cells enter the filter as
small droplets of cytoplasm and can easily cross the filter to the
opposite side, effectively allowing the transmission of cellular information (England, 1969, 1975). Also, as soon as Millipore filters come in
contact with tissue, substances enter the filter which line the pores and
stain as a fibrous layer by TEM (England, 1969). Hay (1977) has shown
that cells positively contact a collagenous substratum. The work reviewed in this chapter supports Hay's view that ECM is necessary for
induction, but that its role is as a template (of a concentration of
fibronectin, collagen, and GAGs) for a cell–collagen interaction which
guides and stimulates the cells into a position of cell–cell interaction.
Support for this interpretation comes from studies implicating cell
surface receptors in induction systems (Tiedemann & Born, 1978;
Sanders & Anderson, 1979), and GAGs shown to be critical for trans-
mitting the inductive signal (Ekblom *et al.*, 1980).

Acknowledgements

My colleagues J. Wakely, D. R. Critchley (University of Leicester) and
L. Kordylewski (Jagiellonian University, Cracow, Poland) have contributed to this chapter with their valuable discussions and preparations.
D. R. Critchley and R. O. Hynes (MIT, Massachusetts, USA) kindly
provided the anti-fibronectin and Dr G. B. Shellswell, Agriculture
Research Council, Meat Research Institute, Langord, Bristol, the anti-
collagen antibodies. Their invaluable help and advice during staining
and photography are gratefully acknowledged.

REFERENCES

ALI, I. U. & HYNES, R. O. (1978). Effect of LETS glycoprotein on cell motility. *Cell*, **14**, 439–46.
BANCROFT, M. & BELLAIRS, R. (1975). Differentiation of the neural plate and neu-
ral tube in the young chick embryo. *Anatomy and Embryology*, **147**, 309–35.
BARD, J. B. L. & HAY, E. D. (1975). The behaviour of fibroblasts from the developing avian cornea. *Journal of Cell Biology*, **67**, 400–18.
BELSKY, E., VASAN, N. S. & LASH, J. W. (1980). Extracellular matrix components
and somite chondrogenesis: A microscopic analysis. *Developmental Biology*, **79**, 159–80.

BERNFIELD, M. R., COHN, R. H. & BANERJEE, S. D. (1973). Glycosaminoglycans and epithelial organ formation. *American Zoologist*, **13**, 1067–83.

CRITCHLEY, D. R., ENGLAND, M. A., WAKELY, J. & HYNES, R. O. (1979). Distribution of fibronectin in the ectoderm of gastrulating chick embryos. *Nature*, **280**, 498–500.

DERBY, M. A. (1978). Analysis of glycosaminoglycans within the extracellular environments encountered by migrating neural crest cells. *Developmental Biology*, **66**, 321–36.

DESSAU, W., VON, DER MARK, H., VON, DER MARK, K. FISHER, S. (1980). Changes in the patterns of collagens and fibronectin during limb bud chondrogenesis. *Journal of Embryology and Experimental Morphology*, **57**, 51–60.

EBENDAL, T. (1976). Migratory mesoblast cells in the young chick embryo examined by scanning electron microscopy. *Zoon*, **4**, 101–8.

EBENDAL, T. (1977). Extracellular matrix fibrils and cells contacts in the chick embryo. The possible roles in orientation of cell migration and axon extension. *Cell and Tissue Research*, **175**, 439–58.

EKBLOM, P., ALITALO, K., VAHERI, A., TIMPL, R. & SAXEN, L. (1980). Induction of a basement membrane glycoprotein in embryonic kidney: Possible role of laminin in morphogenesis. *Proceedings of the National Academy of Sciences, USA*, **77**, 485–9.

ENGLAND, M. A. (1969). Millipore filters studied in isolation and in vitro by transmission electron microscopy and stereoscanning electron microscopy. *Experimental Cell Research*, **54**, 222–30.

ENGLAND, M. A. (1975). Membrane filters do not prevent cell contacts. *Experientia*, **31**, 349–51.

ENGLAND, M. A. (1980). Fibronectin related cell shape changes *in vivo*. *Cell Biology International Reports*, **4**, 801.

ENGLAND, M. A. (1981). Applications of the SEM to the analysis of morphogenetic events. *Journal of Microscopy*, (in press).

ENGLAND, M. A. & COWPER, S. V. (1976). A transmission and scanning electron microscope study of primary neural induction. *Experientia*, **32**, 1578–80.

ERICKSON, C. A., TOSNEY, K. W. & WESTON, J. A. (1980). Analysis of migratory behaviour of neural crest and fibroblastic cells in embryonic tissues. *Developmental Biology*, **77**, 142–56.

GROBSTEIN, C. (1955). Tissue interaction in the morphogenesis of mouse embryonic rudiments in vitro. In *Aspects of Synthesis and Order in Growth*, ed. D. Rudnick, pp. 233–56. Princeton University Press.

GROBSTEIN, C. (1967). Mechanisms of organogenetic tissue interaction. *National Cancer Institute Monographs*, **26**, 279–95.

HAHN, L-H. E. & YAMADA, K. M. (1979). Identification and isolation of a collagen-binding fragment of the adhesive glycoprotein fibronectin. *Proceedings of the National Academy of Sciences, USA*, **76**, 1160–3.

HAMBURGER, V. & HAMILTON, H. L. (1951). A series of normal stages in the development of the chick embryo. *Journal of Morphology*, **88**, 49–92.

HAMILTON, H. L. (1952). *Lillie's Development of the Chick: An Introduction to Embryology*. New York: Henry Holt & Co.

HAY, E. D. (1968). Organisation and fine structure of epithelium and mesenchyme in the developing chick embryo. In *Epithelial Mesenchymal Interactions*, ed.

R. Fleischmajer & R. Billingham, pp. 31–55. Baltimore: Williams & Wilkins.

HAY, E. D. (1973). Origin and role of collagen in the embryo. *American Zoologist*, **13**, 1085–106.

HAY, E. D. (1977). Cell-matrix interaction in embryonic induction. In *Cell to Cell Interactions: International Cell Biology 1976–1977*, ed. K. R. Porter & B. R. Brinkley, pp. 50–7. The Rockefeller University Press.

HAY, E. D. (1978). Fine structure of embryonic matrices and their relation to the cell surface in ruthenium red-fixed tissues. *Growth*, **42**, 399–423.

HAY, E. D. & MEIER, S. (1974). Glycosaminoglycan synthesis by embryonic inductors: neural tube, notochord, and lens. *Journal of Cell Biology*, **62**, 889–98.

KLEINMAN, H. K., MARTIN, G. R. & FISHMAN, P. H. (1979). Ganglioside inhibition of fibronectin-mediated cell adhesion to collagen. *Proceedings of the National Academy of Sciences, USA*, **76**, 3367–71.

LASH, J. W. & VASAN, N. S. (1978). Somite chondrogenesis *in vitro*: stimulation by exogenous extracellular matrix components. *Developmental Biology*, **66**, 151–71.

LAUSCHER, C. K. & CARLSON, E. C. (1975). The development of proline containing extracellular connective tissue fibrils by chick notochordal epithelium *in vitro*. *Anatomical Record*, **182**, 151–68.

LOFBERG, J. & AHLFORS, K. (1978). Extracellular matrix organisation and early neural crest cell migration in the axolotl embryo. *Zoon*, **6**, 87–101.

LOW, F. N. (1967). Developing boundary (basement) membranes in the chick embryo. *Anatomical Record*, **159**, 231–8.

LOW, F. N. (1968). Extracellular connective tissue fibrils in the chick embryo. *Anatomical Record*, **160**, 93–108.

MADRI, J. A., ROLL, F. J., FURTHMAN, H. & FOIDART, J. M. (1980). Ultrastructural localization of fibronectin and laminin in the basement membrane of the murine kidney. *Journal of Cell Biology*, **86**, 682–6.

MANASEK, F. J. (1975). The extracellular matrix: a dynamic component of the developing embryo. In *Current Topics in Developmental Biology*, **10**, 35–102.

MAYER, JR, B. W. & PACKARD, JR, D. S. (1978). A study of the expansion of the chick area vasculosa. *Developmental Biology*, **63**, 335–51.

MEIER, S. & HAY, E. D. (1974). Stimulation of extracellular matrix synthesis in the developing cornea by glycosaminoglycans. *Proceedings of the National Academy of Sciences, USA*, **71**, 2310–13.

MEIER, S. & HAY, E. D. (1975) Stimulation of corneal differentiation by interaction between cell surface and extracellular matrix. I. Morphometric analysis of trans-filter induction. *Journal of Cell Biology*, **66**, 275–91.

OVERTON, J. (1976). Scanning microscopy of collagen in the basement lamella of normal and regenerating frog tadpoles. *Journal of Morphology*, **150**, 805–11.

PINTAR, J. E. (1978). Distribution and synthesis of glycosaminoglycans during quail neural crest morphogenesis. *Developmental Biology*, **67**, 444–64.

PRATT, R. M., YAMADA, K. M., OLDEN, K., OHANIAN, S. H. & HASCALL, V. C. (1979). Tunicamycin-induced alterations in the synthesis of sulfated proteoglycans and cell surface morphology in the chick embryo fibroblast. *Experimental Cell Research*, **118**, 245–52.

REVEL, J. P. (1974). Some aspects of cellular interactions in development. In *The Cell Surface in Development*, ed. A. A. Moscona, pp. 51–65. New York: John Wiley & Sons.

ROSENQUIST, G. C. (1966). A radioautographic study of labelled grafts in the chick blastoderm: development from primitive streak stages to stage 12. *Contributions to Embryology, Carnegie Institute*, **38**, 71–110.

RUOSLAHTI, E. & ENGVALL, E. (1980). Complexing of fibronectin glycosaminoglycans and collagen. *Biochimica et Biophysica Acta*, **631**, 350–8.

SANDERS, E. J. (1979). Development of the basal lamina and extracellular materials in the early chick embryo. *Cell and Tissue Research*, **198**, 527–37.

SANDERS, E. J. & ANDERSON, A. R. (1979). Ultrastructural localization of wheat germ agglutinin-binding sites on surfaces of chick embryo cells during early differentiation. *Journal of Cellular Physiology*, **99**, 107–24.

SAXEN, L., LEHTONEN, E., KARKINEN-JOAS, M., NORDLING, S. & WARTIOVAARA, J. (1976). Are morphogenetic tissue interactions mediated by transmissible signal substances or through cell contacts? *Nature (Lond.)*, **259**, 662–3.

SCOTT, J. E. (1975). Composition and structure of the pericellular environment. *Philosophical Transactions of the Royal Society, London*, **B271**, 235–42.

SOLURSH, M. (1976). Glycosaminoglycan synthesis in the chick gastrula. *Developmental Biology*, **50**, 525–30.

TIEDEMANN, H. & BORN, J. (1978). Biological activity of vegetalizing and neuralizing inducing factors after binding to BAC-cellulose and CNBR-Sepharose. *Wilhelm Roux Archives*, **184**, 285–99.

TIMPL, R., ROHDE, H., GEHRON-ROBEY, P., RENNARD, S. I., FOIDART, J. M. & MARTIN, G. R. (1979). Laminin: a glycoprotein from basement membranes. *Journal of Biological Chemistry*, **254**, 9933–7.

TRELSTAD, R. L. (1973). The developmental biology of vertebrate collagens. *Journal of Histochemistry and Cytochemistry*, **21**, 521–8.

TRELSTAD, R. L. (1977). Mesenchymal cell polarity and morphogenesis of chick cartilage. *Developmental Biology*, **59**, 153–63.

TRELSTAD, R. L., HAY, E. D. & REVEL, J. P. (1967). Cell contact during early morphogenesis in the chick embryo. *Developmental Biology*, **16**, 78–106.

VANROELEN, Ch., VAKAET, L. & ANDRIES, L. (1980) Alcian blue staining during the formation of mesoblast in the primitive streak stage chick blastoderm. *Anatomy and Embryology*, **160**, 361–7.

WAKELY, J. & ENGLAND, M. A. (1979). Scanning electron microscopical and histochemical study of the structure and function of basement membranes in the early chick embryo. *Proceedings of the Royal Society, London*, **B206**, 329–52.

WARTIOVAARA, J., LEIVO, I. & VAHERI, A. (1980). Matrix glycoproteins in early mouse development and in differentiation of teratocarcinoma cells. In *The Cell Surface: Mediator of Developmental Processes. 38th Symposium of The Society for Developmental Biology*, ed. S. Subtelny & N. K. Wessells, pp. 305–24. London: Academic Press.

WESTON, J. A. & BUTLER, S. L. (1966). Temporal factors affecting localization of neural crest cells in the chicken embryo. *Developmental Biology*, **14**, 246–66.

WESTON, J. A., DERBY, M. A. & PINTAR, J. E. (1978). Changes in the extracellular environment of neural crest cells during their early migration. *Zoon*, **6**, 103–13.

WESTON, J. A., PINTAR, J. E., DERBY, M. A. & NICHOLS, D. H. (1977). The morphogenesis of spinal ganglia from neural crest cells. In *Cellular Neurobiology*, ed. Z. Hall *et al*, pp. 217–26. New York: Alan R. Liss.

WYLIE, C. C., HEASMAN, J., SWAN, A. P. & ANDERTON, B. H. (1979). Evidence for substrate guidance of primordial germ cells. *Experimental Cell Research*, **121**, 315–24.

YAMADA, K. M. (1978). Immunological characterization of a major transformation-sensitive fibroblast cell surface glycoprotein. *Journal of Cell Biology*, **78**, 520–41.

YAMADA, K. M. & KENNEDY, D. W. (1979). Fibroblast cellular and plasma fibronectin are similar but not identical. *Journal of Cell Biology*, **80**, 492–8.

YAMADA, K. M., KENNEDY, D. W., KIMOTA, K., & PRATT, R. M. (1980). Characterization of fibronectin interactions with glycosaminoglycans and identification of active proteolytic fragments. *Journal of Biological Chemistry*, **255**, 6055–63.

YAMADA, K. M. & OLDEN, K. (1978). Fibronectins: adhesive glycoproteins of cell surface and blood. *Nature (Lond.)*, **275**, 179–84.

The role of the extracellular matrix in cell movement and guidance

C. C. WYLIE, ALMA P. SWAN AND JANET HEASMAN

Department of Anatomy, St George's Hospital Medical School, Cranmer Terrace, London SW17 ORE

Many general hypotheses exist for explaining cell guidance, and can be grouped under the general headings of guidance by the substrate, and guidance by diffusible molecules from the target. The latter mechanism will not be considered in this chapter, although it has been shown to exist in the guidance of *Dictyostelium* (Konijn, Van de Meene, Bonner & Barkley, 1967), neurons (Menesini Chen, Chen & Levi-Monalcini, 1978), leucocytes (Zigmond, 1978), and the attraction of vascular elements towards a tumour (Ausprunk & Folkman, 1977).

Substrate-mediated guidance may be affected either by the general shape or contours of the substrate (contact guidance) or by the interaction of the cells with specific molecules on the substrate, the cells 'following' the molecules with which they have greatest affinity (haptotaxis). The idea of haptotaxis has arisen from studies of cell migration *in vitro* on substrates to which cells adhere with different affinities (Harris, 1973; Carter, 1967). The extent to which each of these two types of substrate guidance exist *in vivo* has yet to be established both with respect to the molecules involved, and the cell types which might use them for their migration.

This chapter considers various aspects of the relationship between cell migration and the extracellular matrix. We largely confine our attention to the movement of the migratory cells in early embryos. Here, cells may move in sheets (as during gastrulation), in cords (as in gland morphogenesis), or as clumps or single cells (as in the migration of neural crest cells, primordial germ cells, and neurons). It seems to us that only within these shifting populations of cells, whose movements result in morphogenesis, will we find the answers to the problems of cell guidance. The difficulty lies in relating observations on the cells concerned *in vitro* to their corresponding activities within the organism. Details of such correlations exist for only a few cell types. We document examples of these, the extracellular matrix molecules with which they are associated, and the possible ways in which these molecules might affect their movement.

EVIDENCE FOR ORIENTED PATHWAYS FOR CELL MIGRATION IN EMBRYOS

Before a substrate guidance hypothesis can be advanced for cell migration *in vivo*, it has to be demonstrated that such pathways do exist in intact embryos, and that migratory cells do follow them. There is now a considerable body of evidence for both of these facts, with respect to several lineages:

The neural crest

This is a group of cells which detach from the junction of the neural folds and surface ectoderm, at the end of neural tube formation. They are initially arranged as a longitudinal strip of cells, on top of, and partly overhanging, the early neural tube. From here, they migrate laterally until they encounter the developing somites on each side, where they split into two streams, those passing between somite and surface ectoderm (the dorsal pathway), and those passing between somite and neural tube (the ventral pathway) (for reviews see Horstadius, 1950 and Weston, 1970). Neural crest cells migrate extensively through the embryonic body, and contribute to a variety of adult structures. There is now good evidence that the migratory path of these cells constitutes a general pathway along which migratory invasive cells can move. First, grafting experiments, where segments of early neural tube and crest are moved to new axial levels, show that crest cells migrating from the grafts take the route appropriate to the new axial level (Weston, 1970; Noden, 1975; Le Douarin & Teillet, 1974). The pattern of migration is therefore imposed by environmental factors in this case. More recently, injections of cloned crest-derived pigment cells into the neural crest pathway have shown that the pathway ages, supporting progressively less migration of cells of equal migratory potential (Bronner-Fraser & Cohen, 1980; see also Weston & Butler, 1966). These experiments are particularly well controlled, since they take advantage of the fact that the cranial end of the embryo is developmentally older than the caudal end; cells injected into the two ends of the same embryo thus encounter neural crest pathways of different 'ages'. Malignant tumour cells also migrate along the neural crest pathway when injected (Erickson, Tosney & Weston, 1980).

The possibility that the environmental cues for this guided movement may be morphologically visible has been investigated in both the early chick embryo (Bancroft & Bellairs, 1976; Tosney, 1978) and in the axolotl embryo (Lofberg, Ahlfors & Fallstrom, 1980). In both species, a fibrous extracellular matrix (ECM) is visible during crest migration. In the axolotl, this is highly oriented in the direction of migration, i.e., ventrally around the sides of the neural tube, whereas in the chick embryo there is no such obvious polarity. Another constituent of the ECM has been reported, which consists of irregular clumps of fibrogranular material, known as 'interstitial bodies' (Low, 1970; Cohen & Hay, 1971). Both isolated neural crest cells (Davis, 1980) and crest-derived fibroblasts (Bard & Hay, 1975) have been demonstrated to follow guidance cues imposed by collagen fibres in hydrated collagen lattices, and so might be expected to be guided by such features *in vivo*.

Primordial germ cells (PGCs)

These arise early in the development of vertebrate embryos, and migrate from their site of origin to the site of the gonad formation, on the dorsal body wall adjacent to the root of the dorsal mesentery of the gut.

In anuran amphibians, the PGCs arise at the vegetal pole of the blastula (Bounoure, 1934; Blackler, 1958, 1962). Following gastrulation, they lie in the floor of the archenteron where they become incorporated into the developing gut. The PGCs then migrate dorsally within the gut tube (Kamimura, Kotani & Yamagata, 1980), and then through its dorsal mesentery to the dorsal body wall (Blacker, 1958; Whitington & Dixon, 1975; Wylie & Heasman, 1976). This pathway is shown in Fig. 1. The advantage of studying PGCs of anuran amphibians such as *Xenopus laevis* is that they can be isolated during their migratory phase as they pass through the gut mesentery, and so can be studied *in vitro* (Wylie & Roos, 1976). We find that isolated PGCs adhere rapidly to amphibian cells grown *in vitro*, whereas they do not adhere to a variety of artificial substrates, such as glass, plastic or collagen. On cellular substrates, they are both motile and invasive. Furthermore, they elongate in the direction of orientation either of the cells on which they are placed, or the stress fibre cytoskeleton of these cells (Fig. 2) (Wylie *et al.*, unpublished results).

Whilst in the mesentery of the gut, PGCs migrate between the two

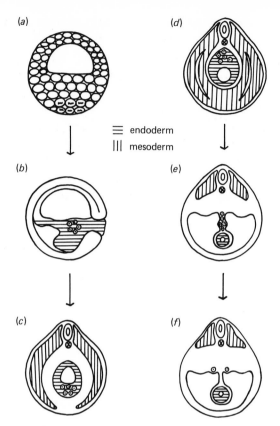

Fig. 1. PGC migration during *Xenopus laevis* embryonic development. (*a*) Blastula stage; the germplasm is incorporated into the cells of the vegetal pole. (*b*) Gastrula stage; the PGCs lie within the endoderm in the floor of the archenteron. (*c*) Stage 20; the PGCs lie within the gut endoderm. (*d*) Stage 35–36; the coelom forms in the lateral plate endoderm. (*e*) Stage 43–44; the PGCs move to the posterior body wall via the dorsal mesentery of the gut. (*f*) Stage 45–46; the gonadal ridges begin to develop on the posterior body wall and are colonised by the PGCs.

layers of simple squamous epithelium which constitute the mesentery at the early stage, i.e. along the basal surfaces of these cells. The shapes of the epithelial cells (CECs), can be established by looking at the coelomic, or apical surface of the epithelium, using the scanning electron microscope. The PGCs can be seen clearly beneath the epithelium, since they are so large that they project through as large lumps. They are always covered over with the epithelium, however, and do not break through into the coelom.

By this technique, we find that CECs are oriented in two directions

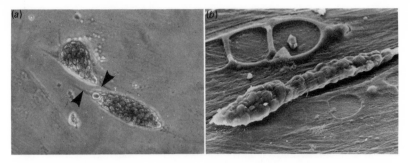

Fig. 2. PGCs elongated on mesentery cell substrate *in vitro*. (*a*) Phase contrast photograph of two adherent PGCs; note the long thin filopodia (arrowed). (*b*) SEM photograph of PGC; the cell is elongated in the direction of the stress fibres in the underlying cells.

whilst in the mesentery. Firstly, nearest the gut, they are found in an oblique orientation, slanting from the gut to dorsal body wall (Fig. 3*a*). Nearer the dorsal body wall, they are found elongated transversely, in a direction outwards from the mesentery, and towards the site of gonad formation (Fig. 3*b*). Junctional areas are also seen where one orientation merges into the other (Fig. 3*c*; Heasman & Wylie, 1981). Whatever the orientation of CECs, however, we always find elongated PGCs oriented in the same direction as the CECs, and this leads us to suppose that PGCs are using the shapes of the overlying epithelial cells for orienting purposes (Heasman & Wylie, 1981). By careful sectioning in either the longitudinal axis, or in the transverse axis of the embryo, we can use the TEM to study the way in which PGC shape is related to the cytoskeleton of the CECs, and any ECM present. Several points emerge from such a study. First, whenever PGCs put out filopodia, they do so in the direction of the long axis of the adjacent CEC (Fig. 4*a*); second, the CECs have a marked stress fibre type of skeleton, with filament bundles running in the long axes of the cells. These filament bundles insert into desmosome-type junctions holding the CECs together (Fig. 4*b*). PGCs conversely, are characterised by a dispersed microfilament system, and organised bundles of filaments are found only in filopodia (Fig. 4*c* & 4*a* inset). Third, the stress fibre orientation of the CECs changes, as might be expected, at sites where changes in polarity of the cells occur. Thus, desmosomes at sites where CECs are at right angles to each other show insertion of filaments at corresponding angles (Fig. 4*d*).

Small adhesion plaques, and areas of close approximation, are found between PGCs and CECs (Fig. 4*c*). We also find similar adhesion plaques

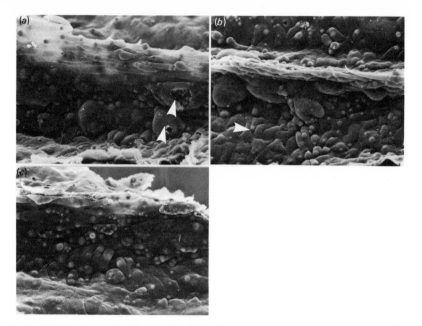

Fig. 3. SEM photographs of stage 43 embryonic mesentery as PGCs migrate through it. (*a*) The CECs in the mesentery proper are elongated in the long axis of the mesentery; PGCs are seen as large bulges (arrowed). (*b*) CECs nearer the body wall are oriented transversely (arrowed) towards the site of gonad formation. (*c*) The junction of transversely and longitudinally oriented CECs can be clearly seen.

when PGCs are *in vitro*, migrating within a matrix of cultured cells (Fig. 4*e*). In both cases, therefore, direct cell-to-cell contacts seem to be important (Heasman & Wylie, 1981). Lastly, we find occasional bundles of ECM fibrils around PGCs *in vivo* (Fig. 4*f*), coaligned with the migrating PGCs. These fibrils are non-banded, and each is 10 nm in diameter.

Occasionally, during preparation for SEM, a dissected specimen of whole mesentery is damaged. If this is fortuitous enough, it permits us to look beneath the coelomic epithelium (Fig. 5). Here we find structures that look like ECM fibrils, in the plane in which PGCs migrate. We are currently continuing to study these.

The anuran amphibian is not the only vertebrate group in which an oriented path can be seen around the PGCs. In the chick embryo, PGCs are found around the cranial end of the primitive streak, at the junction between area opaca and area pellucida (Fujimoto, Ukeshima & Kiyofuji, 1976). Just beneath the surface epiblast at this point there is a crescent-shaped area rich in fibrils, which are oriented primarily in the direction of PGC migration (Critchley, England, Wakely & Hynes, 1979). ECM fibrils are also found at early stages in chick embryogenesis which coalign with the outward migration of mesoderm cells from the primitive streak (Wakely & England, 1979).

Pathways in the developing nervous system

There is evidence both from morphological studies, and from observation of migration from ectopic locations, that oriented pathways exist in the developing nervous system. A now classical piece of research by Cajal (Ramon y Cajal, 1960) demonstrated that pioneer neuron processes can act as a substrate for guidance of axon outgrowth in the cerebellum. Similarly, in the peripheral nervous system, Schwann cells and other cell types can use axons as a substrate for migration (Spiedel, 1964). Conversely, glial cells, once established, can act as substrates for the migration of whole neurons, for example, in the cerebellum (Rakic, 1971), and the cerebral cortex (Schmechel & Rakic, 1979). A second pioneering piece of research in this field is that of Harrison (Harrison, 1904) who demonstrated that in interspecific grafts of *Rana* embryos, the cells destined to form the lateral line system of the graft invaded the host embryo, and faithfully followed its lateral line pathway.

More recently, transplantation of developing eye and hindbrain to ectopic sites in developing frog embryos has shown that outgrowing neuron processes from these transplants can invade the host nervous system. These grow along discrete reproducible pathways in the long axis of the developing spinal cord. Mauthner neurons (from grafted hind brain) pass along the margin of the basal plate, whereas optic neurons grow along the marginal surface of the altar plate (Katz & Lasek, 1979, 1980). Such pathways for neuron migration have been visualised as intercellular spaces visible by light microscopy within the outer margins of the embryonic nervous system (Singer, Nordlander & Egar, 1979; Silver & Sidman, 1980).

The sclerotome

This is a component of the early somite, which subsequently migrates medially and ventrally to surround the notochord, and then the neural tube. Sclerotome cells then differentiate into bone and cartilage of the vertebral column. The initial migration of these cells has been studied in the early chick embryo (Trelstad, Hay & Revel, 1967; Ebendal, 1977). Cells are elongated in the direction of travel, and show close appositions and electron dense plaques with the ECM fibrils which are

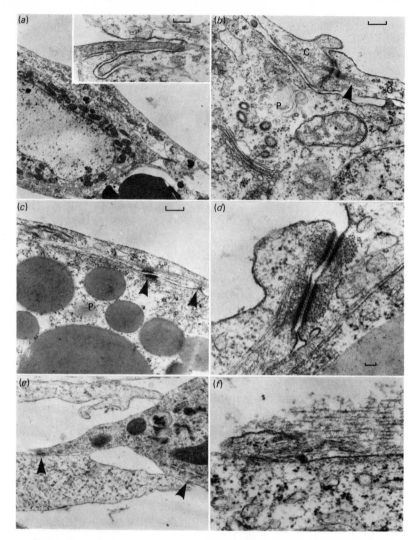

Fig. 4. TEM photographs of sites of cell interaction during PGC migration.

(a) Low-power EM of PGC (cut in TS) covered by a CEC. The PGC has extended a fine straight filopod at one end. Inset: High-power EM of the PGC process, which is associated with an indention in the CEC surface. This PGC process was followed in serial thin sections throughout its width. It was a straight protrusion containing a bundle of longitudinally oriented microfilaments. Bar = 0.2 μm.

(b) Medium-power EM of a specimen cut in LS. A PGC (P) is covered by 2 CECs (C). Note the desmosome-like junction between the CECs and the longitudinally oriented bundle of microfilaments associated with it (arrow). Bar = 0.1 μm.

(c) Medium-power EM of a specimen cut in LS. A cytoplasmic lamella of a CEC covers a PGC (P). Note regions of close approach between the PGC and CEC (arrows) and the desmosome-like junction between the two cells. The microfilaments are very obvious in the CEC, but are not conspicuous in the PGC. Bar = 0.2 μm.

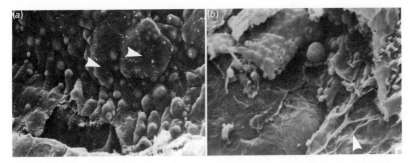

Fig. 5. (*a*) Low-power scanning electron micrograph of section of mesentery damaged during dissection; two PGCs are visible within the mesentery (arrowed). (*b*) High-power scanning electron micrograph of damaged region; beneath the coelomic lining, where the PGCs migrate, can be seen a network of fine fibrils (arrowed).

aligned in the direction of migration. It has been suggested (Ebendal, 1976, 1977) that sclerotome cells migrate, mainly as individuals, along these oriented matrix fibres.

Heart formation

Careful EM observations have demonstrated that cells which will form the endocardial cushion migrate along microfibrils in the acellular cardiac jelly (Markwald, Fitzharris, Bolender & Bernanke, 1979). Furthermore, alignment of the cardiac jelly fibrils becomes obvious immediately before migration of the cushion cells.

Results from all of these developmental systems, taken together, indicate that discrete pathways for cell migration do exist in early embryos, and that their specificity varies (e.g. two different types of neuron choose two different parts of the CNS, whereas several cell types seem capable of using the neural crest pathway). There is a great deal of scope for further work into the cell-type specificity of these pathways.

Of course, the presence of a pathway, unless it is a haptotactic one

(*d*) High-power EM of 2 CECs overlying a PGC at the root of the mesentery (cut in TS). Note the filaments associated with the junctional complex between the two cells. In the mesentery they are cut in TS, while in the CEC of the dorsal body wall, they are longitudinally sectioned. Bar = 0.1 μm.

(*e*) Medium-power EM of PGC filopod cut in LS, protruding between two CECs *in vitro*. Note the dense adhesion plaques (arrowed) between the filopod and one of the CECs.

(*f*) High-power EM of a bundle of extracellular fibrils often seen around PGCs *in vivo*, coaligned with the membranes of migrating PGCs. The fibrils are non-banded and each is about 10nm in diameter.

increasing adhesiveness for example, will only indicate *polarity* to a migrating cell. *Direction* could be superimposed on this in a number of ways. For example, injections of nerve growth factor into the brains of developing rats causes sympathetic axons to abnormally innervate the spinal cord. These axons run to the site of injections by discrete and reproducible channels indicating a possible combination of chemotaxis and orienting pathways (Menesini Chen *et al.*, 1978). Direction could also be imposed by population pressure (i.e. contact inhibition of movement 'against the tide') (Tosney, 1978), or by ageing and dis-assembly of the pathway following passage of the cells (Weston & Butler, 1966; Wylie *et al.*, 1979). Alternatively, direction, once established by any mechanism e.g. haptotaxis (Carter, 1967), may be irreversible, due to intrinsic factors within the migrating cell itself, which is only able to continue in one direction (Gail & Boone, 1970; Albrecht-Buehler, 1977). It is possible that direction in a pathway may only slowly become established, since cells injected early enough into the chick neural crest pathway can pass either way along it, whereas later they can only pass in the conventional direction of crest cell migration (Erickson *et al.*, 1980).

MOLECULES KNOWN TO BE ASSOCIATED WITH EMBRYONIC CELL MIGRATION

Using such migratory cell lineages as those documented above, we can now turn to the way in which known ECM molecules are associated with them, and putative mechanisms by which they act.

Glycosaminoglycans (GAGs)

These are large extracellular polyanions, consisting mainly of repeat-ing disaccharide units. The different species of GAG are distinguished by molecular size, sugar residues, and variable numbers and position of sulphate groups. GAGs are found complexed with proteins, giving rise to large proteoglycan molecules.

Both the type and distribution of GAGs alter during morphogenesis, suggesting that they are in some way causally related to cell move-ments (see reviews by Toole, 1976, 1981). Of particular interest are the changing levels of hyaluronic acid (HA) during embryonic cell migra-tion. An example of this is in the developing cornea of the chick, which is associated with several waves of cell migration (Hay & Revel, 1969; Hay 1980). The last cells to migrate are the corneal fibroblasts, which lay down the complex orthogonal array of collagen fibres characteristic of the adult cornea. These neural crest-derived fibroblasts migrate into an extracellular space greatly expanded by the secretion of HA by the corneal endothelium (Toole & Trelstad, 1971; Trelstad et al., 1974). Subsequent to this migration, hyaluronidase levels rise, the level of HA falls, and the tissue condenses. This same sequence of events occurs in other areas of massive cell migration, e.g. the migration of the chick sclerotome is associated with high levels of HA (Kvist & Finnegan, 1970a, b). Following this migration, HA levels go down, and levels of chondroitin sulphate rise (Toole, 1972). Similarly, the initial neural crest outgrowth takes place in a large space, rich in HA (Pratt, Larsen & Johnston, 1975; Derby, 1978), as does primary mesenchyme formation, and neural fold formation in early rat embryos (Solursh & Morriss, 1977). These observations have led to the general hypothesis that large extracellular spaces rich in GAG are somehow conducive to cell migration. Indeed, in interspecific frog hybrids which cannot release complex carbohydrates into the intercellular spaces, the movements of gastrulation appear to be prevented (Johnson, 1977). However, less is known about migratory cells not surrounded by a large extracellular space, e.g. neural crest cells later on in their migration, or outgrowing nerve processes, GAGs have also been found in adhesion plaques between certain types in vitro and their substrates (Culp, 1976; Culp, Murray & Rollins, 1979). Once again it is suggested that a change in composition of the GAG complex takes place as the cells move.

Collagen

Collagen fibres seem capable of guiding cell movement, though the degree to which this property is specific to the collagen molecule is a vexed question, since it is often found associated with other ECM components, such as GAGs or fibronectin. When collagen fibres are arranged as a lattice in vitro, cells from a variety of sources are found to coalign with them, e.g. fibroblasts (Elsdale & Bard, 1972), neural crest-derived corneal fibroblasts (Bard & Hay, 1975), outgrowing neuron processes (Ebendal, 1976), chick heart fibroblasts (Dunn & Ebendal, 1978), and outgrowing neural crest cells (Davis, 1980). In each case, the collagen fibres play a contact-guiding role, which cannot be due to simple adhesion to the collagen, since guidance is lost when oriented collagen lattices are dried down flat (Dunn & Ebendal, 1978; Davis, 1980). Presumably, therefore, cells are responding to defined shape, rather than simple adhesive interactions. In this role, collagen in vivo

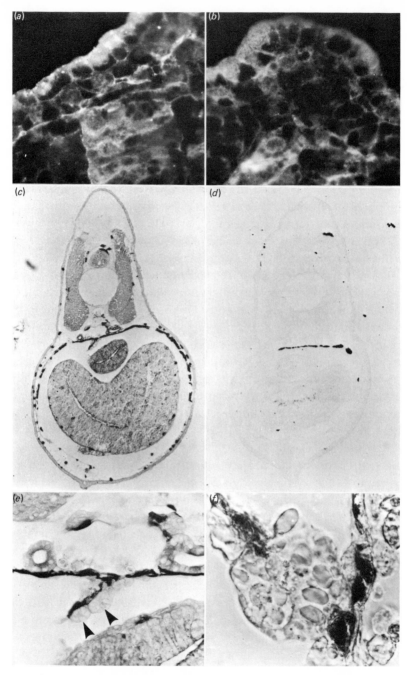

Fig. 6. (*a*) TS *Xenopus laevis* embryo stage through the neural crest. The section was stained with rabbit anti-*Xenopus* fibronectin (FN) and FITC-goat anti-rabbit Ig. There is

may act to limit the options open to a cell, and thus give directional cues (Bard, Hay & Mellor, 1975; Haemmerli & Strauli, 1978). The behaviour of different cell types on collagen matrices derived from basement membranes (Overton, 1979) would tend to support this idea, as would the demonstration that increasing or decreasing the amounts of collagen in tissue aggregates *in vitro* has a corresponding effect on cell migration in the aggregates (Armstrong & Armstrong, 1980).

Fibronectin

This glycoprotein, which is known to be important in cell adhesion, has been shown to play a role in the organisation of the cytoskeleton of transformed cells *in vitro* (Ali, Mautner, Lanza & Hynes, 1977; Willingham *et al.*, 1977). Furthermore, elements of the cytoskeleton coalign with surface fibronectin in many cases *in vitro* (Hynes & Destree, 1978), and when the cytoskeleton is disrupted with cytochalasin B or D, this causes release of surface fibronectin (Mautner & Hynes, 1977; Ali & Hynes, 1977; Kurkinen, Wartiovaara & Vaheri, 1978). These observations, together with the fact that added fibronectin seems capable of stimulating certain cell migration *in vitro* (Ali & Hynes, 1978), suggest that this molecule may play a role in cell migration *in vivo*.

A considerable body of evidence is now accruing that this is so. Fibronectin is found in several sites of cell movement in early embryos. Using immunohistochemical staining, it can be observed in early mouse embryos (Zetter & Martin, 1978; Wartiovaara, Leivo & Vaheri,

some fluorescence around the neural crest cells, which are just beginning to migrate at this stage.

(*b*) As in (*a*) using control (no anti-FN) serum. There is no fluoresence around the neural crest cells.

(*c*) Low-power photomicrograph of TS stage 45 *Xenopus laevis* embryo, stained for fibronectin by the peroxidase-anti-peroxidase method (PAP). Sites of FN localisation show up as dense areas with bright field illumination and are widely dispersed in embryos of this stage.

(*d*) Control for (*c*) using serum containing no anti-FN. The black line seen here is a row of pigment cells around the dorsal body wall.

(*e*) Medium-power photomicrograph of PAP-anti-FN stained TS stage 45 embryos. This shows the mesentery with PGCs migrating through it (arrowed). FN is localized around these PGCs.

(*f*) High-power photomicrographs of PAP-anti-FN stained mesentery containing PGCs, showing fibronectin associated with the mesentery cells around the migrating PGC.

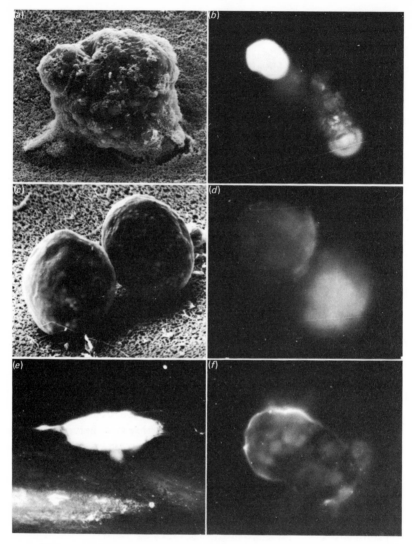

Fig. 7. (a) Scanning electron micrograph of EDTA-isolated PGC on a Millipore filter. PGCs isolated in this way have surfaces contaminated with parts of cells and extra-cellular debris. In this instance there is a whole CEC flattened onto the surface.

(b) Fluorescence micrograph of EDTA-isolated PGCs stained with anti-*Xenopus* fibronectin antibody. The PGCs themselves show some fluorescence, but the very brightly-fluorescent cells are contaminating CECs.

(c) Scanning electron micrographs of PGCs isolated using EDTA and then trypsi-nised. This treatment removes all contamination from the cell surface.

(d) Fluorescence micrograph of PGCs isolated and trypsinised, and stained with anti-*Xenopus* fibronectin. The cells show no fluorescence after trypsinisation.

(e) Fluorescent micrographs of PGC isolated, trypsinised and allowed to recover its

1980), and in early chick embryos in areas of mesoderm migration, primordial germ cell migration, and at the edge of the expanding blastoderm (Critchley et al., 1979). It is also found in chick embryos around neural crest cells as they begin migrating (Mayer, Hay & Hynes, 1981; Newgreen & Thiery, 1980).

We have been studying the distribution and possible role of fibronectin in early amphibian embryos, using an anti-*Xenopus* plasma fibronectin. We find positive staining in transverse sections of early embryos around the neural crest at the beginning of migration (Fig. 6a & b). At later stages, we find fibronectin more widely distributed, particularly in a population of extremely strongly staining cells, widely dispersed in the embryo (Fig. 6c & d). We also find staining around PGCs in the mesentery of the gut (Fig. 6e & f). Since fibronectin appears to be present *in vivo* at the time of PGC migration, we isolated PGCs from the mesentery of the gut, and examined them *in vitro*. We find the following:

(i)   PGCs isolated with EDTA are found to be surrounded by fibronectin-containing cells, shown by SEM and fluorescence in Fig. 7a & b.

(ii)   When treated with 0.05% trypsin at room temperature for four minutes contaminating cells and fibronectin are removed (Fig. 7c & d).

(iii) When the PGCs have recovered their ability to adhere to cultured epithelial cells, they remain fibronectin-negative.

(iv)   However, once stuck to cultured cells, fibronectin is found around PGCs (Fig. 7e & f).

These results suggest that PGCs adhere to fibronectin on the underlying cells *in vitro*, and therefore possibly *in vivo*, but they do not synthesise it themselves. We therefore seeded batches of PGCs, previously trypsinised, and allowed to recover in fibronectin-free medium, onto cultured *Xenopus* mesentery cells, in fibronectin-free medium containing Fab fragments of the anti-*Xenopus* fibronectin. The results are shown graphically in Fig. 8, and demonstrate that adhesion of the PGCs is inhibited by the anti-fibronectin antibody. We therefore

adhesive properties. The cell is stuck to, and elongated on, a mesentery cell substrate *in vitro*, and stained with anti-*Xenopus* fibronectin. This cell is now strongly fibronectin-positive.

(f) Fluorescent micrograph of two PGCs treated as in (e). The PGCs have stuck to and are beginning to elongate on, a mesentery cell substrate *in vitro*. Again, the PGCs are fibronectin-positive.

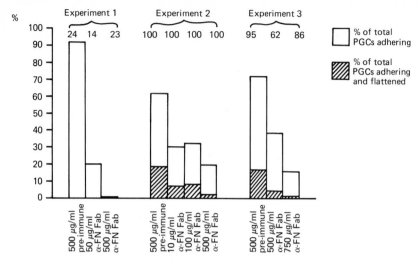

Fig. 8. Adhesion and spreading of PGCs on a mesentery cell substrate, in the presence of Fab fragments of anti-fibronectin. The PGCs were isolated by EDTA, trypsinised and left in fibronectin-free medium for 48 hours to recover their adhesive properties. The numbers at the top of each column indicate the number of cells seeded in each case. The experiment was performed three times, because of the limited number of PGCs obtainable at any one time.

conclude that PGCs do use fibronectin in their *in vitro* adhesion. Since this molecule is also present *in vivo*, it may play here a similar role. This work is currently being submitted for publication. Indeed, fibronectin could act as a common denominator in many of the phenomena described above, since it has been shown to possess binding sites for many other extracellular matrix molecules, as well as cell surfaces (Yamada, Olden & Hahn, 1980).

It is obvious that only by coupling careful morphological observations, with physiological experiments, such as the adhesion blocking experiments described above, with known migratory cells in carefully defined conditions, will we be able to explore the roles of individual ECM components in cell guidance.

Acknowledgements

We would like to thank the Cancer Research Campaign and the Wellcome Trust for financial help in the work described. We are also indebted to Ms Diane Jackson and Ms Claire Varley for expert technical assistance.

REFERENCES

ALBRECHT-BUEHLER, G. (1977). The phagokinetic tracks of 3T3 cells. *Cell*, 11, 395–404.

ALI, I. U. & HYNES, R. O. (1977). Effects of cytochalasin B and colchicine on attachment of a major surface protein of fibroblasts. *Biochimica et Biophysica Acta*, 471, 16–24.

ALI, I. U. & HYNES, R. O. (1978). Effects of LETS glycoprotein on cell mobility. *Cell*, 14, 439–46.

ALI, I. U., MAUTNER, V. M., LANZA, R. P. & HYNES, R. O. (1977). Restoration of normal morphology adhesion, and cytoskeleton in transformed cells by addition of a transformation-sensitive surface protein. *Cell*, 11, 115–26.

ARMSTRONG, M. T. & ARMSTRONG, P. B. (1980). The role of the extracellular matrix in cell motility in fibroblast aggregates. *Cell Motility*, 1, 99–112.

AUSPRUNK, D. H. & FOLKMAN, J. (1977). Migration and proliferation of endothelial cells in preformed and newly formed blood vessels during tumour angiogenesis. *Microvascular Research*, 14, 53–65.

BANCROFT, M. & BELLAIRS, R. (1976). The neural crest cells of the trunk region of chick embryos studied by SEM and TEM. *Zoon*, 4, 73–85.

BARD, J. B. L., HAY, E. D. & MELLOR, S. H. (1975). Formation of the avian cornea. A study of cell movement *in vitro*. *Developmental Biology*, 42, 334–61.

BLACKER, A. W. (1958). Contribution to the study of germ cells in the Anura. *Journal of Embryology and Experimental Morphology*, 6, 491–503.

BLACKLER, A. W. (1962). Transfer of primordial germ cells between two sub-species of *Xenopus laevis*. *Journal of Embryology and Experimental Morphology*, 10, 641–51.

BOUNOURE, L. (1934). Recherches sur la lignee germinale chez la grenouille rousse aux premiers stades du development. *Annales des Sciences Naturelles*, 10^e series, 17(b), 67–248.

BRONNER-FRASER, M. & COHEN, A. M. (1980). Analysis of the neural crest ventral pathway using injected tracer cells. *Developmental Biology*, 77, 130–41.

CARTER, S. B. (1967). Haptotaxis and the mechanisms of cell motility. *Nature*, 213, 256–60.

COHEN, A. M. & HAY, E. D. (1971). Secretion of collagen by embryonic neuroepithelium at the time of spinal cord–somite interaction. *Developmental Biology*, 26, 578–605.

CRITCHLEY, D. R., ENGLAND, M. A., WAKELY, J. & HYNES, R. O. (1979). Distribution of fibronectin in the ectoderm of gastrulating chick embryos. *Nature*, 280, 498–500.

CULP, L. A. (1976). Molecular composition and origin of substrate-attached material from normal and virus-transformed cells. *Journal of Supramolecular Structure*, 5, 239–55.

CULP, L. A., MURRAY, B. A. & ROLLINS, B. J. (1979). Fibronectin and proteoglycans as determinants of cell-substratum adhesion. *Journal of Supramolecular Structure*, 11, 401–27.

DAVIS, E. M. (1980). Translocation of neural crest cells within a hydrated collagen lattice. *Journal of Embryology and Experimental Morphology*, 55, 17–31.

DERBY, M. A. (1978). Analysis of glycosaminoglycans within the extracellular environment encountered by migrating neural crest cells. *Developmental Biology*, **66**, 321–36.

DUNN, S. A. & EBENDAL, T. (1978). Contact guidance of oriented collagen gels. In *Form-shaping Movements in Neurogenesis*, ed. C.-O. Jacobson & T. Ebendal, p. 65. Stockholm: Alunquist & Wiksell.

EBENDAL, T. (1976). The relative rules of contact inhibition and contact guidance in orientation of axons extending on aligned collagen fibrils *in vitro*. *Experimental Cell Research*, **98**, 159–69.

EBENDAL, T. (1977). Extracellular matrix fibrils and cell contacts in the chick embryo. Possible roles of orientation of cell migration and axon extension. *Cell Tissue Research*, **175**, 439–58.

ELSDALE, T. & BARD, J. B. L. (1972). Collagen substrata for studies on cell behaviour. *Journal of Cell Biology*, **54**, 626–37.

ERICKSON, C. A., TOSNEY, K. W. & WESTON, J. A. (1980). Analysis of migratory behaviour of neural crest and fibroblastic cells in embryonic tissues. *Developmental Biology*, **77**, 142–56.

FUJIMOTO, T., UKESHIMA, A. & KIYOFUJI, R. (1976). The origin, migration and morphology of the primordial germ cells in the chick embryo. *Anatomical Record*, **185**, 139–54.

GAIL, M. H. & BOONE, C. W. (1970). The location of mouse fibroblasts in tissue culture. *Biophysical Journal*, **10**, 980–93.

HAEMMERLI, G. & STRAULI, P. (1978). Motility of L5222 leukaemia cells within the mesentery. *Virchows Archiv-Abteilung B. Zellpathologie*, **29**, 167–77.

HARRIS, A. (1973). Behaviour of cultured cells on substrata of variable adhesiveness. *Experimental Cell Research*, **77**, 285–97.

HARRISON, R. G. (1904). Experimentelle Untersuchungen uber die Entwicklung der Sinnersorgane der Seitenlinie bei den Amphibian. *Archiv von Mikroskopisches Anatomie*, **63**, 35–149.

HAY, E. D. (1980). Development of the vertebrate cornea. *International Review of Cytology*, **63**, 263–322.

HAY, E. D. & REVEL, J. P. (1969). Fine structure of the developing avian cornea. *Monographs in Developmental Biology*, vol. 1, ed. A. Wolsky & P. S. Chen. Karger Basel.

HEASMAN, J. & WYLIE, C. C. (1981). The migratory route of the primordial germ cells of the anuran amphibian *Xenopus laevis*, and its role in their guidance. *Philosophical Transactions of the Royal Society of London*, (in press).

HORSTADIUS, S. (1950). *The Neural Crest*. Oxford University Press.

HYNES, R. O. & DESTREE, A. T. (1978). Relationships between fibronectin (LETS protein) and actin. *Cell*, **15**, 875–86.

JOHNSON, K. E. (1977). Extracellular matrix synthesis in blastula and gastrula stages of normal and hybrid frog embryos. I. Toluidine blue and lanthanum staining. *Journal of Cell Sciences*, **25**, 313–22.

KAMIMURA, M., KOTANI, M. & YAMAGATA, K. (1980). The migration of presumptive primordial germ cells through the endodermal cell mass in *Xenopus laevis*. A light and electron microscope study. *Journal of Embryology and Experimental Morphology*, **59**, 1–17.

KATZ, M. J. & LASEK, R. J. (1979). Substrate pathways which guide growing axons in *Xenopus* embryos. *Journal of Experimental Neurology*, **183**, 817–32.

KATZ, M. J. & LASEK, R. J. (1980). Guidance cue patterns and cell migration in multicellular organisms. *Cell Motility*, **1**, 141–57.

KONIJN, T. M., VAN, DE MEENE, J. G. C., BONNER, J. T. & BARKLEY, D. S. (1967). The acrasin activity of adenosine 3′–5′ cyclic phosphate. *Proceedings of the National Academy of Sciences, USA*, **58**, 1152–4.

KURKINEN, M., WARTIOVAARA, J. & VAHERI, A. (1978). Cytochalasin B releases a major surface-associated glycoprotein, fibronectin, from cultured fibroblasts. *Experimental Cell Research*, **111**, 127–37.

KVIST, T. N. & FINNEGAN, C. V. (1970a). The distribution of glycosaminoglycans in the axial region of the developing chick embryo. 1. Histochemical analysis. *Journal of Experimental Zoology*, **175**, 221–40.

KVIST, T. N. & FINNEGAN, C. V. (1970b). The distribution of glycosaminoglycans in the axial region of the developing chick embryo. 11. Biochemical analysis. *Journal of Experimental Zoology*, **175**, 241–58.

LE DOUARIN, N. & TEILLET, M. (1974). Experimental analysis of the migration and differentiation of neuroblasts of the autonomic nervous system and of neuro-ectodermal mesenchymal derivatives, using a biological cell marking technique. *Developmental Biology*, **41**, 162–84.

LOFBERG, J., AHLFORS, K. & FALLSTROM, C. (1980). Neural crest cell migration in relation to extracellular matrix organisation in the embryonic axolotl trunk. *Developmental Biology*, **75**, 148–67.

LOW, F. N. (1970). Interstitial bodies in the early chick embryo. *American Journal of Anatomy*, **128**, 45–56.

MARKWALD, R. R., FITZHARRIS, T. P., BOLENDER, D. L. & BERNANKE, D. H. (1979). Structural analysis of cell: matrix association during the morphogenesis of atrioventricular cushion tissue. *Developmental Biology*, **69**, 634–54.

MAUTNER, V. & HYNES, R. O. (1977). Surface distribution of LETS protein in relation to the cytoskeleton of normal and transformed cells. *Journal of Cell Biology*, **75**, 743–58.

MAYER, B. W., HAY, E. D. & HYNES, R. O. (1981). Immunocytochemical localisation of fibronectin in embryonic chick trunk and area vasculosa. *Developmental Biology*, (in press).

MENESINI CHEN, M. G., CHEN, J. S. & LEVI-MONALCINI, R. (1978). Sympathetic nerve fibres ingrowth into the CNS of neonatal rodents upon intracerebral NGF injection. *Archives Italiennes de Biologie*, (*Pisa*), **116**, 53–84.

NEWGREEN, D. & THIERY, J-P. (1980). Fibronectin in early avian embryos: synthesis and distribution along the migratory pathways of neural crest cells. *Cell and Tissue Research*, **211**, 269–91.

NODEN, D. (1975). An analysis of the migratory behaviour of avian cephalic neural crest cells. *Developmental Biology*, **42**, 106–30.

OVERTON, J. (1979). Differential response of embryonic cells to culture on tissue matrices. *Tissue and Cell*, **11**, 89–98.

PRATT, R. M., LARSEN, M. A. & JOHNSTON, M. C. (1975). Migration of cranial neural crest cells in a cell-free hyaluronic-rich matrix. *Developmental Biology*, **44**, 298–305.

RAKIC, P. (1971). Neuron-glia relationships during granule cell migration in the developing cerebellar cortex. A Golgi and electron microscopic study in *Macacus rhesus. Journal of Comparative Neurology*, **141**, 283–312.

RAMON'Y CAJAL, S. (1960). *Studies on Vertebrate Neurogenesis.* Springfield: Charles C. Thomas.

SCHMECHEL, D. E. & RAKIC, P. (1979). A golgi study of radial glial cells in developing monkey telencephalon: morphogenesis and transformation into astrocytes. *Anatomy and Embryology*, **156**, 115–52.

SILVER, J. & SIDMAN, R. L. (1980). A mechanism for the guidance and topographic patterning of retinal ganglion cell axons. *Journal of Comparative Neurology*, **189**, 101–11.

SINGER, M., NORDLANDER, R. H. & EGAR, M. (1979). Axonal guidance during embryogenesis and regeneration in the spinal cord of the newt: the blueprint hypothesis of neuronal pathway patterning. *Journal of Comparative Neurology*, **185**, 1–22.

SOLURSH, M. & MORRISS, G. M. (1977). Glycosaminoglycans synthesis in rat embryos during the formation of the primary mesenchyme and neural folds. *Developmental Biology*, **57**, 75–86.

SPIEDEL, C. C. (1964). *In vivo* studies of myelinated nerve fibres. *International Review of Cytology*, **16**, 173–231.

TOOLE, B. P. (1972). Hyaluronate turnover during chondrogenesis in the developing chick limb and axial skeleton. *Developmental Biology*, **29**, 321–9.

TOOLE, B. P. (1976). Morphogenetic role of glycosaminoglycans (acid mucopolysaccarides) in brain and other tissues. In *Neuronal Recognition*, ed. S. H. Barondes, p. 275. New York: Plenum Press.

TOOLE, B. P. (1981). Glycosaminoglycans in morphogenesis. In *Cell Biology of Extracellular Matrix*, ed. E. D. Hay. New York: Plenum Press.

TOOLE, B. P. & TRELSTAD, R. L. (1971). Hyaluronate production and removal during corneal developmental in the chick. *Developmental Biology*, **26**, 28–35.

TOSNEY, K. W. (1978). The early migration of neural crest cells in the trunk of the avian embryo: an electron microscopic study. *Developmental Biology*, **62**, 317–33.

TRELSTAD, R. L., HAY, E. D. & REVEL, J-P. (1967). Cell contact during early morphogenesis in the chick embryo. *Developmental Biology*, **16**, 78–106.

TRELSTAD, R. L., HAYASHI, K. & TOOLE, B. P. (1974). Epithelial collagens and glycosaminoglycans in the embryonic cornea: macromolecular order and morphogenesis in the basement membrane. *Journal of Cell Biology*, **62**, 815.

WAKELY, J. & ENGLAND, M. A. (1979). Scanning electron microscopical and histochemical study of the structure and function of basement membranes in the early chick embryo. *Proceedings of the Royal Society of London*, **B206**, 329–52.

WARTIOVAARA, J., LEIVO, I. & VAHERI, A. (1980). Matrix glycoproteins in early mouse development and in differentiation of teratocarcinoma cells. In *The Cell Surface: Mediator of Developmental Processes*, ed. S. Subtelny & N. K. Wessells, pp. 305–28. New York: Academic Press.

WESTON, J. A. (1970). The migration and differentiation of neural crest cells. *Advances in Morphogenesis*, **8**, 41–114.

WESTON, J. A. & BUTLER, S. L. (1966). Temporal factors affecting localisation of neural crest cells in the chicken embryo. *Developmental Biology*, **14**, 246–66.

WHITINGTON, P. McD. & DIXON, K. E. (1975). Quantitative studies of germ plasm and germ cells during early embryogenesis of *Xenopus laevis. Journal of Embryology and Experimental Morphology*, 33, 57–74.

WILLINGHAM, M. C., YAMADA, K. M., YAMADA, S. S., POUYSSEGUR, J. & PASTAN, I. (1977). Microfilament bundles and cell shape are related to adhesiveness to substratum and are dissociable from growth control in cultured fibroblasts. *Cell*, 10, 375–80.

WYLIE, C. C. & HEASMAN, J. (1976). The formation of the gonadal ridge in *Xenopus laevis*. I. A light and transmission electron microscope study. *Journal of Embryology and Experimental Morphology*, 35, 125–38.

WYLIE, C. C., HEASMAN, J., SWAN, A. P. & ANDERTON, B. H. (1979). Evidence for substrate guidance of primordial germ cells. *Experimental Cell Research*, 121, 315–24.

WYLIE, C. C. & ROOS, T. (1976). The formation of the gonadal ridge in *Xenopus laevis*. III. The behaviour of isolated primordial germ cells *in vitro. Journal of Embryology and Experimental Morphology*, 35, 149–57.

YAMADA, K. M., OLDEN, K. & HAHN, L-H. E. (1980). Cell surface protein and cell interactions. In *The Cell Surface: Mediator of Developmental Processes*, ed. S. Subtelny & N. K. Wessells, pp. 43–78. New York: Academic Press.

ZETTER, B. R. & MARTIN, G. R. (1978). Expression of a high molecular weight cell surface glycoprotein (LETS protein) by preimplantation mouse embryos and teratocarcinoma stem cells. *Proceedings of the National Academy of Sciences, USA*, 75, 2324–8.

ZIGMOND, S. H. (1978). Chemotaxis by polymorphonuclear leucocytes. *Journal of Cell Biology*, 77, 269–87.

Growth and migration in endothelial regeneration

ROBERT R. BÜRK

Friedrich Mhescher-Institut, PO Box 273, CH-4002 Basel, Switzerland

In the early mammalian embryo there is rapid proliferation of cells and also migration of cells from one region to another. In the adult as a whole there is only a minimum of proliferation of cells. This can be described as maintenance growth. There is in general no cell migration. However, in particular epithelia there is a high proliferation rate and the functional cells of this proliferation pool all migrate away from their origin, displacing previous cells. It is probably generally true that the maintenance proliferation drives the system. In vascular endothelium it seems that there should be a certain low proliferation rate. The daughters of a division either occupy the space of their mother and displace a decrepit neighbouring cell or one of them is simply lost into the blood stream. This is a 'space-full' response. When the space is full, there is nowhere for one daughter to attach and that daughter is lost.

In contrast to the maintenance situation there is a wound response. When a tissue is disturbed so that some of the cells are separated, the cells can then spread with a minimum of migration back over the gap and close it in a matter of hours. There may be no proliferative response. Endothelial cells in culture have just this property. They grow as a strict monolayer. However, in a confluent monolayer the cell density can vary by a factor of about four. It seems that the cells grow to the higher density so that if a few cells are lost there are enough remaining to spread and cover the gap, thus rapidly restoring the integrity and function of the endothelium. The full wound response only occurs when a part of a tissue is destroyed or removed to leave a larger open space. The remaining cells of the type removed migrate into the open space and proliferate. Here proliferation seems to be an 'open-space' response and is terminated by a 'space-full' situation. Often there is an overshoot in the proliferation leading to hyperplasia.

The proper state of affairs in the adult is a maintenance proliferation that is counterbalanced by a 'space-full' loss. This terminology is appropriate for endothelium, but not for the epithelia described in C. Potten's chapter. The response to a wound in an endothelium is cell migration, followed, when there is an 'open-space', by cell proliferation.

What would happen if control of migration and proliferation failed? The control can fail in two directions. It can fail on or off. Further it can switch on or off improperly. A total failure to switch on proliferation in response to the appropriate situation would presumably be recognized as an acute wasting disease. This type of acute failure seems to be very rare. There is also the possibility that under starvation conditions proliferation is switched on but the necessary nutrients (or more likely the vitamins) are deficient and the response does not occur. Another normal but improper situation occurs during old age. Then, even with adequate nutrition, wound healing is slow. Why? Does the difference lie in the cells themselves or in their environment? Epidermal cells in culture can be maintained for some fifty generations and, if Epidermal Growth Factor is added to the culture medium the cultures can be maintained for one hundred and fifty generations (Rheinwald & Green, 1977). Is therefore the slowing down of wound healing in old age due to the lack of a growth factor, as in the case of cultured epidermal cells? Could wound healing be improved by supplying the appropriate growth factor?

Failure to switch off a proper response to a wound would lead to a tumour-like growth. Hyperplasia, however, seems to be an intrinsic part of the wound response. The hyperplasia can be expected since the cells require a few hours to become committed to cell division and then require a day or two to reach and complete cell division. (Bürk, 1970; Carpenter & Cohen, 1976; Shechter, Hernaez & Cuatrecasas, 1978; Jimenez de Asua, 1980). This lag between the stimulus to divide, which can be received when there is an 'open-space', and the actual division which can then occur in a 'space-full' situation, leads to an overshoot or hyperplasia, until the excess cells are shed. A failure to switch off might be due to the lack of the proper signal but more likely (because if one cell failed to produce the signal others would continue to produce it) the failure would be due to the lack of a receptor mechanism, either the receptor for an inhibitor or, for instance, a lack of junctional communication so that intracellular signal molecules would not be received by a cell. This is the basis for the interest in the effect of phorbol esters on metabolic cooperation. (cf. Newbold, this volume).

If it is assumed that all cells in a mammal contain the same genetic information and that there are growth factors whose concentration properly stimulates maintenance proliferation, then these factors cannot be produced in their target cell. They must be hormones produced elsewhere and the gene must not be expressed in the target cell. If, however, the gene through mutation, transposition or teratogeny becomes expressed ectopically in its target cell then that cell becomes auto-stimulating and will escape the normal growth regulation. We have detected two such growth factors in a hamster tumour cell line, SV28. Both factors stimulate 3T3 cells to migrate and proliferate. It seems that migration is associated with the cell cycle. Only one of the factors is also produced by the parent line of SV28, BHK21/13, and neither is produced by normal secondary cultures of baby hamster kidney fibroblasts. The ectopic auto-stimulation that we have described would be an example of switching on improperly (Bürk, 1980).

It is known (Folkman, 1975) that blood vessels can be induced to form by nearby tumours. Apparently, the tumour releases a substance that diffuses to the neighbouring blood vessels. The substance stimulates the endothelial cells there and these move towards the tumour, proliferating and forming a vessel around a core that later dissolves. The factor that we have isolated from SV28 and BHK21/13 can stimulate the migration and division of aorta endothelial cells. It is therefore a candidate for a Tumour Angiogenesis Factor.

I want to describe now, an attempt to find an example of a chronic failure of maintenance growth and suggest that is likely to be an example of a failure in the migration phase of wound-healing. In parts of the world where wood is plentiful, like in parts of Switzerland, roofs and even the walls of houses are covered with wooden shingles. They form an overlapping covering that keeps out the rain and the wind. The disadvantage of wooden shingles, over slates or asbestos tiles, is that they perish. The wood is not durable. There is a high maintenance cost. They have to be regularly inspected and broken shingles replaced by new ones. If not, the covering becomes weakened and the wind gets underneath and tears off whole areas of shingles. I suggest this is an analogy for a vascular endothelium. The endothelial cells overlap, seal the vessel wall and protect it from the blood flow. If there is inadequate maintenance growth, the endothelium degenerates becoming less and less dense, and more and more fragile, so that eventually when there are insufficient cells to cover the surface in an area of turbulent flow, the blood can get underneath and displace a patch of endothelium. The underlying surface is thrombogenic, so platelets attach and release growth factor. If this process is sustained long enough (about four days) the smooth muscle cells are stimulated to migrate out of the media, through the interna elastica and into the lumen of the vessel. Once this has happened the smooth muscle cells continue to proliferate. If the diet is rich in lipids, these accumulate in the smooth muscle cells giving rise to an atherosclerotic plaque (Ross & Glomset, 1976; Schwartz *et al.*, 1980). A crucial aspect of this hypothesis is that the functional integrity of the endothelium depends on maintenance growth. The Tumour Angiogenesis-like Factor which is produced ectopically in BHK21/13 and SV28 cells is able to stimulate migration and proliferation of aorta endothelial cells. The factor may therefore explain the vascularization of the tumours observed when BHK21/13 or SV28 cells are injected into hamsters. The proper role of this factor could be in the maintenance growth of the endothelium. We therefore wanted to test the hypothesis that there could be a deficiency of such a factor during atherogenesis resulting in a deficiency of endothelial maintenance. With lack of maintenance proliferation the cells in the endothelium would become less densely packed and, if damaged, less able to spread out and close the gaps. Since it is technically easier to test the first phase of wound-healing, i.e. migration, than to test 'maintenance' the migration assay was used.

For the migration assay, cells (Balb/c 3T3 or pig aorta endothelial cells) are allowed to form a confluent quiescent monolayer in a plastic Petri dish. This layer is then wounded by gently pressing a single-edged razor blade down through the cell layer to lightly mark the plastic with a starting line, relaxing and moving the blade to one side to scrape away cells and make a hole in the cell layer. The debris is then removed and the medium replaced with fresh serum-free medium. The test serum is then added and the wounded culture incubated for 22 hours. The cells are then fixed, stained and allowed to dry. The number of cells crossing 1 mm of the starting line are then counted (Bürk, 1973; Bürk, Clopath & Müller, 1979).

Serum was prepared from a pig on a normal diet and stored at $-18°$ C. The diet of the pig was changed to one containing cholesterol, lard and peanut oil. After three weeks a further batch of serum was prepared and stored at $-18°$ C. After 12 weeks of the hyperlipid diet the pig was sacrificed and it was confirmed that the aorta had developed atherosclerotic lesions. The two sera from before and during the diet were thawed, heated to 56° C for 10 minutes and tested in the migration assay. It was found that during the atherogenic diet the serum was deficient in the ability to stimulate cell migration. Mixing experiments showed the presence of an inhibitor rather than a lack of stimulatory activity. It was found that the inhibitor remained when the lipoproteins were removed by centrifugation at a density of 1.21 g cm^{-3}. The inhibitor has subsequently been detected in all of 13 pigs on the same atherogenic diet and not in any of 12 on the normal diet (Bürk *et al.*, 1979). The sera of rats on a hyperlipid diet also showed less migration activity than those from rats on the control diet (Fig. 1). In another experiment where a pig was bled and serum prepared each week, it was found that when the normal diet was changed to the atherogenic, hyperlipid diet, the migration activity of the serum dropped within one week but recovered after two weeks when the normal diet was restored. Consequently, serum was prepared from a pig at two day intervals and the diet changed from normal to atherogenic and back to normal. It can be seen from Fig. 2 that it took more than one, and less than three days for the migration activity to drop and about a week to recover.

The situation in man is similar to that in pigs. There are humans whose sera stimulate cell migration of both 3T3 and pig aorta endothelial cells. There are others whose serum contains an inhibitor of cell migration (Fig. 3). A diet change experiment was performed on one subject. The subject first adopted a prudent diet in which he ate moderately choosing chicken, fish, fruit, vegetables, milk and cheese, but avoiding eggs, animal fat, ice cream and butter. Then for five days he changed to a hyperlipid diet, including six eggs, 50 g of 'speck' (a fat, streaky bacon), two 'Landjaegers' (a salami-like sausage) each day with bread, butter and ice cream where possible. After this he returned to a prudent diet. Sera were prepared from the subject at intervals and stored at −18° C till the end of the experiment. The sera were then thawed and heated to 56° C for 10 minutes before assay. The samples were coded until after the migration activities had been determined. During the hyperlipid bacon, egg and salami diet the migration activity fell, and about a week after resumption of prudent diet the migration activity recovered (Bürk, Clopath, Flammer & Müller, in preparation). Among the various other human sera that have been surveyed one (kindly provided by Dr N. Myant) came from a patient (N.E.) homozyg-

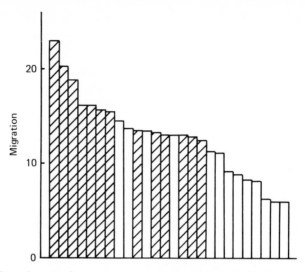

Fig. 1. Effect of serum from rats on a control and hyperlipid diet on Balb/c 3T3 migration. 100 μl serum was used per 2.5 ml Dulbecco's HG medium in a 35 mm dish. Cross-hatched columns, control diet; open columns, hyperlipid diet.

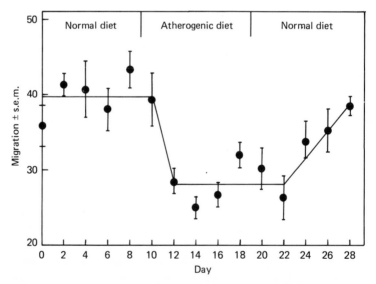

Fig. 2. Effect of diet change on ability of serum to stimulate migration of Balb/c 3T3 cells. A pig was bled every two days and serum prepared and stored at −18° C. On the 9th day the diet was changed to a hyperlipid diet and on the 19th day the diet was changed back to normal. 200 μl serum was used as in Fig. 1. Bars are s.e.m.

ous for Familial Hyperlipidemia (FH). The inhibition observed in the migration assay (Fig. 4) was more extreme than in any other human

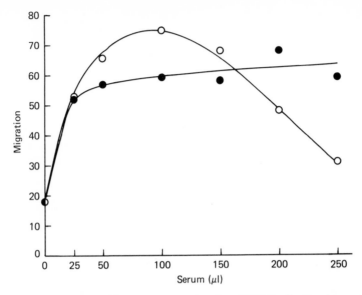

Fig. 3. Effect of two selected human sera on migration of 3T3-B cells. Varying amounts of serum were assayed as in Fig. 1. ●, serum of R.B.; O, serum of K.M.

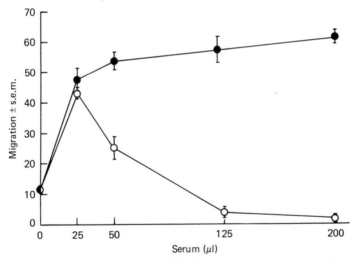

Fig. 4. Effect of serum from a patient (N.E.) homozygous for Familial Hyperlipidemia. Assay as in Fig. 3. ●, foetal calf serum, O, serum of N.E.

serum yet tested. This is of particular interest, because such patients avoid lipids in their diets and yet this serum has the inhibitor. FH homozygotes normally die in their teens as a result of one of the complications of atherosclerosis.

The effect of the inhibitory serum from atherogenic diet-fed pigs on

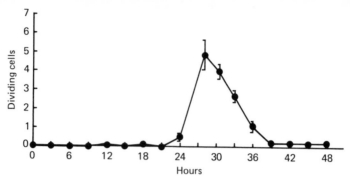

Fig. 5. Time of division of serum-stimulated 3T3-B cells. A culture was set up as for the migration assay and given 10% foetal calf serum at time zero. The number of dividing cells in each of ten × ten fields was determined at subsequent times (Bürk, 1970). Mean ± s.e.m. dividing cells per field, i.e. per 1.51 mm².

cell proliferation has been investigated. The standard migration assay of 22 hours (Bürk, 1973; Bürk *et al.*, 1979) is terminated before the cells, which are quiescent at the beginning of the experiment, start to divide. The first divisions appear at about 24 hours and reach a maximum frequency at about 29 hours (Fig. 5). Thus, an experiment terminated at 22 hours excludes cell division and one terminated at 46 hours includes cell division. When sera from normal and atherogenic pigs were compared at 22 hours and 46 hours the inhibitory effect appeared to have decreased at the late time (Fig. 6*a*). However, when the cell density in the undisturbed part of the cell layer was measured, it was found that with no serum added, it had declined. When serum was added, the layer maintained its initial density for 22 hours. By 46 hours, the cells had divided (as expected) in the highest serum concentrations (Fig. 6*b*). Remarkably, more cells in the serum from atherogenic diet-fed pigs divided than in the serum from normal diet-fed pigs. So not only did the migration inhibitor not inhibit cell division, but the serum of atherogenic diet-fed animals actually stimulated more cells to divide. When the extent of migration was adjusted for cell division (by dividing the number of migrated cells by the relative increase in cell density for that plate, compared to the average for the plates fixed at the beginning of the experiment) it is clear (Fig. 6*c*) that the inhibitory effect on cell migration was still there at 46 hours.

To summarize, an inhibitor of cell migration, but not cell division, has been detected in the serum of atherogenic-fed pigs and rats. A similar inhibitor, appearing with a change to a hyperlipid diet can also

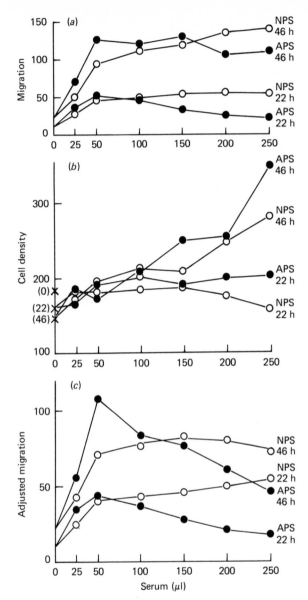

Fig. 6. Effect of serum from normal- and atherogenic-fed pigs on 3T3-B migration and cell density at 22 hours and 46 hours after addition. Sera were assayed as in Fig. 3, except that some plates were kept till 46 hours. ●, serum from pigs fed on atherogenic-hyperlipid diet (APS): O, serum from pigs fed a control diet (NPS). (*a*) Migration as a function of serum concentration at 22 and 46 hours. (*b*) Cell density (per mm^2) as a function of serum concentration at 22 and 46 hours. (*c*) Migration adjusted for cell density as a function of serum concentration at 22 and 46 hours.

be detected in human serum. Further the serum in a Familial Hyper-lipidemia homozygote showed an extreme inhibitory effect.

Since Virchow (1856), it has been thought that atherosclerosis starts with injury to the vascular endothelium. This 'response to injury' hypothesis (Ross & Glomset, 1976) suffers from the difficulty that 'injuries' have not been demonstrated to occur normally. The hypothesis is satisfactory in that experimental injury to the endothelium (e.g. by scratching, ballooning, drying, antibody complexes, homocysteine) does initiate the development of atherosclerotic lesions (Ross, 1979). Recently, Schwartz *et al.* (1980) have proposed that gross injury may not be the initiating event, but rather a loss of integrity. It has further been proposed (Schwartz *et al.*, 1978) that the vascular endothelium has a dynamic integrity which depends on the motility of the cells.

If, and it is by no means certain (cf. Schwartz *et al.*, 1978), it is true that an inhibitor of cell migration, detected in cultured cells growing on plastic in Dulbecco's medium, is also an inhibitor of endothelial cell migration on the vascular wall *in vivo*, then the integrity of the endothelium might be expected to be reduced in the presence of such an inhibitor. One can speculate that the integrity of the vascular endothelium and its strict monolayer structure are the result of a fine balance between factors like the Tumour Angiogenesis Factor and inhibitors that can alter the ability of the cells to migrate and proliferate. While on the one hand migration factors or endothelial growth factors will favour the replacement of decrepit cells, excess of inhibitors will stop the maintenance process. Inhibitors might be produced as a response to external factors such as the nutritional status of the animal or as a consequence of genetic disorder leading to an imbalance in the production of the factors and inhibitors, hence impairing the renewal of the endothelium and causing continuous injury. When one considers that the human aorta endothelium is stretched and allowed to relax some 60–160 times per minute it seems possible that two endothelial cells may occasionally pull apart. If the cells are paralysed (Haudenschild) or frozen (e.g. if the fatty acid composition of their membranes is altered; Shinitzky, 1978) then they may not be able to move together again fast enough to prevent damage, especially in a region of turbulent flow at a fork in the vessel. Even if the cell density of the endothelium were increased in animals on an atherogenic diet, if the cells lack the contractility or fluidity to migrate then the reserve density would not increase the overall integrity of the endothelium and where there is no integrity there will be no integration and where there is no integration the function of the endothelium will fail.

I thank R. Bachmann, P. Clopath and M. Erard for the pig sera, R. Flammer for endothelial cells and R. Cortesi and K. Müller for rat and human serum.

REFERENCES

BÜRK, R. R. (1970). One-step growth cycle for BHK21/13 hamster fibroblasts. *Experimental Cell Research*, **63**, 309–16.

BÜRK, R. R. (1973). A factor from a transformed cell line that affects cell migration. *Proceedings of the National Academy of Sciences, USA*, **70**, 369–72.

BÜRK, R. R. (1980). A progression in the production of growth factors with progression in malignant transformation. In *Control Mechanisms in Animal Cells*, ed. L. Jimenez de Asua et al., pp. 245–57. New York: Raven Press.

BÜRK, R. R., CLOPATH, P. & MÜLLER, K. (1979). Deficiency in the ability of serum from pigs on an atherogenic diet to stimulate the migration of 3T3 cells in culture. *Artery*, **6**, 205–19.

CARPENTER, G. & COHEN, S. (1976). Human epidermal growth factor and the proliferation of human fibroblasts. *Journal of Cellular Physiology*, **88**, 227–38.

FOLKMAN, J. (1975). Tumor angiogenesis. In *Cancer*, vol. 3, ed. F. F. Becker, pp. 355–88. New York: Plenum Press.

JIMENEZ DE ASUA, L. (1980). An ordered sequence of temporal steps regulates the rate of initiation of DNA synthesis in cultured mouse cells. In *Control Mechanisms in Animal Cells*, ed. L. Jimenez de Asua et al., pp. 173–97. New York: Raven Press.

RHEINWALD, J. G. & GREEN, H. (1977). Epidermal growth factor and the multiplication of cultured human epidermal keratinocytes. *Nature*, **265**, 421–4.

ROSS, R. (1979). The pathogenesis of atherosclerosis. *Mechanisms of Ageing and Development*, **9**, 435–40.

ROSS, R. & GLOMSET, J. S. (1976). The pathogenesis of atherosclerosis. *New England Journal of Medicine*, **295**, 369–77, 420–5.

SCHWARTZ, S. M., GAJDUSEK, C. M., REIDY, M. A., SELDEN, S. C. & HAUDENSCHILD, C. C. (1980). Maintenance of integrity in aortic endothelium. *Federation Proceedings*, **39**, 2618–25.

SCHWARTZ, S. M., HAUDENSCHILD, C. C. & EDDY, E. M. (1978). Endothelial regeneration. I. Quantitative analysis of initial stages of endothelial regeneration in rat aortic intima. *Laboratory Investigation*, **38**, 568–80.

SHECHTER, Y., HERNAEZ, L. & CUATRECASAS, P. (1978). Epidermal growth factor: Biological activity requires persistent occupation of high-affinity cell surface receptors. *Proceedings of the National Academy of Sciences, USA*, **75**, 5788–91.

SHINITZKY, M. (1978). An efficient method for modulation of cholesterol level in cell membranes. *FEBS Letters*, **85**, 317–20.

VIRCHOW, R. (1856). *Gesammelte Abhandlungen zur Wissenschaftlichen Medicin*, Teil V, Phlogose und Thrombose im Gefässsystem, pp. 458–638. Frankfurt a.M.: verlag von Meidinger.

Junctional communication in the peripheral vasculature

JUDSON D. SHERIDAN AND DAVID M. LARSON

J.D.S.: Department of Anatomy, 4–135 Jackson Hall, University of Minnesota Medical Schooh,
Minneapolis, MN 55455. D. M. L.: Department of Pathology, Harvard Medical School and
Brigham and Women's Hospital, Boston, MA 02115

Blood vessels are complex organs whose many functions are still incompletely understood. Considerable attention is currently being given to the apparently simplest component of the vessels, the endothelium, which serves as the major dynamic interface between the blood and tissues. The endothelium has many functions (for a review see: Gimbrone, 1976), for example: as a selective barrier, regulating the influx and efflux of nutrients, catabolites, and other molecules; as a primary target for hormones and other vasoactive agents (Buonassisi & Colburn, 1980); and as a participant in various critical protective mechanisms such as vascular repair (Haudenschild, 1980), inflammation, immunological response, and hemostasis (Shepro & D'Amore, 1980).

These functions are not performed independently by single endothelial cells, but by endothelial cells working in concert with each other, with smooth muscle cells, and with pericytes. The interactions of these different cells can occur by a variety of mechanisms, but the one which interests us most involves the direct transfer of small molecules from cell to cell via permeable junctions, or 'junctional communication'. Our interest has led us to investigate a number of basic questions whose answers we feel may give some insight into the general functional significance of permeable junctions as well as further understanding of vascular physiology and pathology. This chapter discusses some of the background for these questions and describes the progress we have made in attempting to answer them.

CULTURED CELLS: SUITABLE MODELS FOR
STUDYING ENDOTHELIAL JUNCTIONS?

Investigations into the physiology of the vascular intima have been greatly aided by the techniques recently developed for growing endothelial cells from various sources in tissue culture (Jaffee *et al.*, 1973; Gimbrone, 1976; Macarak *et al.*, 1977; Folkman *et al.*, 1979; Bürk, this volume). The cultures are amenable to many experiments that are difficult or impossible *in situ* while they retain many of the properties of the intact endothelium, e.g. density-dependent regulation of growth (Haudenschild, 1980; Bürk, this volume), monolayer arrangement of cells (Gimbrone, 1976), hormone responsiveness (Buonassisi & Colburn, 1980), and elaboration of factor VIII antigen (Shepro & D'Amore, 1980).

Cultured endothelial cells would also seem to have many advantages for studies of junctional communication. *In situ*, the cells have prominent gap (and tight) junctions which may be involved in some of the processes known to occur *in vitro*, e.g. wound repair (Schwartz, Haudenschild & Eddy, 1978) and hormone activation (Buonassisi & Venter, 1976). Moreover, the full range of techniques for investigating junctional ultrastructure and permeability are more readily applicable in culture than *in situ*. Junctional changes that result from the adaptation of a three-dimensional tissue to the two-dimensions of monolayer culture are not likely to occur with endothelial cells, which have a two-dimensional arrangement *in situ*. Senescence in culture (Schwartz, 1978), which has been reported to cause junctional changes in other cells (Kelley *et al.*, 1979), should be avoidable by using early passage cells. Nevertheless, culture conditions do not duplicate all of the conditions *in situ*, e.g. the medium is artificial and the cells are growing on a non-physiological substrate. One or more of these differences could lead to junctional changes.

When we began our studies there had been little attention paid to the junctions in endothelial cultures. Therefore, we first had to ask if the junctions were present in culture and what junctional changes, if any, occurred as the cells adapted to cultured conditions. We began with a detailed study of the structure of gap and tight junctions and of the cell-to-cell transfer of small molecules in endothelial cultures. We chose bovine aortic and umbilical vein cells because they are easy to establish in culture and, in the case of the aortic cells, have been extensively studied by other workers (Macarak *et al.*, 1977; &

Schwartz, 1978; see Bürk this volume). As a baseline for comparison we chose cells freshly released in sheets and clumps from collagenase-treated large vessels (in preparation for culturing). In contrast to intact endothelium, these sheets of effluent cells could be handled for ultra-structural and physiological study in a manner directly comparable to that used for the cultures, thus avoiding differences due simply to technique.

As expected from the literature on intact vessels (Simionescu, Simionescu & Palade, 1976), both tight and gap junctions were found in the sheets of aortic and umbilical vein effluent cells (Fig. 1) (Larson & Sheridan, 1979). In thin section, the tight junctions were represented by focal or occasionally broader regions of apposition at which the external laminae of the plasma membranes fused, or even disappeared. In freeze-fracture replicas, the intramembranous components of the tight junctions were seen to follow the typical endothelial pattern which differs from that in most epithelia (see Lane in this volume). Rather than P-face fibrils and clean E-face grooves, the tight junctions had only gentle ridges on the P-faces and the E-face grooves were filled with irregularly distributed particles. The ridges and grooves formed interlacing networks, often more than three elements in width. Gap junctions were also common in thin sections of the effluent cells, displaying the expected, regular apposition of membranes separated by an apparent space of *ca* 2 nm. In freeze-fracture replicas the gap junctions were usually comprised of macular aggregates of large par-ticles seen on the P-faces or pits on the E-faces. The densities of the particles (or pits) showed a pronounced reduction in the centre of most of the larger gap junctions. This rather unusual configuration has not, to our knowledge, been reported for other endothelial gap junctions, and its source and significance are unclear. Since these density dif-ferences were also seen in junctions fixed *in situ* they were evidently not results of the collagenase treatment. It is possible that the different particle arrangements are associated with different channel diameters, as has been suggested in other systems (Peracchia, 1980; Raviola *et al.*, 1980), but we have no direct data relevant to this issue.

The gap junctions were often intercalated within tight junction networks, although occasionally they occurred in isolation. In about half of the interfaces, however, the gap junctions were interconnected by linear strands of uniform, P-face particles. The particle strands were arranged in patterns resembling those of the tight junctions, but the strands and tight junctions were rarely seen together. The significance

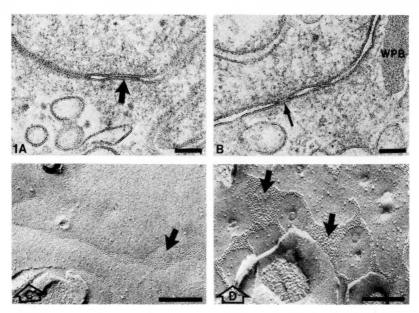

Fig. 1. Junctions between effluent endothelia cells from bovine vessels treated with collagenase. (A) Thin section view of aortic cells showing small gap junction (arrow) with the typical extracellular gap. Bar = 0.1 μm. (B) Thin section view of aortic cells with a short region of apparent membrane fusion (arrow), a probable tight junction. WPB = Weibel-Palade body. Bar = 0.1 μm. (C) Freeze-fracture of umbilical vein cells showing network of E-face pits arranged in linear strands and gap junction aggregates. Arrow points to region of decreased density in the centre of a gap junction. Bar = 0.25 μm. (D) as in (C), but showing P-face view of particle strands and aggregates. Again note decreased densities of particles in the gap junctions (arrows). Bar = 0.25 μm. Shadow angle in C & D indicated by open arrows around labels.

of the strands is not known. They are reminiscent of the early stages of tight junction formation (Porvaznik, Johnson & Sheridan, 1978), but it seems unlikely that such extensive formation would be occurring in the large vessels. Perhaps the strands indicate an incipient breakdown of tight junctions. If so, the cause of the breakdown must be something other than the collagenase treatment since the strands were also seen in intact vessel segments.

When primary endothelial cultures, three to five days old, were studied with thin section and freeze-fracture (Fig. 2), tight and gap junctions were seen, though at an apparently reduced frequency. In thin section the junctions resembled those in the effluent cells. In freeze-fracture replicas, the tight junctions had the characteristic

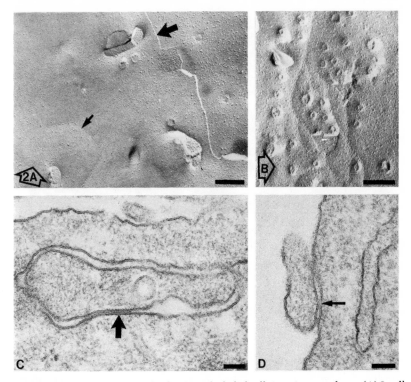

Fig. 2. Junctions between umbilical vein endothelial cells in primary culture. (A) Small gap junction on the P-face and single tight junction groove with particles on the E-face (small arrow). Fracture step apparently occurred along another tight junction strand (large arrow). Bar = 0.25 μm. (B) Unusually complex tight junction array showing primarily E-face grooves associated with micropinocytotic vesicles. Bar = 0.25 μm. (C) Small gap junction (arrow) in thin section. Bar = 0.1 μm. (D) Cross-section of apparent tight junction element (arrow). Bar = 0.1 μm.

endothelial fracture pattern and the gap junctions consisted of macular particle aggregates with central reductions in particle density. The junctions *in vitro*, however, were notably less complex in that fewer than two parallel tight junction elements were generally found in association with small, infrequent gap junctions. Isolated tight and gap junctions were more common than in the effluent cells and there were very few complex arrays of particle strands and aggregates.

As expected from the ultrastructural studies, cultured endothelial cells were found to be capable of the cell-to-cell transfer of potential changes ('electrical coupling') (Fig. 3), dye molecules (fluorescein and Lucifer Yellow) (Fig. 4), and nucleotides (Fig. 5) (Larson & Sheridan, 1979). Detailed quantitative studies of electrical coupling, evaluated

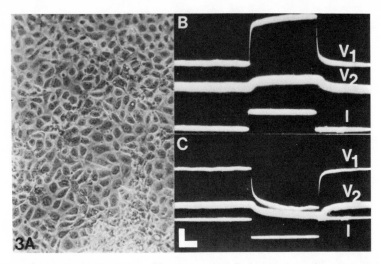

Fig. 3. Electrical coupling between bovine aortic cells in primary culture. (A) Phase-contrast view. (B & C) Intracellular current pulses (I) in depolarizing (B) and hyperpolarizing (C) directions produce voltage deflections in cell injected with current (V_1) and in adjacent cell (V_2). Coupling was usually better than illustrated here. Calibration: vertical, 20 mV, V_1, V_2, 10^{-8} amp, I; horizontal, 20 ms.

according to a two-dimensional model for current spread in a thin sheet (modified from Jongsma and van Rijn, 1972), gave estimates of inter-cellular resistivity (Larson & Sheridan, unpublished). Together with reasonable geometric assumptions, these resistivity values were used to estimate average junctional resistance per cell. Both aortic and umbilical vein cells provided comparable estimates, *ca* 10^6 ohms. For cells having six neighbours, and junctional channels of 10^{10} ohms (Sheridan, Hammer-Wilson, Preus & Johnson, 1978), the estimates further suggest that each interface had *ca* 500 channels (or particles as seen in freeze-fracture). Although our ultrastructural data have not been quantitated on this level, there appear to be no major discrepancies between the sizes of the junctions and the predictions from the electrophysiological measurements.

Because of their irregular size, the effluent cell clumps were unsuitable for detailed, comparative electrophysiological study. However, comparison of the capabilities for dye transfer by effluent and cultured cells was feasible. We found that Lucifer Yellow transferred more rapidly and extensively in the effluent cell clumps than in culture (Larson & Sheridan, unpublished). This difference was confirmed more quantitatively by determining the fall-off of relative

Fig. 4. Transfer of Lucifer Yellow between cultured bovine aortic endothelial cells. Dye was injected with 200 ms current pulses repeated 1/s. Photograph taken after 4 min of injection.

fluorescence intensity as a function of cell diameters away from the injected cell. The distance over which intensity fell to one half the value in the injected cell was one cell diameter in culture and two cell diameters in the effluent cells. Comparable differences were seen when the values were corrected for differences in cell geometry.

Thus, there are both gap and tight junctions in primary endothelial cultures and the cells are capable of transferring inorganic ions, dyes, and nucleotides. However, junctional changes do occur in culture. On the ultrastructural level, the cultured endothelial cells have less complex and possibly smaller gap junctional areas. On the physiological level, the cultures have reduced transfer capabilities, but the differences are small and their significance unclear.

Because the cells we studied had been in culture only a short time, it seemed possible that the changes merely reflected a shift from slow growth *in situ* (Schwartz & Benditt, 1973) to more rapid growth *in vitro* (Schwartz, 1978). Such a simple explanation appears to be insufficient, however. We have recently examined bovine aortic endothelial cultures that had experienced multiple passages and had been confluent and growing at a very slow rate for about a week before fixation (Sheridan, Anderson & Schwartz, unpublished). These cells had even less extensive gap junctions and few if any tight junctions. Physiological studies of such cells will be needed to determine if there is an associated decrease in junctional transfer.

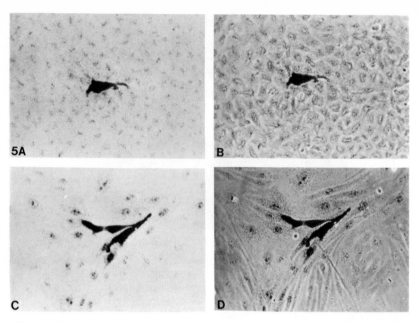

Fig. 5. Uridine-nucleotide transfer between cultured bovine endothelial cells. (A & B) Aortic cells showing extensive transfer over many cell diameters (A = bright-field; B = phase-contrast). Cells co-cultured for 4 h. (C & D) Cloned bovine adrenal capillary cells showing extensive transfer. Note nuclear labelling elongated shape of recipient cells.

We must conclude that further study of junctional communication in endothelial cultures is warranted, but caution must be exercised in generalizing results to the intact vessel, especially regarding quantitative analyses. Ideally, experiments on cultures should be combined with experiments on intact vessels or vessel segments. This operational principle has guided our approach to the next question regarding junctional communication in capillaries and pericytic venules.

CAPILLARIES AND PERICYTIC VENULES: COUPLED OR UNCOUPLED TISSUES?

The endothelia of capillaries and pericytic venules present a totally different problem regarding junctional communication in the vasculature. As elegantly documented by Simionescu *et al.* (1975) and confirmed by other workers (Schneeberger & Karnovsky, 1976), these endothelia have tight junctions, but apparently lack gap junctions. In other systems, a lack of recognizable gap junctions has been associated with alteration in certain processes, such as regulation of cell proliferation, suggesting a functional dependence on junctional communication (Loewenstein, 1979). Yet neither capillaries nor venules show such functional defects. Therefore, either junctional communication is not as critical for these processes as previously believed or the cells can still communicate using structures other than gap junctions.

These considerations have prompted us to directly test the ability of endothelial cells in capillaries and pericytic venules to engage in junctional communication (Sheridan, 1980). We have used both intact and cultured endothelia in our investigations. The intact vessels have been studied primarily in explants of rat omentum, which has microvessels that are readily accessible to microelectrodes and in which the lack of gap junctions in capillaries and pericytic venules has been most extensively documented (Simionescu, Simionescu & Palade, 1975). To date we have restricted our attention to the transfer of the fluorescent dye Lucifer Yellow because it can be localized in tissue section and thus permits identification of the interacting cells.

Our results on capillary endothelia are quite clear: the cells typically transfer dye over many cell diameters within a few minutes (Fig. 6). This transfer does not occur via the extracellular space since dye within the vessel lumen or outside the vessel wall is not taken up by the cells. Thus, permeable junctions are involved.

The results on pericytic venules are less clear. Transfer has been seen (Fig. 6) in a number of instances, but often there is no transfer even after a long injection period. The negative results may reflect the lack of gap junctions, but there are other complicating factors. The venular endothelial cells are more difficult to impale than the capillary cells. This is partly due to the enveloping sheath of pericytes (which are often impaled instead of the endothelial cells, see below) and partly due to the greater amount of associated collagen fibers. Furthermore, the cells failing to transfer commonly have a distorted, retracted shape which might indicate a response to histamine released as a result of trauma. Venular endothelium is especially sensitive to histamine, which has been reported to produce endothelial cell contraction (Majno *et al.*, 1969; but see Hammersen, 1980).

Thus, we can conclude that endothelial cells in capillaries and in some pericytic venules are capable of junctional communication. Whether other endothelial cells in pericytic venules are truly not coupled or are uncoupled by damage and/or histamine effects will be

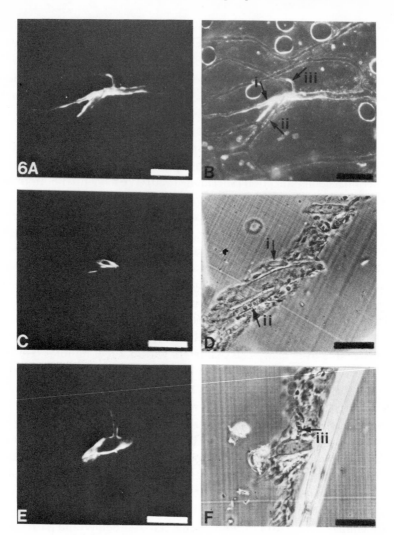

Fig. 6. Transfer of Lucifer Yellow in capillary and venule endothelium. (A & B) Dark-field fluorescence and white light images of an injected vessel. An endothelial cell in the pericytic venule was injected and transfer to two smaller branches, one a definite capillary, was observed by the end of 7 min injection period. (C & D) En face sections of the vessel are seen with epifluorescence and phase-contrast. At this level, the section catches a large branch and two smaller branches (i and ii; see B). Note dye in one of the smaller branches (i) and the bright nucleus lying between i and the larger branch. This nucleus can be seen in (D) and is clearly not from an endothelial cell. (E & F) A more superificial section shows the pattern of transfer in part of the venule and a capillary (iii; see B). A single row of RBC can be seen in iii (see F). Bar = 100 μm, (A & B); 50 μm, (C–F).

determined only by further work.

The strong implication from these studies is that some structures besides gap junctions are responsible for the dye coupling in capillaries. If so, an attractive candidate is the tight junction, which provides intimate associations between apposed cell membranes and contains intramembranous subunits that could form transmembrane channels. This idea derives some rather circumstantial support from previous studies. Tight junction-like elements may electrically couple certain cultured fibroblasts (Larsen, Azarnia & Loewenstein, 1977). Developing tight junctions have linear strands of intramembranous particles that are indistinguishable from gap junction subunits (Porvaznik, Johnson & Sheridan, 1978). There are even some data suggesting that tight junctions and gap junctions draw upon a common pool of precursors (Elias & Friend, 1976). It is perhaps significant that the venules, which may have more labile coupling, also have rather primitive tight junctions containing fewer intramembranous particles (Simionescu *et al.*, 1975).

A role for endothelial tight junctions in coupling is appealing, but another possibility should be considered. While it is unlikely that even very small, infrequent aggregates of junctional particles would have been missed in the detailed studies of capillaries in a variety of tissues, there may be unaggregated gap junction particles that form patent channels across the extracellular space. If the junctional particles were well dispersed, they would probably not be distinguishable from the many large non-junctional particles in the membrane. There is no precedent for coupling by way of totally dispersed particles, but it has been suggested that particles in loose clusters may have permeable channels (Johnson, Hammer, Sheridan & Revel, 1974; Raviola *et al.*, 1980). Without some specific probe, e.g. an antibody to junctional protein, the possibility of coupling via isolated gap junctional particles is difficult to test directly. However, this alternative would become less likely if coupling could be shown to change in direct relation to the amount of tight junction particles.

Manipulation of tight junctions *in situ* would be difficult, but might be possible in culture. Accordingly, we have begun a study of cultured capillary cells obtained from two sources, bovine adrenal gland (from B. Zetter; Folkman *et al.*, 1979) and rat pancreas. The cloned bovine cells grow as monolayers (Fig. 5) and are easily passaged. The pancreatic cells, derived by collagenase treatment of whole pancreas or isolated islets, form small islands (Fig. 7) that show minimal growth.

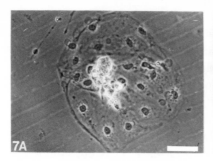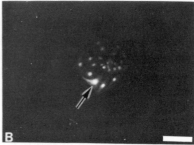

Fig. 7. Lucifer Yellow transfer in endothelial island from rat pancreas. (A) Phase-contrast and (B) dark-field fluorescence (fields reversed). Island injected for 3–4 min. Bar = 50 μm, (A); 100 μm, (B).

We have shown that dye transfer occurs in nearly all cases with the capillary-like endothelial islands from rat pancreas (Fig. 7) and in better than half of the cases with the bovine adrenal cells. The lower incidence of dye coupling in the bovine cultures is still hard to interpret because the cells are more difficult to impale than cultured bovine aortic cells or the rat cells. Thus, uncoupling as a result of damage is a definite possibility. More extensive coupling in the bovine capillary cultures has been observed with nucleotide transfer techniques (Fig. 5). Thus, the physiological studies on the cultured cells have been encouraging.

Our initial studies on the structure of the junctions in bovine capillary cultures have yielded some unexpected results. Freeze-fracture replicas have revealed gap as well as tight junctions. In fact the gap junctions are comparable in size and number to those in bovine aortic cultures. This finding raises a number of possibilities. It is possible, but rather unlikely, that adrenal capillaries have gap junctions and differ from other capillaries. Perhaps the cells did not originally arise from capillaries, but instead from other microvessels, e.g. arterioles or muscular venules, that have endothelial gap junctions *in situ* (Simionescu, Simionescu & Palade, 1975). This possibility, however, also seems unlikely in view of the careful selection technique used in establishing the cultures (Folkman *et al.*, 1979) as well as the many capillary-like properties expressed by the cells *in vitro*, e.g. growth dependence on angiogenesis factor, and the capacity to form capillary-like cellular tubes (Folkman & Haudenschild, 1980). One intriguing possibility is that some factor or condition *in vitro* causes capillary cells to express an otherwise latent ability to form gap junctions. Such a change might, for example, mimic the change that must occur when capillaries develop into larger vessels during embryogenesis or neovascularization (Wagner, 1980). Perhaps something as simple as the increase of surface covered by the endothelial cells *in vitro* is the signal for the junctional change. If so, the gap junctions might again disappear when the capillary cells form *in vitro* tubes (Folkman & Haudenschild, 1980), an idea that is easily tested. Alternatively, perhaps the change occurs after multiple subculturing, which apparently causes the cells to gradually lose their growth dependency on angiogenesis factor (Folkman, Haudenschild & Zetter, 1979; B. Zetter, personal communication). In this case we might expect gap junctions to be absent from earlier cultures.

Although the culture story remains unclear, the possibility of coupling in the absence of gap junctions *in vivo* has important implications for the general field of junctional communication. We have come to associate coupling with gap junctions, largely on the basis of circumstantial evidence. Yet it appears that the junctional permeability and typical gap junction structure can be dissociated: coupling can be abolished without removing gap junctions (Peracchia, 1980), and, as our results and those of others (Larsen, Azarnia & Loewenstein, 1977) suggest, coupling can occur when gap junctions are apparently absent. Consequently it is risky on pure morphological grounds to map the distribution of coupling pathways or to make inferences about the quantitative effectiveness of junctional communication. Nevertheless, the dissociation of coupling and gap junctions does not argue against the dependence of coupling on intercytoplasmic channels. In some cases those channels apparently close while the gross structure containing them, the gap junction, undergoes only minor morphological change. In the other cases, the channels must still be present, but apparently in some other form than gap junctions, e.g. unaggregated subunits or tight junctions.

The detailed functional role of the coupling between endothelial cells remains to be established. It is fair to assert that the endothelium, like other tissues, should be considered as an integrated unit rather than as a set of independent cells, at least for processes depending on electrical potentials or intracellular concentrations of small molecules. That is true even for the capillaries, and perhaps the pericytic venules, which apparently lack gap junctions.

One process in which such a functional integration may be critical is the regulation of cell proliferation. This possibility has been discussed in various forms by a number of authors (e.g. Loewenstein, 1968, 1979; Sheridan, 1976). The typical model begins with the premise that small molecules promoting progression through the G_1 phase of the cell cycle are distributed from cell to cell via gap junctions (Loewenstein, 1968, 1979). According to the model, as the cell density increases in a population of cells dividing asynchronously, more cells will form junctions, and transient increases in the concentration of growth-stimulating molecules will be dampened by their spread to adjacent cells. The ultimate result will be an arrest of cell proliferation at some critical cell density. Reinitiation of growth in an inhibited population might occur simply by closing off (or losing) the junctions provided that the endogenous signal molecule continues to be produced periodically while the cells are inhibited (Burton, 1971). Alternatively, cells might require some outside stimulus to raise the signal above threshold. In this case, the junctions could serve to distribute excess intracellular signal molecules from initially more responsive cells to their less responsive neighbours, thereby ensuring a more uniform reaction from the population.

Despite the conceptual appeal of this model, it has as yet rather little direct experimental support, most of which has come from studies of transformed cells that have lost growth regulation in apparent association with junctional defects (Loewenstein, 1979). There are some reasons to believe that endothelial cells may provide a good system for testing the growth-modulatory effects of junctions. As described elsewhere in this volume (Bürk), endothelial cells show a conspicuous density-dependent inhibition of cell division both in culture and *in situ*. Like other density inhibited cells, the endothelial cells are stimulated to divide when the monolayer is interrupted, or 'wounded': DNA synthesis is initiated both in cells that migrate into the wound and in cells near the wound edge (Schwartz *et al.*, 1978). However, there are fewer complicating factors in the wounding response by endothelial cells. In contrast to the behavior of other cultured cells, confluent endothelial cells do not re-enter the cell cycle when exposed to increased serum concentrations, added growth factors, or fresh growth medium (Haudenschild, Zahniser, Folkman & Klagsbrun, 1976).

According to the model previously discussed, we might explain the wounding response of endothelial cells in two ways involving junctional communication. Wounding might cause cells to lose or close off their junctional channels, preventing the junctions from continuing to dampen periodic intracellular growth signals. Alternatively, the junctions might transmit a wound-induced mitogenic signal from the cells nearest the wound to adjacent and more distant cells. The first explanation, wound-induced closure of the junctional channels, is consistent with the known effect of cell damage in reducing junctional permeability (Loewenstein, 1976). However, in order for this explanation to account for the mitogenesis of cells far from the wound edge, the transfer of another signal to close off junctions would have to be postulated. Moreover, for the 'shut-down' signal to move via the junctions, particularly over many cell diameters, there would have to be a suitable delay between the arrival of the signal and the closure of the junctions.

Evidence against the first explanation has been obtained for certain non-endothelial cells in culture. Stoker (1975) found no change in metabolic cooperation by cells in the mitogenic zone near a wound edge *in vitro*. The sensitivity of the method, however, especially its temporal resolution, may not have been great enough to detect a transient junctional alteration. More direct studies, for example with dye transfer, are needed.

In regard to the second explanation, it is interesting that DNA synthesis is initiated farther away from the wound edge *in situ* than in culture (Schwartz, Haudenschild & Eddy, 1978). This difference might result from the apparent decrease in junctional transfer (and probably junctional size) we have found in culture.

The role of junctions implied by the second explanation is consistent with the results obtained recently by Selden, Rabinovitch & Schwartz (1981). They found that vinblastine acts as a potent mitogen when applied briefly to confluent, growth-inhibited endothelial cells. Many of the vinblastine-treated cells retract from each other, but the retraction can be dissociated from mitogenesis. Stimulation of thymidine incorporation occurs in non-retracted cells, and cytochalasin D reduces the percentage of cells entering S-phase but does not prevent cell retraction. The effect of cytochalasin D might result either from blocking the production of a mitogenic signal associated with cell movement, as suggested by the authors (Selden *et al.*, 1981), or by reducing junctional transfer as reported for cytochalasin B (Stoker, 1975).

HETEROLOGOUS COUPLING: A MECHANISM FOR FUNCTIONAL INTEGRATION OF VASCULAR CELLS?

With the possible exception of some capillaries, endothelial cells are always associated with other fixed cell types, particularly pericytes, fibroblasts, and smooth muscle cells. This association can involve cell contact and/or junction formation as shown originally by Rhodin (1967, 1968). Myoendothelial junctions in thin section, for example, resemble gap junctions while pericyte–endothelial contacts have been less well characterized. These intriguing ultrastructural findings raise the important possibility that the endothelial and non-endothelial cells are coupled.

Rhodin was impressed by the implications of this possible heterologous coupling, which he discussed with a great deal of insight in his classic papers (Rhodin, 1967, 1968). Elaborating upon an earlier idea of Fawcett's (1959), Rhodin suggested that signals generated in the pericyte by the action of vasoactive agents might be transferred to the endothelial cells via pericyte–endothelial junctions. The pericyte network would act in essence as a greatly expanded receptor surface, magnifying endothelial reactions to such molecules as histamine and serotonin. Rhodin also suggested that signals generated in endothelial cells by blood-borne agents might transfer to smooth muscle cells via myoendothelial junctions. Such a mechanism might be particularly important for the precapillary sphincters where the frequency of myoendothelial junctions is especially high.

Despite these interesting possibilities, there have been no direct physiological studies of heterologous coupling between vascular cells, nor in fact have the junctional structures been fully characterized with freeze-fracture. Consequently, we have begun an investigation of this problem again using both intact vessels and culture systems.

Our first, and most dramatic, finding (Sheridan, 1980) was that pericytes in rat omentum are effectively dye-coupled to each other and to some fibroblasts, but not to other fibroblasts or to mast cells (Fig. 8). This result was not anticipated, though in retrospect it is not particularly surprising if one chooses to view pericytes as some modified form of fibroblast (Rhodin, 1968). However, pericytes may be related to smooth muscle cells (Rhodin, 1968) and, if so, the heterologous transfer may be even more significant. Whichever point of view we take, our results suggest that the extent of the cell population that potentially influences the endothelium is greater in terms of cell number and

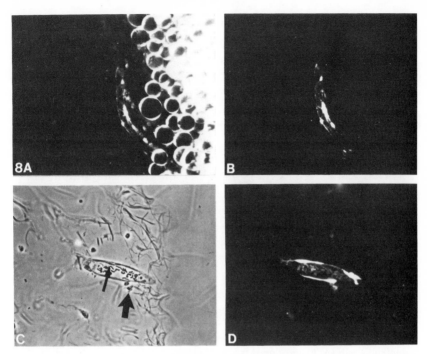

Fig. 8. Dye transfer involving pericytes. (A & B) White-light and dark-field fluorescence pictures, respectively, taken a few minutes after injection of a pericyte. (C & D) Phase-contrast and epifluorescence micrographs of section of vessel. Transfer between pericytes and to one or more fibrolasts (large arrow in C) but not to endothelial cell (small arrow in C). More extensive transfer was usually seen; here there were few nearby fibroblasts or other cells.

diversity than even Rhodin imagined.

A more critical part of Rhodin's scheme, however, is the proposed coupling between pericytes and endothelial cells and here the results have been mixed. A few definitive cases of such transfer have been seen, always associated with inter-endothelial transfer (Fig. 7). However, equally clear examples without detectable transfer to endothelial cells have been found, chiefly in pericytic venules (Fig. 8). The variability may be real, but the problem already considered for the endothelium of pericytic venules (see page 133) must be kept in mind. If these endothelial cells are uncoupled from each other as a result of histamine and trauma, they might well uncouple from the pericytes. Once again, a resolution of this problem awaits further work.

Our studies of endothelial-smooth muscle coupling so far have been restricted to culture systems. We have shown that bovine aortic en-

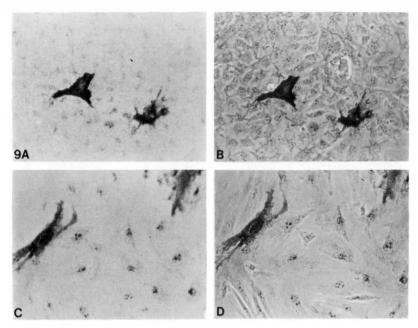

Fig. 9. Uridine-nucleotide transfer from aortic smooth muscle donors to bovine aortic endothelial cells (A, B) and cloned bovine capillary cells (C, D). Format as in Fig. 5.

dothelial and smooth muscle cells become coupled rather indiscriminately *in vitro* as tested either with dye injection or nucleotide transfer (Fig. 9) techniques. More recently we have found that the bovine capillary cells also form permeable junctions with the smooth muscle cells (Fig. 9). These coculture experiments demonstrate that there is no intrinsic specificity restricting the ability of endothelial and smooth muscle cells to form permeable junctions, but they do not prove that the heterologous junctions form *in situ*. This conclusion may seem obvious for the capillary endothelial cells, which rarely if ever come into contact with smooth muscle cells *in situ*. However, the possibility, discussed above, that the cultured capillary cells can differentiate into a form more like the endothelium of larger, smooth muscle-containing vessels adds more meaning to the myoendothelial transfer *in vitro*. If direct studies of intact vessels confirm the presence of functional myoendothelial junctions, the coculture systems may prove useful for physiological studies of the interactions between the coupled cells.

CONCLUSION

We have seen that blood vessels have the structural and functional machinery to carry out junctional communication within the endothelium and between the endothelium and at least some of the associated cells. The basic elements of this communication system are retained in culture, although certain subtle modifications are evident. On the most general level, the interconnected cells have the potential of acting as functional units in a diversity of processes that depend upon electrical events and/or intracellular concentration of small molecules. Whether and by what means the vessels realize their potential for communication will remain major questions for future work.

REFERENCES

BUONASSISI, V. & COLBURN, P. (1980). Hormone and surface receptors in vascular endothelium. In *Advances in Microcirculation*, vol. 9 ed. B. M. Altura, pp. 76–94. Basel: S. Karger.

BUONASSISI, V. & VENTER, J. C. (1976). Hormone and neurotransmitter receptors in an established vascular endothelial cell line. *Proceedings of the National Academy of Sciences, USA*, **73**, 1612–16.

BURTON, A. C. (1971). Cellular communication, contact inhibition, cell clocks, and cancer: the impact of the work and ideas of W. R. Loewenstein. *Perspectives in Biology and Medicine*, **14**, 301–18.

ELIAS, P. M. & FRIEND, D. S. (1976). Vitamin-A-induced mucous metaplasia: an *in vitro* system for modulating tight and gap junction differentiation. *Journal of Membrane Biology*, **34**, 39–54.

FAWCETT, D. W. (1959). *The Microcirculation*. Urbana, Illinois: University of Illinois Press.

FOLKMAN, J., HAUDENSCHILD, C. C. & ZETTER, B. R. (1979). Long-term culture of capillary endothelial cells. *Proceedings of the National Academy of Sciences, USA*, **76**, 5217–21.

FOLKMAN, J. & HAUDENSCHILD, C. C. (1980). Angiogenesis *in vitro*. Nature, **288**, 551–6.

GIMBRONE, M. A. JR (1976). Culture of vascular endothelium. In *Progress in Hemostasis and Thrombosis*, vol. III, ed. T. H. Spaet, pp. 1–28. New York: Grune & Stratton.

HAMMERSEN, F. (1980). Endothelial contractility – does it exist? In *Advances in Microcirculation*, vol. 9, ed. B. M. Altura, pp. 95–134. Basel: S. Karger.

HAUDENSCHILD, C. C. (1980). Growth control of endothelial cells in atherogenesis and tumor angiogenesis. In *Advances in Microcirculation*, vol. 9, ed. B. M. Altura, pp. 226–251. Basel: S. Karger.

HAUDENSCHILD, C. C., ZAHNISER, D., FOLKMAN, J. & KLAGSBRUN, M. (1976). Human vascular endothelial cells in culture. Lack of response to serum growth factors. *Experimental Cell Research*, **98**, 175–83.

JAFFEE, E. A., HOYER, L. W. & NACHMAN, R. L. (1973). Synthesis of anti-hemophilic factor antigen by cultured human endothelial cells. *Journal of Clinical Investigation*, **52**, 2757–64.

JOHNSON, R. G., HAMMER, M., SHERIDAN, J. & REVEL, J. P. (1974). Gap junction formation between reaggregated Novikoff hepatoma cells. *Proceedings of the National Academy of Sciences, USA*, **71**, 4536–40.

JONGSMA, H. J. AND VAN RIJN, H. E. (1972). Electronic spread of current in monolayer cultures of neonatal rat heart cells. *Journal of Membrane Biology*, **9**, 341–60.

KELLEY, R. O., VOGEL, K. G., CRISSMAN, H. A., LUJAN, C. J. & SKIPPER, B. E. (1979). Development of the aging cell surface. *Experimental Cell Research*, **119**, 127–43.

LARSEN, W. J., AZARNIA, R. & LOEWENSTEIN, W. R. (1977). Intercellular communication and tissue growth. IX. Junctional membrane structure of hybrids between communication-competent and communication-incompetent cells. *Journal of Membrane Biology*, **34**, 39–54.

LARSON, D. M. & SHERIDAN, J. D. (1979). Structure of gap and tight junctions and transfer of small molecules from cell to cell in cultured bovine endothelial cells. *Anatomical Record*, **193**, 599.

LOEWENSTEIN, W. R. (1968). Communication through cell junctions. Implications in growth control and differentiation. *Developmental Biology*, (suppl. 2), 151–83.

LOEWENSTEIN, W. R. (1976). Permeable junctions. *Cold Spring Harbor Symposium of Quantitative Biology*, **40**, 49–63.

LOEWENSTEIN, W. R. (1979). Junctional intercellular communication and the control of growth. *Biochimica et Biophysica Acta*, **560**, 1–65.

MACARAK, E. J., HOWARD, B. V. & KEFALIDES, N. A. (1977). Properties of calf endothelial cells in culture. *Laboratory Investigations*, 36, 62–7.

MAJNO, G., SHEA, S. M. & LEVENTHAL, M. (1969). Endothelial contraction induced by histamine-type mediators. *Journal of Cell Biology*, **42**, 647–72.

PERACCHIA, C. (1980). Structural correlates of gap junction permeation. *International Review of Cytology*, **66**, 81–146.

PORVAZNIK, M., JOHNSON, R. G. & SHERIDAN, J. D. (1978). Tight junction formation between cultured hepatoma cells: possible stages in assembly and enhancement with dexamethasone. *Journal of Supramolecular Structure*, **10**, 13–30.

RAVIOLA, E., GOODENOUGH, D. A. & RAVIOLA, G. (1980). Structure of rapidly frozen gap junctions. *Journal of Cell Biology*, **87**, 273–9.

RHODIN, J. A. G. (1967). The ultrastructure of mammalian arterioles and precapillary sphincters. *Journal of Ultrastructure Research*, **18**, 181–223.

RHODIN, J. A. G. (1968). Ultrastructure of mammalian venous capillaries, venules, and small collecting veins. *Journal of Ultrastructure Research*, **25**, 425–500.

SCHNEEBERGER, E. E. & KARNOVSKY, M. J. (1976). Substructure of intercellular junctions in freeze-fractured alveolar-capillary membranes of mouse lung. *Circulation Research*, **38**, 404–11.

SCHWARTZ, S. M. (1978). Selection and characterization of bovine aortic endothelial cells. *In Vitro*, **14**, 966–80.

SCHWARTZ, S. M. & BENDITT, E. P. (1973). Clustering of replicating cells in aortic endothelium. *Proceedings of the National Academy of Sciences, USA,* **73**, 651–3.

SCHWARTZ, S. M., HAUDENSCHILD, C. C. & EDDY, E. M. (1978). Endothelial regeneration. I. Quantitative analysis of initial stages of endothelial regeneration in rat intima. *Laboratory Investigation,* **38**, 568–80.

SELDEN, S. C., III, RABINOVITCH, P. S. & SCHWARTZ, S. M. (1981). Effects of cytoskeletal disrupting agents on replication of bovine endothelium. (In Press).

SHEPRO, D. & D'AMORE, P. A. (1980). Endothelial cell metabolism. In *Advances in Microcirculation,* vol. 9, ed. B. M. Altura, pp. 161–205. Basel: S. Karger.

SHERIDAN, J. D. (1976). Cell coupling and cell communication during embryogenesis. In *The Cell Surface in Animal Embryogenesis,* ed. G. Poste & G. L. Nicholson. Amsterdam: Elsevier.

SHERIDAN, J. D. (1980). Dye transfer in small vessels from the rat omentum; homologous and heterologous junctions. *Journal of Cell Biology,* **87**, 61a.

SHERIDAN, J. D., HAMMER-WILSON, M., PREUS, D. & JOHNSON, R. G. (1978). Quantitative analysis of low-resistance junctions between cultured cells and correlation with gap junctional areas. *Journal of Cell Biology,* **76**, 532–44.

SIMIONESCU, M., SIMIONESCU, N. & PALADE, G. E. (1975). Segmental differentiations of cell junctions in the vascular endothelium: the microvasculature. *Journal of Cell Biology,* **67**, 863–85.

SIMIONESCU, M., SIMIONESCU, N. & PALADE, G. E. (1976). Segmental differentiations of cell junctions in the vascular endothelium: arteries and veins. *Journal of Cell Biology,* **68**, 705–23.

STOKER, M. P. (1975). The effects of topoinhibition and cytochalasin B on metabolic cooperation. *Cell,* **6**, 253–7.

WAGNER, R. C. (1980). Endothelial cell embryology and growth. In *Advances in Microcirculation,* vol. 9, ed. B. M. Altura, pp. 45–75. Basel: S. Karger.

Spatial inter-relationships in surface epithelia: their significance in proliferation control

C. S. POTTEN, S. CHWALINSKI AND M. T. KHOKHAR

Paterson Laboratories, Christie Hospital and Holt Radium Institute, Manchester M20 9BX

INTRODUCTION

All surface epithelia are clearly highly polarized tissues having well-defined functional and proliferative 'poles'; see Fig. 1 where three examples that have been studied closely within our laboratory are schematically illustrated. These can be used as general model systems for all surface epithelia and in this paper we should like to consider one of these, the small intestine, in some detail. We shall use this site as an example to illustrate what are believed to be some fairly general basic principles. Some aspects of the spatial inter-relationships between cells in the other two regions have been discussed elsewhere (Hume, 1980, 1981; Hume & Potten, 1976; Potten, 1976, 1980, 1981; Potten & Allen, 1975). In all these rapidly replacing tissues the functional cells become senescent and die, or are removed through mechanical interaction with the environment, after a functional life-span characteristic of the tissue. This loss is, under steady state conditions, precisely balanced by cell proliferation elsewhere in the tissue i.e. at the pole opposite to the functional one. The consequence of this is a continual movement of maturing cells along a well-defined migratory pathway. The velocity of migration can be measured and varies from tissue to tissue or site to site. A major fundamental question, that remains at present largely unresolved, is what is the driving force for this migration and how is it and the related cell proliferation controlled? Various observations suggest that migration is controlled quite independently of proliferation.

Present addresses: S. C., Department of Patophysiology, Institute of Rheumatology, M. Spartanska 1, Warsaw, Poland; M. T. K., Department of Biology as Applied to Medicine, The Middlesex Hospital Medical School, London WIP 6DB.

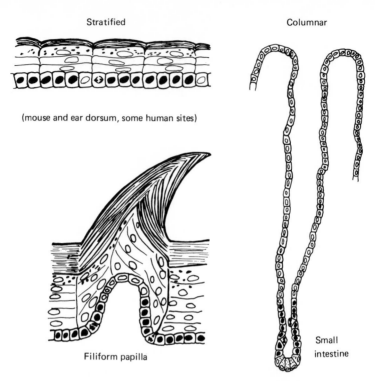

Fig. 1. Schematic representation of three regions of surface epithelium with their proliferative and differentiated functional 'poles'; a simple stratified epithelium, a complex stratified epithelium and a simple columnar epithelium. Proliferative cells are shown as having solid nuclei while post-mitotic cells are shown with 'empty' (unshaded) nuclei.

SMALL INTESTINE – GENERAL STRUCTURE AND CELL HIERARCHY

Some aspects of these inter-relationships are illustrated for small intestine in Fig. 2. Here the functional component (pole) is the villus. Senescent columnar cells are lost from the entire villus surface but this loss is particularly pronounced at the villus tip where the 'oldest' cells are seen (see villus I). The replacement of these lost cells is restricted to the 'hidden' bags of cells encircling each villus base, the crypts. Growing evidence has led to a general acceptance of the idea that cell replacement in this, and probably all replacing tissue is achieved via an hierarchical organization of proliferative cells. The origins of the hierarchies or cell lineages are the cells ultimately responsible for the hierarchy, the stem cells. In the crypts the precise number of these stem

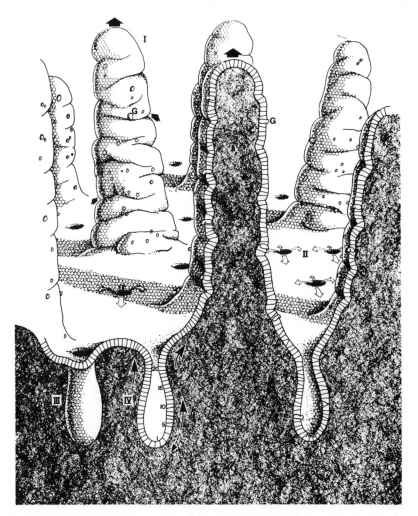

Fig. 2. Diagrammatic representation of the three-dimensional inter-relationships between crypts and villi in the small intestine showing the flow patterns (migrational
pathways) of the epithelial cells from the crypts onto the villi and eventually off the
villi into the lumen of the intestine. G = goblet cells; P = Paneth cells. Both crypts
and villi are shown in section and surface view.

cells remains obscure but they could be very few (1–15), equivalent to
the number of cells in the circumference of the crypt (15–20), or as
many as 80 but less than the number of proliferative (stem-cell derived)
transit cells (150 per crypt). Clearly if the number of stem cells is as high
as 80 then these must 'overlap' with some of the rapidly cycling pro-

liferative cells. The number of cell divisions in the transit population is likely to be 3–8 depending on the number of stem cells and hence the number of lineages.

SPATIAL RELATIONSHIP OF CRYPTS AND VILLI

The ratio of crypts to villus varies slightly from site to site but for mouse jejunum is reported to be between 9 and 14 (Hagemann, Sigdestad & Lesher, 1970; MacDonald & Ferguson, 1978) although a value roughly half this has recently been reported (Smith & Jarvis, 1979). For B6D2Fl mice the relative numbers of crypts and villi per unit area (as determined from whole mount preparations) and hence the crypt to villus ratios are shown in Table 1. The numbers in the rat (438g) are generally about three times higher (Clarke, 1972).

The spatial inter-relationships between crypts and villi are illustrated in Fig. 3 where the epithelium has been removed by hyaluronidase treatment to reveal the position of the otherwise hidden, crypts. The central villus in this case is surrounded by up to 10–12 crypts in this particular case. It is difficult to be precise about the number around any particular villus since all crypts have a variable proportion of the circumference of their orifices directed towards any particular villus. The figures shown in Table 1 represent average figures taken over a large area.

CELL MIGRATION FROM CRYPTS

At present there is no evidence to suggest that significant amounts of cell loss occur on the inter-crypt 'table'. Therefore each crypt would tend to have its own characteristic asymmetry with regard to cell production and cell flow (Fig. 4). Some may be largely unidirectional (e.g. crypts 6 and 10), others bidirectional (e.g. crypts 1, 2, 4 and 8) and yet others could in theory be tri- or multi-directional i.e. it is unlikely that the flow of cells out of a crypt is equal on all sides (see also Smith & Jarvis, 1979). Nor is it likely that the average total flow from one crypt necessarily equals the average total flow from a neighbouring crypt. These patterns of crypts associated with particular villi imply that cell migration from the crypts may be concentrated at particular points on the circumference of the crypt 'mouth'. If migration was, in some way,

Table 1. *Numbers of crypts and villi per mm² at various positions along the intestinal tract of B6D2F1 mice*

Position	Crypts/mm²	Villi/mm²	Crypts/villus
1 Duodenum	546 ± 23	50 ± 3	10.1
2 Jejunum	501 ± 14	50 ± 2	10.0
3 Ileum	516 ± 33	78 ± 5	6.6
4 Ileum	458 ± 30	74 ± 8	6.2
5 Ileum	473 ± 56	78 ± 8	6.1
6 Ileum	506 ± 38	84 ± 7	6.0
7 Ileum	535 ± 7	75 ± 4	7.1
8 Caecum	563 ± 74	—	
9 Colon	354 ± 16	—	
10 Colo-rectum	557 ± 22	—	

The entire intestine from the stomach to anus was divided into 10 roughly equal portions.

directly dependent on cell production the conclusion would be that cell proliferation in the crypts would be similarly asymmetrical. Although the distribution in the crypt of proliferating cells is non-random (being restricted to a band of about 150 cells across the middle of the crypt), there is no evidence for overall asymmetry from one side of the crypt to the other. It is curious to speculate as to the fate and behaviour of the epithelial cells on the non-migrating side of a crypt. Do these cells age and become senescent or are they gradually moved round into the migratory region having been only temporarily in the non-migrating 'backwater'?

It seems likely that migration is a process under an independent set of migration controls. It certainly appears to continue in the absence of mitotic activity e.g. after irradiation (Potten, Hendry, Moore & Chwalinski, 1981). Since the mechanisms involved in migration remain unknown it is impossible to speculate as to how this process is controlled in other than very general terms. Migration *out of the crypt* is an exclusive property of non-proliferative cells. However, it is unclear whether these represent cells that have exhausted an inherent limited division potential, or those which have received a migratory stimulus while in cycle and moved into an area (microenvironment) out of the crypt that is incompatible with further cell cycle activity.

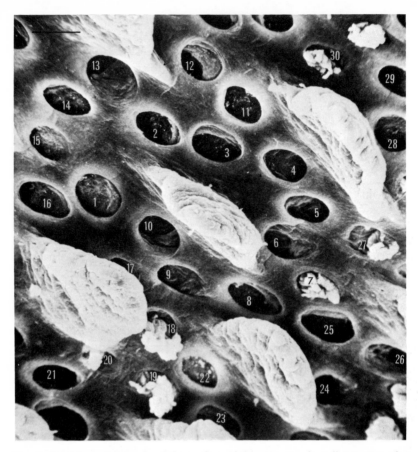

Fig. 3. Scanning micrograph of the surface of the mucosa of small intestine after removal of the epithelium with hyaluronidase. Hence, what is seen here is the connective tissue 'bed' on which the epithelium sits i.e. the holes once occupied by crypts and the connective tissue core of the villi. This permits the otherwise hidden crypts (numbered 1–30) to be seen around the bases of the villi. The connective tissue core of a villus (centre picture) can be seen viewed from directly above. This is surrounded by 10 crypts (1–10) plus two others 14 and 15 that probably contribute cells to this villus. Scale bar = 4 μm. I am grateful to Dr T. Allen for his help in preparing this picture.

CELL LOSS FROM THE VILLUS

The cells reach the villus tip about two days after their last cell division in the crypt (Quastler & Sherman, 1959; Sigdestad, Hagemann & Lesher, 1970; Tsubouchi & Matsuzawa, 1973) but this may take longer under some circumstances e.g. in germ-free mice (Tsubouchi & Matsuzawa, 1973).

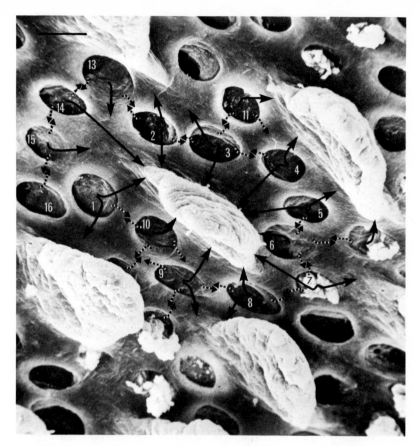

Fig. 4. The same scanning micrograph as shown in Fig. 3, with the probable migratory pathways for cells out of the crypts onto the central villus. Major flow paths are shown by solid arrows while minor or negligible flow is shown by dashed arrows. Some crypts are essentially unidirectional e.g. 6 and 10 while others are bidirectional e.g. 2, 4, 5, 8 etc. Scale bar = 4 μm.

Again, little is known about the process of loss at the end of this 2–3 day life-span. Some ultrastructural aspects have been studied (Potten & Allen, 1977). Do the cells become senescent and die at the end of their defined life-span (i.e. according to some inherent programme)? Are they removed by external forces e.g. abrasion, while still healthy cells when they reach the villus tip? Or, are they passively removed by a lack of space, and finite amount of connective tissue on which to sit, or the constant 'pushing' of new migrating cells from below? Changes in the constitution of the diet and the microbiological status of the gut can affect the villus length (Clarke, 1975).

CHANGES IN THE CRYPT POPULATION:
CRYPT FISSION

It would be interesting to know whether the total cell production rate per crypt is lower for a crypt like crypt 10 in Fig. 4 than for a crypt like crypt, 4 or 8. Examples can be seen of crypts with even less apparent direct contact with villi than is shown by crypt 10. Certainly the size and overall levels of tritiated thymidine labelling vary from crypt to crypt by a factor of about two for the extremes. However, there is no clear tendency for large or small crypts to be high or low in proliferation. There are indications from multiple hydroxyurea treatments of intestine for the existence of some very resistant crypts perhaps because they contain some very slowly cycling stem cells (Potten, Hendry, Moore & Chwalinski, 1981).

The number of villi per unit area in the rat reaches adult values early in life and may decline slightly with age while the number of crypts increases during early life and reaches a plateau later in life (Clarke, 1972; Maskens, 1977). The consequence of this is that the crypt to villus ratio increases slightly with age (Clarke, 1977). Increases in crypt number are probably achieved by a crypt fission process (Maskens, 1977; Hattori & Fujita, 1974) that is also thought to account for the increase in crypt numbers after radiation depletion (Cairnie & Millen, 1975). This crypt fission is believed to originate from the crypt base and be recognizable in sectioned material as bifurcating crypts (see Fig. 5). In the mouse the number of such bifurcating crypts in section is greatest in young mice but a significant number persist throughout adult life (Table 2). There is also an approximately two-fold greater incidence in fission in the plane parallel to the long axis of the intestine (i.e. in longitudinal sections of gut, parallel to the long axis of the elliptical profile each crypt has in cross-section; see Fig. 3). The long axis of the ellipse is about 25% greater than the short axis.

Since the number of crypts in an adult animal is roughly constant the presence of 2% that are budding at any time suggests that some crypts are lost and subsequently replaced. It would be interesting to know if the budding crypts were positioned in any particular way in relation to the villi or whether they represented slow or rapidly proliferating crypts. Could these represent a set of 'master crypts' that play a small role in the day-to-day cell replacement but are responsible for the maintenance of crypt numbers (i.e. the replacement of damaged or transitory crypts) possibly those situated immediately beneath a villus such as seen occasionally in section (unpublished) or in whole mount preparations (Smith & Jarvis, 1979)?

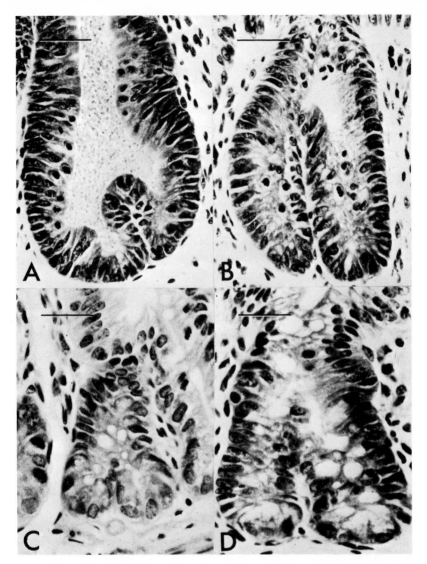

Fig. 5. Photomicrographs of crypts showing evidence of fission at various times after a cytotoxic insult. (A) 7 days after 12Gy ^{137}Cs γ-rays (B) 7 days after 12Gy ^{137}Cs γ-rays (C) 4 days after 0.4 g/kg hydroxyurea (D) 3 days after 50 mg/kg Myleran. Scale bar in A & B = 30 μm; in C & D = 20 μm.

CONTROL OF CELL PROLIFERATION WITHIN THE CRYPT

The total cell output from a crypt is likely to be the consequence of controls acting on the following processes: (1) the rate of progression of

Table 2. *The percentage of bifurcating crypts* ($\pm$ *s.e.*) *seen in transverse* *(TS) and longitudinal (LS) section of small intestine*

Age of mouse (weeks)	TS	LS
3–4	2.5 $\pm$ 0.8	4.4 $\pm$ 1.8
7–8	1.5 $\pm$ 0.2	3.8 $\pm$ 0.6
13–14	0.4 $\pm$ 0.1	1.1 $\pm$ 0.2
16–17	0.8 $\pm$ 0.3	2.4 $\pm$ 0.4
52–53	1.1 $\pm$ 0.4	2.6 $\pm$ 0.8
Mean (overall)	1.26	2.86
(13–53 weeks)	0.77	2.03

A bifurcating crypt was one with a cleft of more than two cells deep and less than nine cells (crypts with a cleft of nine or more cells were regarded as having split). At least 1000 sections (showing Paneth cells and a lumen) were scored from each of four mice.

stem cells through the cell cycle; (2) the rate of removal of cells from the stem to the transit compartments (this represents a differentiation step and may be linked in some way to the control of cell cycle progression and mitosis); (3) the rate of progression of transit cells through the cell cycle; (4) the number of 'amplifying' cell divisions in the transit population. This may be determined by the rate of removal from the transit population which could involve the stimuli triggering migration. These are illustrated in Fig. 6. Controls (1) and (2) are likely to involve on/off switches, stimulators and inhibitors, operating in most cases in G_1 or controlling the entry and exit from G_0. Many of these controls are known to operate in the haemopoietic system (Lord, 1979).

There have, over the years, been several suggestions that the cellular output from crypts is controlled by a negative feedback from the villus, i.e. the functional cells produce and secrete an inhibitor the concentration of which reaching the crypt controls progression through the cycle of crypt cells. This seems unlikely for the following reasons:
(1) If the inhibitor is released to diffuse freely then increasing concentrations would be expected 'downstream' in the intestine. Considering the large number of villus cells and the complex three-dimensional inter-relationships between villi and crypts it is difficult to see how a delicate control could be achieved (see below).
(2) It is often quoted that post-irradiation regeneration studies support this hypothesis since the crypt cell cycle times are reduced when the villus is largely absent. However, careful analysis shows that (a) clonal

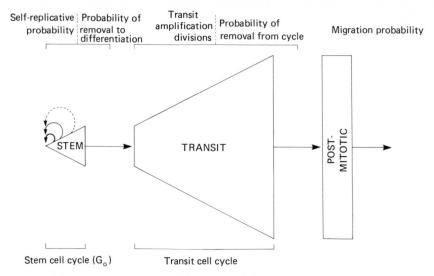

Fig. 6. Schematic representation of the crypt cellular hierarchy with a small stem cell compartment that may itself be hierarchical, a large amplifying transit population and a post-mitotic migrating compartment. The sites of the major theoretical control systems are indicated.

regeneration begins within a few hours of irradiation at a time when the villus is unaffected (Potten & Hendry, 1975), (b) cell cycle progression continues in most transit cells throughout the post-irradiation period and is not noticeably more rapid at times when the villus is completely absent and (c) a low dose of radiation (300-R) which causes little detectable change to the villus results in some rapid (within 6 hours) and fairly dramatic changes in the length of G_1 over the first 2 days post-irradiation (Lesher, 1967).

(3) From the spatial pattern of crypts and villi (Fig. 7) the 'field of influence' (diffusion contour?) for each villus could result in the sort of cell migration patterns seen in Fig. 4 but it is harder to see how this could directly control cell cycle progression. Some crypts would effectively receive a 'double dose' e.g. numbers 2, 8, 16, 27 etc. while others might receive less than a single dose (e.g. number 14 and to a lesser extent number 1). It is not known whether the latter are more proliferatively active than the former.

(4) Since movement onto the villus appears to be a process that can continue in the absence of mitotic activity the villus might be more likely to control this process.

Consequently the control of cell proliferation within the crypt is

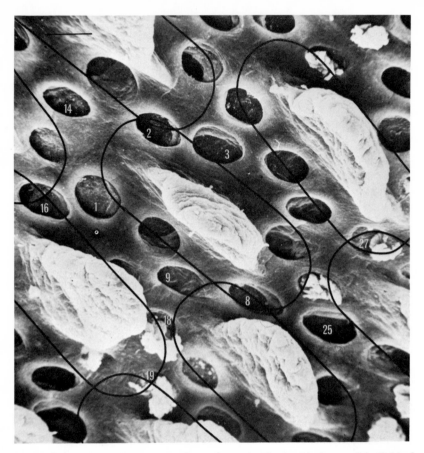

Fig. 7. The same scanning micrographs as shown in Fig. 3 with the possible 'field of influence' of each villus shown. These are contours drawn at an arbitrary equidistance from the base of each villus and could be taken to represent theoretical diffusion concentration contours. In some regions these overlap and the crypts in these over-lapping regions might be exposed to higher levels of a hypothetical control factor e.g. crypts 2, 8, 16, 27 and possibly 18. In other regions crypts can be seen which have less contact with the contour lines of 1, 3, 14 and 25. This figure does not represent any actual concentration contours but merely is used to illustrate the possible complexity of a villus-derived diffusable control factor. Scale bar = 4 μm.

more likely to be a local process restricted to the crypt itself. The movement of cells out of the crypt might result in a fall in local inhibitors through a reduction in inhibitor-producing cells in the crypt which may result in a triggering of G_1 cells into S and subsequently into M in the upper regions of the proliferative compartment. This type of control would operate at a cell to cell level i.e. at the level of extremely local microenvironments. The entire system could be controlled by a cascade type of mechanism initiated at the level of the upper crypt by the movement of cells onto the villus with this movement itself influenced by the general status of the villus.

SPATIAL ORGANIZATION WITHIN THE CRYPT

It is not yet fully clear how the cells of the proliferative hierarchy are spatially organized within the crypt. It is clear that the oldest cells with the lowest probability of more than one further cell division are located near the top of the crypt while the cells with the greatest division potential, the ancestors of the hierarchy (stem cells), must be located near the base. However, a branching family tree is difficult to accommodate spatially within the crypt unless some branches of the tree are arranged into vertical columns. This was considered from a theoretical point of view originally by Quastler & Sherman (1959). Since the branches of the tree represent classes of cells with a common ancestry they might be expected to show some signs of cell cycle synchrony. There have been reports both where evidence in support of synchrony is lacking (Cairnie, Lamerton & Steel, 1965) and where direct evidence in support of synchrony has been presented (Sawicki, Blaton & Pindor, 1977). Fig. 8 shows the results from five experiments where the frequency distribution for the number of mitoses per crypt (or half crypt) section have been determined. These, in most cases, fit the Poisson distribution very closely suggesting that the distribution of mitoses per section is random. There is a deviation from the Poisson distribution in two cases but in neither case was it significant. A more detailed analysis of the distribution of mitoses according to their position and possible 'clumping' indicated a random distribution *along* the crypt (Chwalinski et al., unpublished data). A recent analysis of the spatial distribution of labelled cells using (1) two by two contingency tables for the frequency of labelled or unlabelled adjacent pairs of cells and the chi-squared test as applied by Mantel-Haenzel (1959) and (2) a modified version of Siegel's runs test (1956) has indicated that labelled cells are not randomly distributed in the vertical plane of the crypt ($p < 0.0001$ and < 0.01 respectively). This indicates that there is a tendency for a labelled cell to have other labelled cells above or below and similarly unlabelled cells tend to be flanked by other unlabelled cells which indicates some level of synchronization. This partial synchronization may be such that it cannot be detected by a 'window' as narrow as that

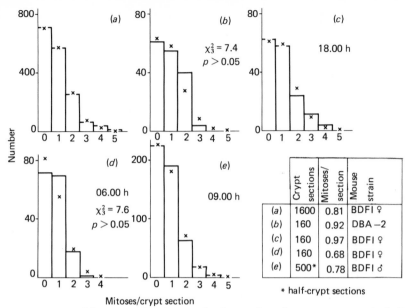

Fig. 8. A series of frequency distributions for the number of mitoses per crypt (or half crypt) section from five different experiments. The points represent the Poisson distribution calculated using the mean value of mitoses per crypt section in each case.

for mitosis (1 hour) but can by that for the S-phase (6–7 hour). Thus within the crypt there are indications that the hierarchies have specific spatial organizations which may have implications for the cascade type of control outlined above. A particular branch of the cellular family tree may behave independently of neighbouring branches. A vertical column, representing a particular generation of the overall family tree, may migrate, and divide thus altering the hierarchical status within that column and hence the 'age' of the micromilieu for the cells situated beneath.

This type of organization suggests that cell replacement is achieved through a highly ordered process where the behaviour of individual cells is determined by their position within the tissue, the microenvironment, and their position within a cellular hierarchy or cell lineage. Cells appear to be removed from the top positions of the crypt (and thus the top of the hierarchy) by a well controlled process that ensures that each villus receives just the right amount of cells from *a part* of the crypts with which it is associated. This removal may alter the local microenvironment at the top of the crypt (by lowering the local levels of inhibitors) which may start a cascade of controlled cell divisions down the cellular hierarchy.

Acknowledgements

This work was supported by the Cancer Research Campaign.

REFERENCES

CAIRNIE, A. B., LAMERTON, L. F. & STEEL, G. G. (1965). Cell proliferation studies in the intestinal epithelium of the rat. I. Determination of the kinetic parameters. *Experimental Cell Research*, **39**, 529–38.

CAIRNIE, A. B. & MILLEN, B. H. (1975). Fission of crypts in the small intestine of the irradiated mouse. *Cell and Tissue Kinetics*, **8**, 189–96.

CLARKE, R. M. (1972). The effect of growth and of fasting on the number of villi and crypts in the small intestine of the albino rat. *Journal of Anatomy*, **112**, 27–33.

CLARKE, R. M. (1975). Diet, mucosal architecture and epithelial cell production in the small intestine of specified-pathogen-free and conventional rats. *Laboratory Animals*, **9**, 201–9.

CLARKE, R. M. (1977). The effects of age on mucosal morphology and epithelial cell production in rat small intestine. *Journal of Anatomy*, **123**, 805–11.

HAGEMANN, R. F., SIGDESTAD, C. P. & LESHER, S. (1970). A quantitative description of the intestinal epithelium of the mouse. *American Journal of Anatomy*, **129**, 41–52.

HATTORI, T. & FUJITA, S. (1974). Fractographic study on the growth and multiplication of the gastric gland of the hamster. The gland division cycle. *Cell and Tissue Research*, **153**, 145–9.

HUME, W. J. (1980). Proliferative organisation in mouse tongue epithelium. Ph. D. Thesis. University of Manchester.

HUME, W. J. (1981). Stem cells in oral epithelia. In *The Identification and Characterisation of Stem Cells*, ed C. S. Potten. Edinburgh: Churchill-Livingstone (in press).

HUME, W. J. & POTTEN, C. S. (1976). The ordered columnar structure of mouse filiform papillae. *Journal of Cell Science*, **22**, 149–160.

LESHER, S. (1967). Compensatory reactions in intestinal crypt cells after 300 roentgens of cobalt-60 gamma irradiation. *Radiation Research*, **32**, 510–19.

LORD, B. I. (1979). Proliferation regulators in haemopoiesis. *Clinics in Haematology*, **8**, 435–51.

MACDONALD, T. T. & FERGUSON, A. (1978). Small intestinal epithelial cell kinetics and protozoa infection in mice. *Gastroenterology*, **74**, 496–500.

MANTEL, N. & HAENZEL, W. (1959). Statistical aspects of the analysis of data from retrospective studies of disease. *Journal of the National Cancer Institute*, **22**, 719–48.

MASKENS, A. P. (1977). Histogenesis of colon glands during postnatal growth. *Acta Anatomica*, **100**, 17–26.

POTTEN, C. S. (1976). Identification of clonogenic cells in the epidermis and the structural arrangement of the epidermal proliferative unit (EPU). In *Stem Cells of Renewing Cell Populations*, ed. A. B. Cairnie, P. K. Lala & D. G. Osmond, pp. 91–102. New York: Academic Press.

POTTEN, C. S. (1980). Proliferative cell populations in surface epithelia: Biological models for cell replacement. *Biomathematics*, **38**, 23–35.

POTTEN, C. S. (1981). Cell replacement in epidermis (keratopoiesis) via discrete units of proliferation. *International Review of Cytology*, **69**, 271–318.

POTTEN, C. S. & ALLEN, T. D. (1975). Control of epidermal proliferative units (EPUs). An hypothesis based on the arrangement of neighbouring differentiated cells. *Differentiation*, **3**, 161–5.

POTTEN, C. S. & ALLEN, T. D. (1977). Ultrastructure of cell loss in intestinal mucosa. *Journal of Ultrastructure Research*, **60**, 272–7.

POTTEN, C. S. & HENDRY, J. H. (1975). Differential regeneration of intestinal proliferative cells and cryptogenic cells after irradiation. *International Journal of Radiation Biology*, **27**, 413–24.

POTTEN, C. S., HENDRY, J. H., MOORE, J. V. & CHWALINSKI, S. (1981). Cytotoxic effects in gastrointestinal epithelium (as exemplified by small intestine). In *Cytotoxic Insult to Tissue*, ed. C. S. Potten & J. H. Hendry. Edinburgh: Churchill-Livingstone (in press).

QUASTLER, H. & SHERMAN, F. G. (1959). Cell population kinetics in the intestinal epithelium of the mouse. *Experimental Cell Research*, **17**, 420–38.

SAWICKI, W., BLATON, O. & PINDOR, M. (1977). Spatial distribution of DNA-synthesizing cells in colonic crypts of the guinea pig. *American Journal of Anatomy*, **148**, 417–26.

SIEGEL, S. (1956). *Non Parametric Statistics for the Behavioural Sciences*. New York: McGraw-Hill.

SIGDESTAD, C. P., HAGEMANN, R. F. & LESHER, S. (1970). A new method for measuring intestinal cell transit time. *Gastroenterology*, **58**, 47–8.

SMITH, M. W. & JARVIS, L. G. (1979). Use of differential interference contrast microscopy to determine cell renewal times in mouse intestine. *Journal of Microscopy*, **118**, 153–9.

TSUBOUCHI, S. & MATSUZAWA, T. (1973). Correlation of cell transit time with survival time in acute intestinal radiation death of germ-free and conventional rodents. *International Journal of Radiation Biology*, **24**, 389–96.

Metabolic cooperation in tumour promotion and carcinogenesis

Chemical Carcinogenesis Division, Institute of Cancer Research, Chalfont St Giles, Bucks HP8 4SP, UK

An understanding of the mechanism of tumour promotion is important in cancer research because it should reveal how potentially tumorigenic ('initiated') cells can remain dormant within a tissue treated with a sub-threshold dose of carcinogen. This knowledge could have applications in cancer prevention and therapy as well as furthering our insight into fundamental cancer mechanisms. Recent work in several laboratories has shown that phorbol ester tumour promoters are potent inhibitors of metabolic cooperation in certain established mammalian cell lines in culture. Moreover, the potency of each phorbol ester in this respect correlates closely with its tumour-promoting activity. On the basis of these findings, it is tempting to speculate that inhibition of metabolic cooperation is an important step in tumour promotion. However, caution should be exercised in formulating such a hypothesis until convincing evidence is obtained that promoters produce the same effect *in vivo*. If this is found to be the case then the corollary of the hypothesis, viz. that the phenotype of an initiated cell can be suppressed by junctional communication with its normal counterparts, would become an attractive possibility.

THE CONCEPT OF TWO-STAGE CARCINOGENESIS

It is now over sixty years since experimental carcinogenesis was first demonstrated in laboratory animals by two Japanese workers (Yamagiwa & Ichikawa, 1918) who repeatedly applied coal tar condensate to the ears of rabbits. Subsequent advances in experimental tumour production led naturally to a search for the specific compounds responsible for the carcinogenicity of tars and in the 1930s the first pure chemical carcinogens, polycyclic aromatic hydrocarbons (PAH) were identified by E. L. Kennaway's group at the Institute of Cancer Research in London (reviewed by Haddow, 1974). As these investigations were extended, it soon became clear that carcinogenic activity is not confined to PAH, and today hundreds of compounds of diverse chemical structure are known to possess this property. Many of these compounds, including PAH, require metabolism to derivatives which will react with cellular macromolecules before they can exert their biological effects.

In most of the early animal experiments, which were designed to investigate the relationship between the structure of a chemical and its carcinogenicity, it was necessary to administer either a single large dose or repeated smaller doses of a carcinogen (by subcutaneous injection or by skin painting) to produce tumours. However, the existence of substances which reinforce the action of carcinogens but are not in themselves carcinogenic was soon recognised. In 1941, Berenblum, investigating the role of irritation in carcinogenesis, found that croton oil (a complex lipid mixture obtained from the seeds of the shrub *Croton tiglium*) caused a marked augmentation of benzo(α)pyrene carcinogenesis, but that two other irritants, turpentine and xylene, were inactive in this respect. The induction of skin tumours in mice using a sub-threshold dose of a carcinogen ('initiation') followed by repeated treatment with a non-carcinogenic promoter such as croton oil ('promotion') was first described by Mottram in 1944. Thus, there gradually emerged the concept that the application of a carcinogen to a susceptible tissue leaves an indelible effect on it such that subsequent stimuli may evoke neoplasia. Some carcinogens, such as PAH, are known as 'complete' carcinogens because at high doses they possess both initiating and promoting activity, while others are pure initiators (for example, urethane) and usually require the application of a promoter before they can induce tumours.

Chemical carcinogens are therefore valuable tools in fundamental cancer research because they can be used to follow the cancer process under controlled experimental conditions. Moreover, series of compounds are available which are closely-related chemically but which differ widely in carcinogenic potency. The action of such compounds on a variety of biological systems has been the subject of intensive study, in the hope that clues may arise as to the particular properties responsible for cancer initiation.

The hypothesis that DNA is the critical cellular target for carcinogens has received much support in recent years. This hypothesis has arisen as a result of the finding that most carcinogens damage DNA in some way and, moreover, in the case of certain classes of chemical carcinogen, that there is a quantitative correlation between carcinogenicity and the extent to which a compound will react with this macromolecule (Brookes & Lawley, 1964). More specifically, of the three major consequences to the mammalian cell of carcinogen-induced DNA modification, viz.: cell death, gross chromosome damage and gene mutation, the last named appears to be the major determinant of carcinogenic potency. Highly potent carcinogens possess the additional property of high mutagenic efficiency when tested in mammalian cells in culture, in that they are able to induce large numbers of gene mutations at doses which produce low levels of cytotoxicity. (Newbold *et al.*, 1980; Brookes & Newbold, 1980).

The results of research outlined above strongly suggest (but do not of course prove) that there is a connection between mutagenesis and carcinogenesis, and from this has arisen the idea that initiation is the result of one or more mutations induced in a somatic cell capable of division. The fact that initiation of mouse skin is a rapid process and essentially irreversible (despite the fact that mouse epidermis is a constantly proliferating tissue with a replacement time of 8–10 days) is consistent with such a notion.

The mechanism by which tumour promoters act is far less clear. In contrast to initiation, promotion is largely reversible and repeated applications of the promoter are required to produce tumours. In mouse skin the first tumours to appear are papillomas, which can regress if treatment with the promoter is withdrawn. Only after continuous promotion over many weeks are malignant tumours observed. It is not known if these develop from papilloma cells by progression or whether they arise independently.

The most potent known tumour promoter for mouse skin is 12-*O*-tetradecanoyl-phorbol-13-acetate (TPA) the active principle of croton oil, isolated and characterised by Hecker (1968) and Van Duuren (1969); because of its potency, this has been the most extensively studied promoter. The structure of TPA is shown in Fig. 1. The molecule has both a lipophilic portion (a long fatty acid chain) and a hydrophilic portion, both of which are required for promoting activity; this amphipathic (hydrophilic–hydrophobic) nature of the TPA molecule gives it an affinity for cell membranes. Recently, evidence has been presented that specific receptors exist for TPA on the surface of avian and mammalian cells (Driedger & Blumberg, 1980; Shoyab &

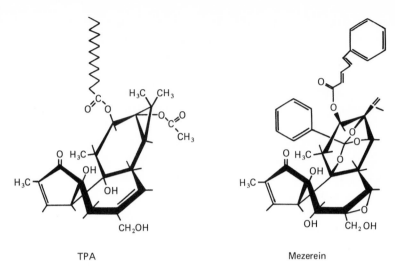

TPA   Mezerein

Fig. 1. Structure of TPA and the related diterpene ester, mezerein.

Todaro, 1980). While TPA is metabolised by cellular enzymes in many cell types, it is generally accepted that metabolism is not required for promoting activity. Moreover, TPA has not been found to be mutagenic in any bacterial or mammalian cell system so far examined.

Phorbol diester promoters, exemplified by TPA, have been shown to have an enormous number of biological and biochemical effects on mammalian tissues and mammalian cells in culture. Some of these are listed in Table 1 but the reader is referred to recent reviews on the subject for further information (Scribner & Suss, 1978; Diamond, 1980). Initially, ideas on the mechanism of tumour promotion were based on the observation that promoters are potent inducers of hyperplasia and inflammation. However, while these effects may be necessary they are clearly insufficient for promotion, since a number of potent irritant and hyperplasiogenic chemicals are inactive as promoters.

Of the many effects on mammalian cells previously ascribed to TPA *in vivo* or *in vitro*, those which involve the induction of synthesis of specific proteins and those which modulate differentiation have been most extensively studied (Diamond, 1980). This has led to the proposal of mechanisms for promotion based, for example, upon increased expression of modified genes in initiated cells, or on blockage of differentiation and subsequent amplification of a pluripotent initiated cell pool. However, it is because phorbol diester promoters produce such pleiotropic effects that those properties specifically related to

Table 1. *Some biological and biochemical effects*
of TPA in mammalian systems

Mouse skin	Mammalian cells in culture
Erythema, irritation & inflammation	Altered cell morphology
Increase in mitotic activity of basal layer (hyperplasia)	Induction of anchorage independent growth
Induction of 'dark' cells	Stimulation of glucose transport and metabolism
Increased keratinisation of upper epidermis	Induction of ornithine decarboxylase
Induction of DNA synthesis	Induction of plasminogen activator
Synthesis of 'new' proteins	Stimulation of DNA synthesis
Induction of ornithine decarboxylase	Modulation of cell differentiation
	Inhibition of metabolic co-operation
Tumour promotion	
	Enhancement of malignant transformation

For further information see reviews by Diamond (1980) and Scribner & Suss (1978).

promotion are so difficult to identify and therefore why, despite recent progress, the mechanism of promotion remains obscure.

VALUE OF UNDERSTANDING THE MECHANISM OF PROMOTION

The persistence of latent malignant cells in mouse epidermis for up to one year following a single application of an initiating dose of carcinogen has been clearly demonstrated after promotion with croton oil (Berenblum & Shubik, 1949) or TPA (Van Duuren et al., 1975). Just how these initiated (premalignant) cells are kept in check or retain a normal phenotype under these conditions, and how promoters can alter this stable state, are obviously important and intriguing problems, a solution to which could contribute significantly to our understanding of cancer mechanisms. In addition, it is conceivable that, armed with this knowledge, we would be in a better position to devise new approaches to cancer prevention and therapy. The fact that human exposure to environmental mutagens is, in general, likely to be more akin to initiation rather than complete carcinogenesis (with the possible exception of rare, heavy occupational exposure) suggests that promotion plays a role in human cancer. Indeed, it may be that, for a variety of political and scientific reasons, there are practical limitations to the extent to which initiation can be controlled through a reduction of levels of mutagens in the environment. Since promotion is a reversible process requiring repeated exposure to the promoting agent, it could in theory be more easily controlled than initiation which is rapid and irreversible.

Although it is not clear what exact roles initiators and promoters play in the genesis of human lung cancer by cigarette smoke, the fact that the risk of contracting the disease is at least partly reversible after discontinuing the habit indicates that promotion could play a strong part. This is supported experimentally by demonstrations of the promoting activity of cigarette smoke condensate. (Wynder, *et al.*, 1978).

Before the idea can be taken too seriously that promotion is important in human cancer, it is of course essential that two-stage carcinogenesis is shown to be a general phenomenon rather than a peculiarity of mouse skin treated with phorbol esters. While it is true that the bulk of information on the subject does derive from the mouse skin–TPA model, there is now considerable evidence that the two-stage system is not totally tissue or species specific. The promoting effect of phenobarbital on rat liver previously initiated by 2-acetylaminofluorene (Peraino, Fry & Staffeldt, 1971) and that of saccharin on methylnitrosourea-initiated rat bladder (Hicks, Wakefield & Chowaniec, 1975) are two noteworthy examples. It has also been claimed that two-stage malignant transformation of mammalian cells in culture can be achieved using a variety of carcinogens as initiators and TPA as a promoter (Heidelberger, Mondal & Peterson, 1978).

TUMOUR PROMOTERS AS INHIBITORS OF METABOLIC COOPERATION

Metabolic cooperation between mammalian cells was first recognised when it was discovered that the defective phenotype of cultured mutant cells lacking the purine salvage enzyme hypoxanthine-guanine phosphoribosyl transferase (HPRT; EC 2.4.2.8) could be corrected by intimate contact with normal cells (Subak-Sharpe, Burk & Pitts, 1969). The most likely reason for this phenomenon is that a molecule of low molecular weight, presumably a nucleotide, is transferred directly from wild-type (HPRT$^+$) cells to the HPRT$^-$ mutants. Narrow hydrophilic channels, approximately 2 nm in diameter, which traverse structures known as gap junctions, are thought to permit the passage of ions and small molecules between cells in contact (Weinstein, Merk & Alroy, 1976; Pitts & Simms, 1977). Gap junctions are formed between most cells *in vivo* and *in vitro*; their precise function in non-excitable tissues is at present unclear, although it has been proposed that they could play an important role in development and tissue homeostasis (Wolpert, 1978; Loewenstein, 1979).

This form of intercellular communication has a specific consequence *in vitro* for those attempting to quantitate frequencies of spontaneous or induced mutations at the HPRT locus, in that it severely hampers the selection with purine analogues (for example, 8-azaguanine) of rare HPRT$^-$ mutants in crowded cultures. HPRT$^-$ cells in contact with HPRT$^+$ cells become sensitive to the cytotoxic effects of 8-azaguanine because they can acquire lethal amounts of 8-azanucleotides from their normal counterparts (Fig. 2). The result is that mutation frequencies can only be measured reliably at cell densities where intercellular contact is minimal (Newbold *et al.*, 1975).

In 1977, Trosko and coworkers found that TPA increased the frequency of HPRT$^-$ mutants induced in a Chinese hamster cell line (V79) by ultraviolet light; the enhancement was ascribed tentatively to an effect of TPA on gene repression. However, this explanation was later shown to be incorrect by Trosko and, independently, by two other groups. Convincing experimental evidence was produced in all three laboratories that TPA can cause a marked inhibition of metabolic cooperation between mammalian cells and thus is able to increase the recovery of HPRT$^-$ mutants in crowded cultures of wild-type cells. I will briefly review this evidence.

In order to distinguish between an effect of TPA on mutant recovery and effects on mutagenesis and gene expression, Trosko's group (Yotti, Chang & Trosko, 1979) cocultured a small number of HPRT$^-$ mutant V79 cells with various numbers of HPRT$^+$ cells in medium containing a purine analogue (in this case, 6-thioguanine). As expected, the recovery of HPRT$^-$ cells as colonies was reduced considerably in the presence of wild-type cells ($> 1 \times 10^5$/9-cm petri dish). However, low doses of TPA ($\geqslant 1$ ng/ml of medium) were able to reverse this effect completely, allowing virtually 100% recovery of mutants. Moreover, when a variety of phorbol-related compounds were tested in this system, a good correlation was observed between the documented promoting activity of each phorbol ester on initiated mouse skin and its ability to increase mutant recovery (Fig. 3). These workers recognised

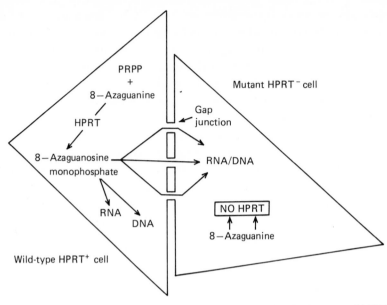

Fig. 2. Schematic representation of metabolic cooperation between mutant (HPRT⁻) and wild-type (HPRT⁺) mammalian cells cultured in the presence of a purine analogue (8-azaguanine). HPRT⁻ cells are normally resistant to the cytotoxic effects of 8-azaguanine because they cannot convert it to the nucleotide. They can, however, acquire lethal amounts of the nucleotide analogue through gap-junctional communication with HPRT⁺ cells.

that the newly-discovered property of TPA could explain the phenomenon of tumour promotion and also proposed that their system could be used as a screening test for other promoters. In later publications, Trosko showed that several non-phorbol promoters, including saccharin (Trosko, Dawson, Yotti & Chang, 1980) phenobarbital, anthralin and deoxycholic acid (Trosko, Yotti, Dawson & Chang, 1981) were also active at enhancing mutant recovery.

A more direct method for measuring the effect of TPA on metabolic cooperation was used by Murray & Fitzgerald (1979). Cultures of mouse epidermal cells (the HEL/37 cell line) were prelabelled by incubation for 2–3 h with [³H]uridine to produce high levels of intracellular [³H]uridine nucleotides. These 'donor' cells were then washed with unlabelled medium to remove remaining free [³H]uridine and then cocultured for 4 h with unlabelled mouse (3T3) 'recipient' cells. Cell membranes are normally impermeable to nucleotides so their passage from donor to recipient requires the presence of junctions. Metabolic cooperation between HEL/37 and 3T3 cells in contact, revealed autoradiographically as a transfer of label, was effectively blocked by the tumour-promoting phorbol esters TPA and phorbol-12,13-didecanoate in the culture medium, but not by non-promoting phorbol derivatives. Murray & Fitzgerald proposed a mechanism for tumour promotion based on their findings and suggested that their results warranted a more detailed examination of the effects of tumour promoters on intercellular communication.

Newbold & Amos (1981) conducted their study on V79 cells but determined the effect of TPA on metabolic cooperation in three ways in the same cell system: (a) by measuring the recovery *in situ* of rare HPRT⁻ mutants induced by the carcinogen methylnitrosourea (MNUA) at long expression times and therefore in the presence of large numbers of wild-type cells, (b) by measuring the recovery of HPRT⁻ mutants in reconstruction (coculture) experiments and (c) by observing [³H]uridine nucleotide transfer between cells in contact, using autoradiography. TPA was shown to be highly potent at inhibiting metabolic cooperation by all three methods. Maximum recovery of mutants in coculture experiments was achieved with a dose of 1 nM TPA. A series of related compounds was tested and, as in the case of the two studies above, the magnitude of the response reflected the potency of each agent as a tumour promoter (Fig. 3).

One interesting diterpene ester examined by Newbold & Amos (1981) was mezerein (structure shown in Fig. 1). Mezerein (an ester of 12-hydroxydaphnetoxin) is a useful compound in this type of study because, although it is a weak promoter, it is equivalent in potency to TPA in many other respects. For example, it is as potent an inhibitor of differentiation in mouse leukaemia cells, mouse neuroblastoma cells and chick embryonic cells, as an inhibitor of the binding of epidermal growth factor to Hela cells, and at inducing hyperplasia, inflammation and ornithine decarboxylase in mouse skin; mezerein is also approximately five times as effective as TPA at inducing plasminogen activator in chick embryo fibroblasts. On the basis of these findings it has been proposed that TPA must accomplish an as yet unidentified event which is essential for tumour promotion but which is not exhibited to the same degree by mezerein (Mufson *et al.*, 1979). In their coculture experiments, Newbold & Amos (1981) found that mezerein was relatively weak at enhancing the recovery of HPRT⁻ mutants, which is in keeping with its tumour-promoting activity (Fig. 3). A similar result for mezerein has been obtained more recently by Murray's group (Guy, Tapley & Murray, 1981) using a modified cell rescue method for

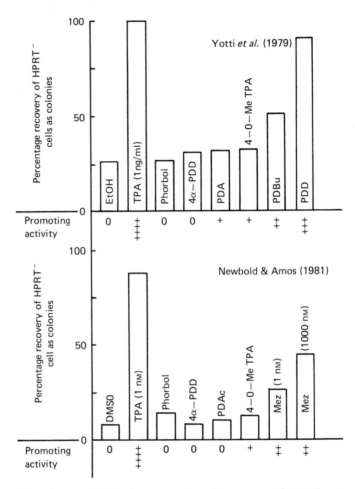

Fig. 3. Effect of TPA and related compounds on the recovery of HPRT⁻ mutant V79 cells in crowded cultures of wild-type (HPRT⁺) cells: relationship between the degree of enhancement of mutant recovery and tumour-promoting activity (rated from O→ + + + +). Data obtained from Yotti *et al.* (1979) and Newbold & Amos (1981).

Abbreviations: EtOH, ethanol; DMSO, dimethylsulphoxide; TPA, 12-O-tetra-decanoyl-phorbol-13-acetate; 4α-PDD, 4α-phorbol-12, 13-didecanoate; PDA, phorbol-12, 13-diacetate; 4-O-MeTPA, 4-O-methyl TPA; PDBu, phorbol-12, 13-dibutyrate; PDD, phorbol-12, 13-didecanoate; PDAc, phorbol-13, 20-diacetate; Mez, mezerein. Compounds were added to the culture medium to give a final concentration of 1 ng/ml (Yotti *et al.* 1979) and 1 nM (Newbold & Amos, 1981); mezerein was also tested at 1 μM in the latter study.

determining metabolic cooperation.

The above investigations have demonstrated unequivocally that

TPA and other promoters have a potent inhibitory effect on metabolic cooperation between certain mammalian cells growing as monolayers on plastic substrates. However, I must emphasise that it is by no means clear whether the phenomenon is a general one which applies to all mammalian cells *in vitro* and *in vivo*, or whether TPA merely affects the way in which a few cell types come into contact and form junctions in culture. In fact, Pitts (personal communication) has recently cast doubts on the universality of the effect on cultured cells; for example, he was unable to detect any inhibition of cooperation in pure cultures of hamster BHK21-C13 or mouse 3T3 cells by the uridine nucleotide-transfer method. If some cell types are truly refractory to TPA in this respect (rather than, for instance, requiring a longer exposure to the phorbol ester) then one would be justified in questioning somewhat any assumption that TPA has a direct effect on gap junctions. Even so, it should be remembered that there are variations in the responsiveness of different tissues and species towards the promoting effects of TPA, and therefore if inhibition of cell communication is important in promotion, one might not expect all cell types to respond similarly in culture. Although Murray & Fitzgerald (1979) have shown that TPA is able to block metabolic cooperation between established mouse epidermal cells and 3T3 cells in coculture, it is now obviously of paramount importance to establish whether TPA disturbs junctional communication in functioning mouse epidermis.

Until the questions posed above are answered satisfactorily, it is not really worth speculating at length about a possible mechanism by which TPA could interfere with gap junction assembly or permeability. However, one factor thought by many to be of overriding importance in regulating junctional communication is the intracellular concentration of free calcium ions (Peracchia, 1980). There is some indication that TPA can alter Ca^{2+} levels in cells. Fisher & Weinstein (1981) have recently observed a small transient increase in total Ca^{2+} levels in rat embryo cell treated with TPA, and Whitfield (Whitfield, MacManus & Gillan, 1973; Boynton, Whitfield & Isaacs, 1976) found that TPA induces Ca^{2+}-dependent DNA synthesis in rat thymic lymphocytes, but only when Ca^{2+} is present in the medium. It would be a relatively simple task to examine Ca^{2+} levels in various TPA-sensitive and TPA-insensitive cells before and after TPA treatment, and relate these to Ca^{2+} levels required to shut off junctional communication.

A NEW WORKING HYPOTHESIS FOR TUMOUR PROMOTION AND LATENT MALIGNANCY

If an effect of tumour promoters on intercellular communication is confirmed in mouse skin, an alternative hypothesis for the mechanism of tumour promotion, which does not depend primarily upon an alteration of the state of differentiation of initiated cells or modified gene expression, could be considered seriously. Promoters might permit the expression of a premalignant phenotype which had previously been masked by communication with surrounding normal cells, possibly by preventing the passage of growth regulatory factors from the normal cells to their initiated counterpart. As a consequence of the existence of gap junctions between cells, many mammalian tissues behave as functional units (syncytia) at least as far as intracellular low-molecular weight molecules (800–1000 daltons) are concerned (Finbow & Pitts, 1981). Non-toxic (initiating) doses of potent carcinogens, albeit highly mutagenic (Brookes & Newbold, 1980) would not be expected to produce the kind of tissue disruption necessary to reduce intercellular communication and therefore it is likely that, under these conditions, considerable restriction would be imposed on the ability of an initiated cell to proliferate autonomously. This state of affairs is likely to persist until communication is impeded by the application of a promoter and the initiated cell is effectively isolated from its neighbours. In conjunction with induced local hyperplasia, the promoter could thus allow the expansion of a clone of initiated cells, some of which (perhaps those on the inside of the clone) would be protected from the growth-controlling influences of normal cells. Such cells could then divide further and progress towards full malignancy by repeated selection of fitter variants. A hypothetical scheme for promotion based on these ideas is shown in Fig. 4.

The above model relies on a temporary loss of junctional communication between normal and initiated cells. Interestingly, a similar loss occurs in regenerating rat liver following partial hepatectomy (Yee & Revel, 1978). Partial hepatectomy is known to be a powerful cocarcinogenic stimulus in the livers of rats treated with a single systemic dose of carcinogen such as dimethylnitrosamine (Craddock, 1978). The hypothesis does not, however, require that there be a permanent loss of junctions in cancer cells. In fact, while a survey of a large number of tumours has revealed that junctional abnormalities are fairly common, lack of communication is by no means the rule (Weinstein *et al.*, 1976).

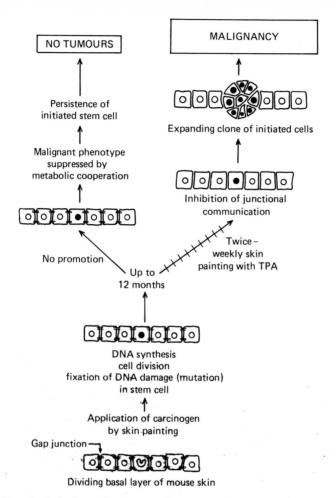

Fig. 4. Hypothetical scheme for latent malignancy and tumour promotion in mouse skin based on metabolic cooperation and its inhibition. Note that the 'initiated' state will persist unexpressed for at least 1 year after treatment with the carcinogen.

One prediction of the hypothesis is that agents which, in contrast to TPA, improve junctional communication between cells, should act as inhibitors of carcinogenesis. In this connection, Elias et al. (1980) have recently shown that retinoic acid produces a substantial proliferation of gap junctions in human basal cell carcinoma. Since retinoids have been shown to impede the development of epithelial malignancies and induce regression of human basal cell epitheliomas, these authors have proposed that an increase in gap-junctional communication may relate to their anti-neoplastic activity. Retinoids are also known to be potent inhibitors of the promotional phase of mouse skin tumorigenesis (Verma *et al.*, 1980); their effects on gap junctions in this tissue and on metabolic cooperation between cultured cells therefore deserve investigation.

CONCLUSION

Despite our lack of knowledge about the role junctional communication plays in cellular growth control, recent observations that tumour promoters are powerful inhibitors of metabolic cooperation under certain conditions, have led to the formulation of a rather attractive working hypothesis for tumour promotion and latent malignancy. However, as I have stressed throughout this chapter, a great deal more work needs to be done to add substance to these ideas. In particular, it is essential that the aforesaid property of promoters be shown to be more than an indirect effect on a few cell types in culture, and that the precise mechanism by which cooperation is blocked be defined. In addition, we need to know much more about the ways in which normal cells can alter the behaviour of counterparts at various stages in the process of transformation to malignancy, and whether their influence can be reduced or abolished by tumour-promoting agents such as the phorbol esters. The availability of a representative cell culture system for this purpose could, in theory, facilitate identification of the important molecules transferred. However, it first remains to be seen whether the hypothesis can stand the test of further, more rigorous experiment.

Acknowledgements

The author would like to thank Ms J. Amos and Miss H. Anton for their help in preparing the manuscript.

The work was supported by the National Cancer Institute USA (Grant No. RO1-CA2580701) and by grants to the Institute of Cancer Research from the Medical Research Council and Cancer Research Campaign.

REFERENCES

BERENBLUM, I. (1941). The cocarcinogenic action of croton resin. *Cancer Research*, 1, 44–8.

BERENBLUM, I. & SHUBIK, P. (1949). The persistence of latent tumour cells induced in the mouse's skin by a single application of 9 : 10-dimethyl-1 : 2-benzanthracene. *British Journal of Cancer*, 3, 384–6.

BOYNTON, A. L., WHITFIELD, J. F. & ISAACS, R. J. (1976). Calcium-dependent stimulation of BALB/C 3T3 mouse cell DNA synthesis by a tumour-promoting phorbol ester (PMA). *Journal of Cellular Physiology*, **87**, 25–32.

BROOKES, P. & LAWLEY, P. D. (1964). Evidence for the binding of polynuclear aromatic hydrocarbons to nucleic acid of mouse skin: relation between carcinogenic power and their binding to DNA. *Nature (London)*, **202**, 781–4.

BROOKES, P. & NEWBOLD, R. F. (1980). Biological effects of specific hydrocarbon-DNA reactions. In *Carcinogenesis: Fundamental Mechanisms and Environmental Effects*, ed. P. Pullman, P.O.P. Ts'o & H. Gelboin, pp. 81–90. London: D. Reidel.

CRADDOCK, V. M. (1978). Cell proliferation and the induction of liver cancer. In *Primary Liver Tumours*, ed. H. Remmer, H.M. Bolt, P. Bannasch & H. Popper, pp. 377–83. Lancaster: MTP Press.

DIAMOND, L. (1980). Tumour promoters and the mechanism of tumour promotion. *Advances in Cancer Research*, **12**, 1–74.

DRIEDGER, P. E. & BLUMBERG, P. M. (1980). Specific binding of phorbol ester tumour promoters. *Proceedings of the National Academy of Sciences, USA*, **77**, 567–71.

ELIAS, P. M., GRAYSON, S., CALDWELL, T. M. & McNUTT, N. S. (1980). Gap junction proliferation in retinoic acid-treated human basal cell carcinoma. *Laboratory Investigation*, **42**, 469–74.

FINBOW, M. E. & PITTS, J. D. (1981). Permeability of junctions between animal cells. *Experimental Cell Research*, **131**, 1–13.

FISHER, P. B. & WEINSTEIN, I. B. (1981). Enhancement of cell proliferation in low calcium medium by tumour promoters. *Carcinogenesis*, **2**, 89–95.

GUY, G. R., TAPLEY, P. M. & MURRAY, A. W. (1981). Tumour promoter inhibition of intercellular communication between cultured mammalian cells. *Carcinogenesis*, **2**, 223–7.

HADDOW, A. (1974). Sir Ernest Laurence Kennaway FRS, 1881–1958: Chemical causation of cancer then and today. *Perspectives in Biology and Medicine*, **17**, 543–88.

HECKER, E. (1968). Cocarcinogenic principles from the seed oil of *Croton tiglium* and from other Euphorbiaceae. *Cancer Research*, **28**, 2338–49.

HEIDELBERGER, C., MONDAL, S. & PETERSON, A. R. (1978). Initiation and promotion in cell cultures. In *Carcinogenesis: Mechanisms of Tumour Promotion and Cocarcinogenesis*, ed. T. J. Slaga, A. Sivak & R. K. Boutwell, pp. 197–202. New York: Raven Press.

HICKS, R. M., WAKEFIELD, J. & CHOWANIEC, J. (1975). Evaluation of a new model to detect bladder carcinogens or co-carcinogens; results obtained with saccharin, cyclamate and cyclophosphamide. *Chemico-Biological Interactions*, **11**, 225–33.

LOEWENSTEIN, W. R. (1979). Junctional intercellular communication and the control of growth. *Biochimica et Biophysica Acta*, **560**, 1–65.

MOTTRAM, J. C. (1944). A developing factor in experimental blastogenesis. *Journal of Pathology and Bacteriology*, **56**, 181–7.

MUFSON, R. A., FISCHER, S. M., VERMA, A. K., GLEASON, G. L., SLAGA, T. J. & BOUTWELL, R. K. (1979). Effects of 12-0-tetradecanoyl-phorbol-13-acetate and mezerein on epidermal ornithine decarboxylase activity, isoproterenol-stimulated levels of cyclic AMP and induction of mouse skin tumours *in vivo*. *Cancer Research*, **39**, 4791–5.

MURRAY, A. W. & FITZGERALD, D. J. (1979). Tumour promoters inhibit metabolic cooperation in cocultures of epidermal and 3T3 cells. *Biochemical and Biophysical Research Communications*, **91**, 395–401.

NEWBOLD, R. F. & AMOS, J. (1981). Inhibition of metabolic cooperation between mammalian cells in culture by tumour promoters. *Carcinogenesis*, **2**, 243–9.

NEWBOLD, R. F., BROOKES, P., ARLETT, C. F., BRIDGES, B. A. & DEAN, B. (1975). The effect of variable serum factors and clonal morphology on the ability to detect HPRT-deficient variants in cultured Chinese hamster cells. *Mutation Research*, **30**, 143–8.

NEWBOLD, R. F., WARREN, W., MEDCALF, A. S. C. & AMOS, J. (1980). Mutagenicity of carcinogenic methylating agents in associated with a specific DNA modification. *Nature (London)*, **283**, 596–9.

PERACCHIA, C. (1980). Structural correlates of gap junction permeation. *International Review of Cytology*, **66**, 81–146.

PERAINO, C., FRY, R. J. M. & STAFFELDT, E. (1971). Reduction and enhancement by phenobarbital of hepatocarcinogenesis induced in the rat by 2-acetylamino-fluorene. *Cancer Research*, **31**, 1506–12.

PITTS, J. D. & SIMMS, J. W. (1977). Permeability of junctions between animal cells. *Experimental Cell Research*, **104**, 153–63.

SCRIBNER, J. D. & SUSS, R. (1978). Tumour initiation and promotion. *International Review of Experimental Pathology*, **18**, 137–98.

SHOYAB, M. & TODARO, G. J. (1980). Specific high activity cell membrane receptors for biologically active phorbol and ingenol esters. *Nature (London)*, **288**, 451–5.

SUBAK-SHARPE, J. H., BURK, R. R. & PITTS, J. D. (1969). Metabolic cooperation between biochemically marked mammalian cells in tissue culture. *Journal of Cell Science*, **4**, 353–67.

TROSKO, J. E., CHANG, C. C., YOTTI, L. P. & CHU, E. H. Y. (1977). Effect of phorbol myristate acetate on the recovery of spontaneous and ultraviolet light-induced 6-thioguanine and ouabain-resistant Chinese hamster cells. *Cancer Research*, **37**, 188–93.

TROSKO, J. E., DAWSON, B., YOTTI, L. P. & CHANG, C. C. (1980). Saccharin may act as a tumour promoter by inhibiting metabolic cooperation between cells. *Nature (London)*, **285**, 109–10.

TROSKO, J. E., YOTTI, L. P., DAWSON, B. & CHANG, C. (1981). *In vitro* assay for tumour promoters. In *Short-Term Tests for Chemical Carcinogens*, ed. H. Stich. New York: Spring-Verlag, in press.

VAN DUUREN, B. L. (1969). Tumour-promoting agents in two-stage carcinogenesis. *Progress in Experimental Tumour Research*, **11**, 31–68.

VAN DUUREN, B. L., SIVAK, A., KATZ, C., SEIDMAN, I. & MELCHIONNE, S. (1975). The effect of aging and interval between primary and secondary treatment in two-stage carcinogenesis on mouse skin. *Cancer Research*, **35**, 502–5.

VERMA, A. K., SLAGA, T. J., WERTZ, P. N., MUELLER, G. C. & BOUTWELL, R. K. (1980). Inhibition of skin tumour promotion by retinoic acid and its metabolite 5,6-epoxyretinoic acid. *Cancer Research*, **40**, 2367–71.

WEINSTEIN, R. A., MERK, F. B. & ALROY, J. (1976). The structure and function of intercellular junctions in cancer. *Advances in Cancer Research*, **23**, 23–79.

WHITFIELD, J. F., MACMANUS, J. P. & GILLAN, D. J. (1973). Calcium-dependent stimulation by a phorbol ester (PMA) of thymic lymphoblast DNA synthesis and proliferation. *Journal of Cellular Physiology*, **78**, 355–68.

WOLPERT, L. (1978). Gap Junctions: channels for communication in development. In *Intercellular Junctions and Synapses*, ed. J. Feldman, N. Gilula & J. D. Pitts, pp. 83–96. London: Chapman and Hall.

WYNDER, E. L., HOFFMANN, D., McCOY, G. D., COHEN, L. A. & REDDY, B. S. (1978). Tumour promotion and cocarcinogenesis as related to man and his environment. In *Carcinogenesis: Mechanisms of Tumour Promotion and Cocarcinogenesis*, ed. T. J. Slaga, A. Sivak & R. K. Boutwell, pp. 59–77. New York: Raven Press.

YAMAGIWA, K. & ICHIKAWA, K. (1918). Experimental study of the pathogenesis of carcinoma. *Journal of Cancer Research*, **3**, 1–21.

YEE, A. G. & REVEL, J. P. (1978). Loss and reappearance of gap junctions in regenerating liver. *Journal of Cell Biology*, **78**, 554–64.

YOTTI, L. P., CHANG, C. C. & TROSKO, J. E. (1979). Elimination of metabolic cooperation in Chinese hamster cells by a tumour promoter. *Science*, **206**, 1089–91.

Tissue organisation and neoplasms

CHARLES ROWLATT

Imperial Cancer Research Fund, Lincoln's Inn Fields, London WC2A 3PX

INTRODUCTION

In this chapter the information which the variety of neoplasms possible in an organ can give about its tissue-organising mechanisms is examined. A general rigorous behavioural definition of neoplasms is derived which is compatible with both histopathological and experimental observations. The necessary behaviour arises from three broad classes of change: random alteration of genetic material, inappropriate activation of parts of the genome and insertion of new information in the genome. While behaviour resulting from the last two will be relatively uniform, random alteration will eventually be expressed in every relevant organising mechanism. As the structure of neoplasms follows from the residual behaviour of its cells, the range of neoplasms arising in each organ by this class of alteration reflects the key tissue organising mechanisms.

It is now generally recognised (e.g. Murray, 1908; Law, 1952; Klein & Klein, 1956; Strong, 1958; Cairns, 1975) that the process of tumour development involves cellular alteration leading to escape from regulation in the organism. When considering the biological status of a neoplasm at a particular time, two concepts must be distinguished: the basis for the actual structural and behavioural properties of that neoplasm, and the relative instability of these properties with the passage of time. Clinical malignancy is a very broad concept which includes both.

The basis for the actual properties of a neoplasm is the particular pattern of cell and tissue organising mechanisms which are altered at that time. By defining neoplasm in these terms we will see how structure and behaviour are a reflection of those mechanisms which are abnormal. Although single 'behavioural' abnormalities may provide the basis for particular criteria of clinical malignancy (e.g. invasiveness) in general the more classes of behaviour disturbed the more disorganised and clinically malignant is the tumour. The second concept, the risk of any part of the neoplasm acquiring further relevant properties, allows a biological distinction to be made between clinically benign neoplasms, which are all stable, and unstable neoplasms which are all already or potentially clinically malignant. The apparent paradox that not all stable tumours are clinically benign and not all clinically malignant tumours are unstable is avoided by separating malignancy from the idea of risk of further change.

In this chapter the change of properties with time is discussed briefly, as it affects choice of suitable material. The main discussion concerns the properties of the neoplasm at a particular moment.

DEFINITION OF NEOPLASM

Derivation

An authoritative histopathologist's definition is given by Willis (1967):

> A tumour is an abnormal mass of tissue, the growth of which exceeds and is uncoordinated with that of the normal tissues, and persists in the same excessive manner after cessation of the stimuli which evoked the change.

This essentially descriptive definition mentions causative stimuli but does not define the underlying process.

A limited definition of neoplasm can be derived from two general assumptions: that normal dividing cells can undergo spontaneous or induced mutation, and that organisms have a variety of factors which regulate the behaviour of the component cells.

'Normal cells' when grown in culture in adverse chemical or physical conditions reveal the presence of genetic variants resistant to the adverse conditions. These occur spontaneously both in bacterial (Lederberg & Lederberg, 1952) and mammalian cell systems (Sanford, 1965). Physical or chemical mutagens increase the yield. Although these mutant clones have escaped a common inhibitory effect, different intermediate stages, controlled by different genes, may have been changed in each clone (Meager, Ungkitchanukit & Hughes, 1976). Similar mutation is assumed to occur *in vivo*, either spontaneously (Burnet, 1974) or as the result of an external agent (Auerbach, 1978).

The cells in an organism are also regulated by extracellular extrinsic factors. Each cell is a member of some cell lineage, usually classified empirically from the terminally differentiated form. While the stability of each step of restriction and the final determination of cells in a lineage appears to be due to intracellular interactions (Gurdon, 1967; Davidson, 1976; Wessells, 1977) the particular state of differentiation in 'determined' cells is maintained by external stimuli (Sengel, 1976; Wessells, 1977). Many of these stimuli are derived from other differentiated cell types, of the same or different lineages. Sequential orderly embryonic development is achieved by temporal and spatial apposition of appropriate inducing and competent reacting tissues (Waddington, 1966; Sengel, 1976; Wessells, 1977).

On these assumptions neoplasms would form when mutations which avoid the normal restraining mechanisms in the organism occur in individual cells. However firm evidence (Stevens, 1967) suggests that epigenetic as well as genetic processes are involved in some cases. Therefore the basic phenomenon in cancer is one of stable behavioural change in cells and the mechanism need not be the same in every case. The definition of neoplasm proposed here follows Willis's pattern but includes both genetic and epigenetic mechanisms as alternatives providing the necessary abnormal behaviour.

A *neoplasm* is a mass of tissue generated by cells capable of division which have acquired either permanent expressible heritable change or stable epigenetic change so that the same or other cells no longer respond appropriately to one or more normal tissue organising stimuli, chemical or physical, intracellular or extracellular, in the organism in which it occurs.

The formation of neoplasms is the gross consequence of the neoplastic process. This process of *neoplasia* may be considered to start when the generating cell or cells become altered. Although we have specified the cellular process necessary to form a neoplasm, we do not assume that a detectable neoplasm is an inevitable consequence of the process. *Carcinogenesis* may be considered as the demonstration of neoplasia by the production of detectable neoplasms.

Discussion

This definition of neoplasm specifies the actual properties: provided the conditions are fulfilled, a neoplasm will have formed. Any transmissible stable alteration in behaviour so that there is escape from any tissue-organising mechanism which can cause a tissue mass will initiate neoplasia. If the alteration to genetic material is focal and random, we conclude that:

(i) The range of neoplasms found in any organ reflects the tissue-organising mechanisms in that organ.

(ii) A defect in every relevant organising mechanism will be represented, alone or in combination, in some neoplasms of that organ.

The alteration of neoplasms with time which allows some to acquire new and often more malignant properties is known as progression (Foulds, 1954). The stepwise nature of progression (Foulds, 1954; Rous & Beard, 1935; Greene, 1940) suggests that it is predominantly a consequence of genetic damage (Law, 1952; Strong, 1958) occurring in successive episodes, providing further selective advantage to subclones. We may therefore also conclude that:

(iii) When a neoplasm arises as a consequence of induced random damage, the higher the level of damage, the greater will be the proportion of clinically malignant to clinically benign neoplasms.

The most important malignant behavioural characters of neoplasms are whether cells invade locally and whether fragments separate to form metastatic deposits. There is a variety of ways in which cells or groups of cells may be distributed through the body, which depend both on the anatomical location of the primary neoplasm and on the altered behaviour of the cells. Cells with appropriate behaviour will tend to follow the simplest route (Willis, 1967; Poste, Doll, Hart & Fidler, 1980).

The histopathologist supports the term malignant with a description of this abnormal structure, behaviour or distribution of cells of the tissue of origin. Direct evidence for an unstable neoplasm is given by foci of altered cells within an otherwise uniform specimen. He may use the term potentially malignant when he sees evidence of instability in tissues without gross abnormality. Evidence of this increased risk may be an increased number of cells in division or evidence of increased instability in intracellular mechanisms (Comings, 1973; Busch, 1978) suggested by nuclear irregularity or abnormal chromosomes (Benedict, 1977). Unstable is the appropriate term when these abnormal cytological indications are found in exfoliate cells, or in lesions such as carcinoma-in-situ in which overt malignant behaviour is not found.

APPROPRIATE MATERIAL

However, not all the alterations are based on focal and random change and evidence exists for a variety of classes of change. The insertion of new genetic material by, for example viruses (Rapp, 1976; Crawford, 1980; Duesberg, 1980; Sharp, 1980) producing particular behaviour in the cell which gives rise to neoplasms is well established.

Evidence for inappropriate activation of part of the genome has been drawn from several fields. The direct epigenetic effect by transposing rapidly differentiating tissue to an alternative site and disrupting the process to produce neoplasms (Stevens, 1967) has been mentioned above. That cells from these neoplasms can apparently differentiate normally has been shown by reintroducing them into an earlier stage of development and obtaining normal development (Illmensee & Mintz, 1976). Traditional histological descriptions using the term 'dedifferentiated' implied expression of inappropriately early stages of differentiation, and this has been confirmed recently by the demonstration of foetal antigens on the surface of neoplastic cells (Uriel, 1979). However this can also occur in certain inflammatory conditions, suggesting that it is a general reaction. Abnormalities of chromatin structure and of the karyotype indicate that instability of the genetic apparatus is an important process in malignancy (Comings, 1973; Benedict, 1977; Busch, 1978).

Whether random focal alterations to the genetic material occur with physical and chemical carcinogens is still debated, although information about the interactions between DNA and both physical (Maher & McCormick, 1979) and a variety of chemical carcinogens (Miller, 1978; Coombs, 1980) is accumulating. It has been maintained that species lifespan is governed by the effectiveness of the correction of random genetic errors (Burnet, 1974; Cutler, 1975) and that stochastic errors occur at all levels of organisation during ageing. If cellular damage accumulated during ageing is focal and random (irrespective of whether it is primary or secondary) neoplasms arising from ageing alone would be predicted and should, on the present hypothesis, tend to be benign and distributed widely among organs. In contrast, neoplasms arising from specific carcinogens would tend to be malignant and their distribution would reflect the physical, chemical or biological path of the carcinogen through the organism.

Each class of change can be expected to have characteristic features. Viral tumours, in which a specific extrinsic programme is introduced, will show this similar behaviour, and may often express common antigens. Tumours arising by inappropriate expression will also tend to show a limited range of behaviour. Only in the case of random alteration can the full range of possible behavioural abnormalities be seen. When these changes occur frequently, multiple changes in behaviour may be superimposed producing very complex behaviour, possibly associated with clinical malignancy. It is the isolated random changes at low frequency which may be expected to demonstrate the loss of particular organising mechanisms most clearly. This suggests that the smallest neoplasms induced by the most gentle methods will provide the best information about tissue-organising mechanims.

HISTOPATHOLOGICAL EXAMPLE

Statement

The general hypothesis that tissue-organising mechanisms and neoplasms of those tissues are complementary phenomena has a predictive value. If a model of these mechanisms in an organ is available and if the proposed definition of neoplasm is valid, the simple naturally occurring neoplasms in the organ may be predicted if we assume that single mechanisms become disturbed.

In a very simple model of skin (for discussion Sengel, 1976; Foulds, 1975), renewal occurs continuously from dividing cells in the basal layer. Cells may either differentiate and be shed from the surface without further division, or they may remain as basal cells, retaining their capacity for division and properties which ensure that this division only occurs near stromal tissues. In this model (Fig. 1) cell division is controlled by (i) a negative feedback mechanism from differentiated skin cells (ii) a constraint limiting division to the basal zone, and (iii) systemic factors modulating diurnal rhythm etc. (Bullough, 1962). Stromal influences (Sengel, 1976) are not included in this simplified model.

Let us assume (A) that differentiation in a clone of dividing cells is permanently altered so that the negative feedback is lost (Fig. 1a). Undifferentiated skin cells will accumulate, and more division will occur. However as division can still only occur in the basal layer it is still limited by the rate and extent to which the basal layer may expand. The resulting growth is steady with basal cell compression; the risk of further mutation remains similar to that in normal skin.

Now let us assume (B) that a clone of basal cells becomes per-

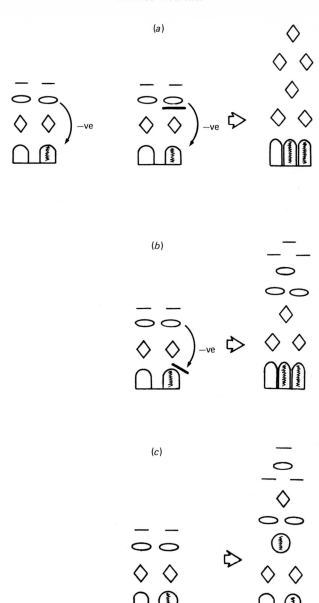

Fig. 1. Diagrams to show the structural effects of heritable defects in three parts of the homeostatic model for skin outlined in the text. The normal balance is represented at the top left, with cells in or between division confined to the basal zone.and cells in successive stages of differentiation forming strata above. (*a*) Defect A, a block in differentiation. (*b*) Defect B, insensitivity to the negative feedback control of division. (*c*) Defect C, loss of the constraint to divide in the basal zone.

manently insensitive to the negative feedback (Fig. 1*b*). As with (A) some extra division will occur, also limited to the basal zone. Differentiated cells will continue to be shed from the surface; the size and growth of any resulting neoplasm (B) will depend on the discrepancy between cell input and cell loss.

Finally, let us assume (C) that a clone of cells becomes permanently released from the constraint of dividing in the basal layer near the stroma (Fig. 1*c*). Dividing cells will be found among the differentiating cells. The number of these abnormal cells will remain small for a long time because the negative feedback and systemic controls are unaffected. The orientation of the skin layers will be disturbed by replenishment from the middle strata. Any increase in the rate of division will cause an increase in abnormal cell mass, although much of this may be shed in the continuing cell flux. If adjacent normal basal cells repopulate the basal layer, they may *displace all abnormal cells* and the incipient neoplasm (C_1) will disappear as a direct consequence of the initial lesion.

Alternatively, (Neoplasm C_2), the cell mass may increase, retaining terminally differentiated cells. Division is no longer confined to a single layer and the number of dividing cells will increase exponentially, although slowly at first. Loss of the negative feedback mechanism will increase the rate of cell division, bringing further increased risk of mutation and progression. Change allowing extension into the underlying tissues may occur at any stage and, depending on the secondary properties acquired during progression, will provide foci capable of metastasis to form secondary deposits.

The structure and behaviour of these neoplasms is very similar to those arising in skin (Table 1, Fig. 2). Neoplasm A resembles basal cell carcinoma (Fig. 2*a*), neoplasm B resembles benign papillomas (Fig. 2*b*), neoplasm C_1 resembles intra-epidermal carcinoma (Fig. 2*c*) which may fail to develop and neoplasm C_2 resembles the varieties of squamous cell carcinoma (Fig. 2*d* & 2*e*).

Discussion

Skin was chosen because carcinoma of the epidermis falls into two well-recognised classes. The model probably does show how abnormalities in single parts of a complex growth control mechanism may account for some of their properties. But as the basis for each behavioural mechanism is complex, tumours with the same structure and behaviour need not necessarily have identical cytological (Weinstein,

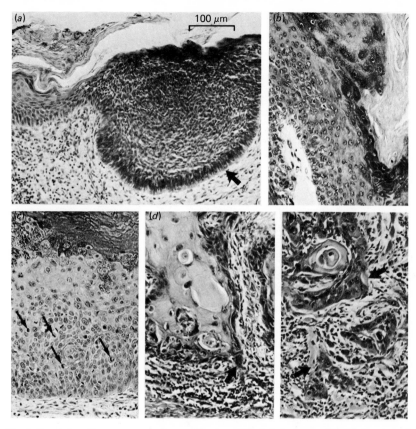

Fig. 2. Examples of common neoplasms of human skin and their postulated behavioural defects; standard processing, haematoxylin and eosin stain. (*a*) Small basal cell carcinoma with adjacent normal skin on left (TRC 2252); note accumulation of immature cells, and tight packing ('palisading') in the basal layer (arrow): possibly defect A. (*b*) Part of a verruca senilis (senile wart) (TRC 100): note presence of all layers in excess: possibly defect B. (*c*) Region of intra-epidermal carcinoma (TRC 1569): note the skin thickening and the abnormal distribution of dividing cells (arrows), although all stages of differentiation are present: possible defect C. (*d*) Part of a squamous cell carcinoma (TRC 4949): most stages of differentiation can be recognised although the pattern is abnormal. Penetration of the basement membrane by an isolated cord of cells is occurring (arrow): multiple defects. (*e*) Part of an invasive squamous cell carcinoma (TRC 1543): note the separation of degenerate muscle cells (arrows) by clumps of cells which retain differentiated characters: multiple defects.

Table 1. *Examples of well-recognised forms of neoplasms of skin[a] with an assessment of their growth behaviour*

Name	Probable origin	Characteristic structure	Characteristic growth	Characteristic spread
seborrhoeic keratosis (verruca senilis, hard pigmented mole)	epidermis	papilloma in which a thickened, hyperkeratotic epithelium lines elongated dermal papillae	normally benign with intradermal growth limited by palisaded basal cells	malignant spread very rare
basal cell carcinoma (rodent ulcer)	epidermis or its adnexa	large or small ramifying basophilic tumour cell masses surrounded by a 'palisade' of basal cells	slow continuous direct growth	distant spread very rare indeed
intra-epidermal carcinoma (*carcinoma-in-situ*)	epidermis	areas of thickened epithelium, keratinised but with abnormal stratification and cell division at all levels	no penetration of the basement membrane; when this occurs it becomes a squamous cell carcinoma	
squamous cell carcinoma	epidermis or its appendages	tumour of 'prickle cells', with varying degrees of differentiation, spreading into dermis	growth varies from *slow* usually with differentiation to *fast* with little differentiation and basal separation of small groups of cells	differentiated forms tend to show distant metastasis late undifferentiated forms show early lymphatic spread

[a]summarised from Ashley (1978).

Merk & Alroy, 1976) or biochemical (van Potter, 1973; Peterson, 1976; Pitot, 1977) defects. Variants within the major classes should represent defects in additional control mechanisms.

The model itself is grossly oversimplified. Although the negative feedback postulated could be mediated by locally acting chalones (Bullough, 1975; Rytömaa, 1976) these have not been demonstrated unequivocally (Montagna & Parakkal, 1974). Here the actual mechanism is not specified, and could, for example, be physical. The requirement that cells divide in and only on the basal zone may be too rigorous as divisions occur in a broader band (Foulds, 1975). The exclusion both of other cell types within the epidermis and of all dermal effects also oversimplifies the model. Dermal tissues in particular are known to affect skin differentiation (Bullough, 1975; Sengel, 1976), probably by positive factors.

When the predicted appearances are compared with those of skin neoplasms and other lesions, the importance of dermal–epidermal interactions is emphasised. Disturbances of both elements are described (Ashley, 1978; Lever & Schaumburg-Lever, 1975) in both basal cell and squamous cell neoplasms and other lesions. Loss of this interaction in invasive squamous cell carcinoma reflects this interdependence.

This model of the tissue interactions is clearly too simple an approximation of the cellular interactions, and must be improved. However, it does have two interesting implications.

Firstly, if an agent which effects the negative feedback mechanism is isolated (such as chalone: Bullough, 1962) it will control basal cell carcinomas which arise from an absent factor but not squamous cell carcinomas which arise by insensitivity to the factor. Therapeutic trials of such agents should be planned to distinguish between these mechanisms. An intriguing generalisation is that whenever a simple neoplasm shows blocked differentiation a replaceable negative feedback factor may be missing.

Secondly, loss of polarisation of the stratified epithelium is only one (necessary) criterion for recognising carcinoma-in-situ; nuclear abnormality is also important. The model emphasises that one behavioural defect could have important consequences as a self-limiting process for the lesion (Walton, 1976; Kinlen & Spriggs, 1978).

In other systems similar models have been proposed. By characterising the surface differentiation antigens in a range of leukaemias it has been possible to allocate some of them to particular stages in differentiation (Greaves, Janossy, Francis & Minowada, 1978). In these cases systemic chalones may affect the total cell mass (Rytömaa, 1976). One classification of non-Hodgkin lymphomas using similar immunological techniques suggests that positive antigen stimulation of the tissue, without the negative feedback of terminal differentiation, generates the neoplasms (Habeshaw, 1979).

CONCLUSION

The purpose of this chapter has been to indicate the ways in which the failure of normal functional integration of cells in tissues may give rise to neoplasms. The whole range of neoplasms arising in any tissue reflects the range of possible interactions which may be involved. Provided that it is appreciated both that the proportions of particular types of neoplasm in an organ may be biased by the process of induction, and also that particular defects in behaviour may present in random combinations as well as singly, useful classes of neoplasms can be selected. Rarer forms may provide important information. It should then be possible to show how single or multiple defects in a homeo-static model of the organ generate these neoplasms.

REFERENCES

ASHLEY, D. J. B. (Ed.) (1978). *Evan's Histological Appearances of Tumours*, 3rd edn. Edinburgh: Churchill Livingstone.

AUERBACH, C. (1978). A pilgrim's progress through mutation research. *Perspectives in Biology and Medicine*, **21**, 319–34.

BENEDICT, W. (1977). The importance of chromosomal changes in the expression of malignancy. In *The Molecular Biology of the Mammalian Genetic Apparatus*, vol. 2, ed. P. Ts'o, pp. 229–39. North Holland: Elsevier.

BULLOUGH, W. S. (1962). The control of mitotic activity in adult mammalian · tissues. *Biological Reviews*, **37**, 307–42.

BULLOUGH, W. S. (1975). Mitotic control in adult mammalian tissue. *Biological Reviews*, **50**, 99–127.

BURNET, M. (1974). *Intrinsic Mutagenesis: A Genetic Approach to Ageing*. Lancaster: MTP.

BUSCH, H. (1978). The relation of gene control mechanisms to cancer. *Zeitschrift für Krebsforschung*, **92**, 123–35.

CAIRNS, J. (1975). Mutation selection and the natural history of cancer. *Nature*, **255**, 197–200.

COMINGS, D. E. (1973). A general theory of carcinogenesis. *Proceedings of the National Academy of Sciences, USA*, **70**, 3324–8.

COOMBS, M. M. (1980). Chemical carcinogenesis: a view at the end of the first half-century. *Journal of Pathology*, **130**, 117–46.

CRAWFORD, L. V. (1980). Transforming genes of DNA tumor viruses. *Cold Spring Harbor Symposia on Quantitative Biology*, **44**, 9–11.

CUTLER, R. G. (1975). Evolution of human longevity and the genetic complexity governing ageing rate. *Proceedings of the National Academy of Sciences, USA*, **72**, 4664–8.

DAVIDSON, E. H. (1976). *Gene Activity in Early Development*, 2nd edn. New York: Academic Press.

DUESBERG, P. H. (1980). Transforming genes of retroviruses. *Cold Spring Harbor Symposia on Quantitative Biology*, **44**, 13–29.

FOULDS, L. (1954). The experimental study of tumour progression. *Cancer Research*, **14**, 327–39.

FOULDS, L. (1975). *Neoplastic Development 2*. London: Academic Press.

GREAVES, M., JANOSSY, G., FRANCIS, G. & MINOWADA, J. (1978). Membrane phenotypes of human leukemic cells and leukemic cell lines: clinical correlates and biological implications. In *Differentiation of Normal and Neoplastic Hemopoietic Cells*, ed. B. Clarkson, P. A. Marks & J. E. Till, pp. 823–41. Cold Spring Harbor Laboratory.

GREENE, H. S. N. (1940). Familial mammary tumors in the rabbit IV. The evolution of autonomy in the course of tumor development as indicated by transplantation experiments. *Journal of Experimental Medicine*, **71**, 305–24.

GURDON, J. B. (1967). On the origin and persistence of a cytoplasmic state inducing nuclear DNA synthesis in frogs' eggs. *Proceedings of the National Academy of Sciences, USA*, **58**, 545–52.

HABESHAW, J. A. (1979). Hypothesis: non-Hodgkin lymphomas are abnormal immune responses. *Cancer Immunology and Immunotherapy*, **7**, 37–42.

ILLMENSEE, K. & MINTZ, B. (1976). Totipotency and normal differentiation of single teratocarcinoma cells cloned by injection into blastocysts. *Proceedings of the National Academy of Sciences, USA*, **73**, 549–53.

KINLEN, L. J. & SPRIGGS, A. I. (1978). Women with positive cervical smears but without surgical intervention. *Lancet*, ii, 463–5.

KLEIN, G. & KLEIN, E. (1956). Conversion of solid neoplasms into ascites tumours. *Annals of New York Academy of Sciences*, **63** Art. 5, 640–61.

LAW, L. W. (1952). Origin of the resistance of leukaemic cells to folic acid antagonists. *Nature*, **169**, 628–9.

LEDERBERG, J. & LEDERBERG, E. M. (1952). Replica plating and indirect selection of bacterial mutants. *Journal of Bacteriology*, **63**, 399–406.

LEVER, W. F. & SCHAUMBURG-LEVER, G. (1975). *Histopathology of the Skin*, Philadelphia: J. B. Lippincott Company.

MAHER, V. M. & McCORMICK, J. J. (1979). DNA repair and carcinogenesis. In *Chemical Carcinogenesis and DNA*, vol. II, ed. P. L. Grover, pp. 133–58. Florida: CRC Press Inc.

MEAGER, A., UNGKITCHANUKIT, A. & HUGHES, R. C. (1976). Variants of hamster fibroblasts resistant to Ricinus communis toxin (Ricin). *Biochemical Journal*, **154**, 113–24.

MILLER, E. C. (1978). Some current perspectives on chemical carcinogenesis in humans and experimental animals: presidential address. *Cancer Research*, **38**, 1479–96.

MONTAGNA, W. & PARAKKAL, P. F. (1974). *The Structure and Function of Skin.* New York: Academic Press.

MURRAY, J. A. (1908). Spontaneous cancer in the mouse. Histology, metastasis, transplantability and relations of malignant new growths to spontaneously affected animals. *Imperial Cancer Research Fund Scientific Report*, 3, 69–115.

PETERSON, J. A. (1976). Clonal variation in albumin messenger RNA activity in hepatoma cells. *Proceedings of the National Academy of Sciences, USA*, 73, 2056–60.

PITOT, H. C. (1977). Natural history of neoplasia. *American Journal of Pathology*, 89, 402–11.

POSTE, G., DOLL, J., HART, I. R. & FIDLER, I. J. (1980). In vitro selection of murine B16 melanoma variants with enhanced tissue-invasive properties. *Cancer Research*, 40, 1636–44.

POTTER, R. VAN (1973). Biochemistry of cancer. In *Cancer Medicine*, ed. J. F. Holland & E. Frei, pp. 178–92. Philadelphia: Lea & Febiger.

RAPP, F. (1976). Viruses as an etiologic factor in cancer. *Seminars in Oncology*, 3, 49–63.

ROUS, P. & BEARD, J. S. (1935). The progression to carcinoma of virus-induced rabbit papillomas. *Journal of Experimental Medicine*, 62, 523–48.

RYTÖMAA, T. (1976). The chalone concept. *International Review of Experimental Pathology*, 16, 155–206.

SANFORD, K. K. (1965). Malignant transformation of cells in vitro. *International Review of Cytology*, 18, 249–311.

SENGEL, P. (1976). *Morphogenesis of Skin.* Cambridge University Press.

SHARP, P. A. (1980). Summary: Molecular biology of viral oncogenes. *Cold Spring Harbor Symposia on Quantitative Biology*, 44, 1305–22.

STEVENS, L. C. (1967). Origin of testicular teratomas from primordial germ cells in mice. *Journal of the National Cancer Institute*, 38, 549–52.

STRONG, L. C. (1958). Genetic concept for the origin of cancer: historical review. *Annals of New York Academy of Sciences*, 71 Art. 6, 810–38.

URIEL, J. (1979). Retrodifferentiation and the fetal patterns of gene expression in cancer, *Advances in Cancer Research*, 29, 127–74.

WADDINGTON, C. H. (1966). Mendel and the study of development. *Proceedings of the Royal Society (London)*, B164, 219–29.

WALTON REPORT (1976). Cervical cancer screening programs: Epidemiology and natural history of carcinoma of the cervix. *Canadian Medical Association Journal*, 114, 1003–12.

WEINSTEIN, R. S., MERK, F. B. & ALROY, J. (1976). The structure and function of intercellular junctions in cancer. *Advances in Cancer Research*, 23, 23–89.

WESSELLS, N. K. (1977). *Tissue Interactions and Development.* Menlo Park, California: W. A. Benjamin Inc.

WILLIS, R. A. (1967). *Pathology of Tumours.* London: Butterworths.

Tumour formation: the concept of tissue (stroma-epithelium) regulatory dysfunction

GISELE M. HODGES

Tissue Interaction Laboratory, Imperial Cancer Research Fund, London WC2A 3PX, UK

1. INTRODUCTION

Tumour formation is a pathological phenomenon reported for nearly all animal phyla (Huxley, 1958) and expressed probably in most tissues or organs with constituent cells capable of division; and, is one of the major concerns of present-day medicine. The purpose of this article is not to review the entire field of tumour genesis but rather, to consider within the range of epigenetic mechanisms implicated in the activation (and repression) of tumour formation the involvement of stromal-epithelial regulatory mechanisms. Such an approach may be rationalized on the basis that the great majority of spontaneous human and animal tumours arise from epithelia. The more general considerations relating to the terminology, aetiology and processus of carcinogenesis have been the subject of many reviews and the reader is referred to a selection of presentations for coverage of these areas (see for example Smithers, 1962; Prehn, 1972, 1976; Willis, 1973; Epstein, 1974; German, 1974; Pierce, 1974; Becker, 1975; Hecker, 1976; Searle, 1976; Ahuja & Anders, 1977; Cahan & Schottenfeld, 1977; Koprowski, 1977; Knudson, 1977; Pitot, 1977a, b; Temin, 1977; Clark, Griffen & Shaw, 1979; Klein, 1979; Coombs, 1980; Farber & Cameron, 1980).

2. STROMAL-EPITHELIAL INTERACTION

A. Basic concepts

Many of the complex epithelio-mesenchymal structured organs which arise during later embryogenesis are the consequence of tissue interactions between mesenchymal cells derived from the mesoderm and an epithelium derived from any one of the three germ layers. During the past thirty years a large body of literature has accumulated on the stromal-epithelial interactive processes from which have emerged basic concepts on the regulation of such phenomena as proliferation, differentiation, death, and migration of cells leading to the development of structures of different organs. There remains, however, a lack of information on the substances participating in these interactive processes. Nevertheless, it is possible to classify, at least in part, stromal-epithelial regulatory function according to mechanisms of action during definite stages of embryological development. This section is intended to systemize the information as a whole, but without specific discussion, for which purpose reference may be made to a series of reviews in the literature including Fleischmajer & Billingham (1968), Thomas (1970), Slavkin & Greulich (1975), Sengel (1976), Karkinen-Jääskeläinen *et al.* (1977), Lash & Burger (1977), Poste & Nicolson (1977), Wessells (1977), Cunha *et al.* (1980). These document that orderly embryonic differentiation is dependent both on precisely timed and localized cellular interactions and on appropriate temporal and spatial application of inducing and effector tissues, with that between stroma and epithelium being central to normal organogenesis of many embryonic rudiments.

Isolated culture or coculture in various combinations of separated epithelial and stromal tissues, under *in vitro* or *in vivo* conditions, has allowed analysis of the interactive effects of either cell type (epithelial or stromal) on the growth, function or development of the other, and provided evidence of locally generated inductive factors emanating from one or other. From such analysis of combinations between stromal and epithelial tissues of different chronological age (hetero-chronic), regional origin (heterotropic), species (heterospecific), genetic (heterogenetic) or pathological (heteropathic – see Section 2C) constitution have been identified various inductive influences of stromal cells. This has led stroma to be seen as possessing certain qualities that generate specific genetic expression and regional specialization from the epithelial repertoire, with the epithelium seen as containing all the structures common to it.

In the absence of stroma, epithelia show in general a limited capacity for survival and cytodifferentiation, failure to undergo morphogenesis, and loss of histotypic organization. Following exposure to stroma, an input of stromal signals can be distinguished on the basis of the output or response of the epithelium. Such stromal effects can be broadly categorized into:

(a) A growth-promoting activity whose interplay with other epithelial control mechanisms results in a stabilization of mitotic activity in the epithelium: the informational signal is seen as permissive and non-specific.

(*b*) A relatively non-specific morphogenetic signal which can act across species lines and permit phenotypic expression, i.e. cellular and functional differentiation and morphogenesis of epithelia: this is seen as a permissive effect in that the signal is understood and expressed by heterotypic epithelia but the response is specified by the responding epithelium.

(*c*) A specific morphogenetic signal that contains determinate information which controls the local characteristics of epithelia, and which can redirect the determined state of an epithelium: this type of interaction presumes some control of selectivity in gene usage and the effect is considered as instructive.

Epithelial competence, or response, to inductive cues fom the stroma can be variable in relation to type and age of the epithelium and may range from a general to a strict mesenchymal specificity though the mechanistic basis for this remains to be established. The stroma may also act as the target tissue for certain hormones or vitamins. These substances exert their action through a tissue interactive process with the stroma complementing or mediating their effect on the epithelium. Furthermore, variation in the quantitative mass of stroma can contribute to altered epithelial morphogenesis and cytodifferentiation. That cells of a common embryonic type must attain a certain critical mass before they can effect their developmental capabilities or influence, and that this critical mass may vary according to the tissue type, has been well recognized in relation to early stages of embryogenesis (Deuchar, 1975).

Stromal inductive effects have been designated, in general terms, as either *instructive* (or directive) when a choice of determined state is made, i.e. when the induction affects the state of determination by directing the effector tissue into a pathway of development which it would not have otherwise followed; or, *permissive* when expression of the prospective developmental fate of a tissue is allowed, i.e. when the induction affects the extent to which an already determined cell can express that state by differentiation along its particular pathway of development. The permissive role has also been seen as one of metabolic complementarity between cells of different type. Many studies of stromal-epithelial interaction have tended to concentrate on the inductive action of the stromal substratum on the epithelial cover. As a consequence, there has been a tendency to consider such inductions as unidirectional processes with the stroma seen as the inductor (active) tissue and the epithelium as the responsive or effector (passive) tissue. There is, however, clear evidence that for at least a number of embryonic organs, inductive interactions are far more complex with involvement of varying sequences of reciprocal permissive or restrictive actions of epithelial and stromal tissues one on the other (see Thomas, 1970; Sengel, 1976). The developmental response of interacting tissues may be specified, therefore, by either the responding or the inducing tissue depending in part on the level of the inductive stimulus and on the responsiveness of interacting tissue, both of which may vary temporally or spatially. Though, in many cases, which tissue takes the initial lead in the morphogenetic interaction has yet to be fully defined.

The basic mechanism(s) of stromal-epithelial inductive interactions and their molecular basis remains still largely conjectural. Three alternative mechanisms, originally proposed by Grobstein (1955), suggested either a long-range transfer of labile informational molecule(s) between the tissues; direct cell–cell mediation of the inductive signal between heterotypic populations; or, a matrix-mediated interaction at the interface of the inductor and responding tissues. More recently, these proposals have been incorporated into two projected major systems of interaction (Saxén, 1977, 1979) viz. long-range transmission of inductive signals over distances of the order of 50 000 nm (which would include the earlier concepts of free diffusion and of matrix interaction); and, short-range transmission over an interspace of some 5 nm (which would include concepts of short-range diffusion, interaction of surface-associated molecules, and transfer of signal molecules through intercellular channels). Present theory suggests that probably no single mechanism is involved, and that certain aspects of morphogenesis may be dependent upon different mechanisms with possibly different modes of interaction mediated by different types of signal molecules (Saxén, 1979).

Over the last decade, particular attention has focussed on the structure and function of the basal lamina located at the interface between interacting tissues (Kefalides, 1979). This has evolved from the increasing evidence of basal lamina involvement in the regulation of developmental events, with various constituents of this extracellular matrix, especially collagen and glycosaminoglycans, clearly important factors for the morphogenesis and cytodifferentiation of different epithelia (Slavkin & Greulich, 1975; Hay, 1977; Kefalides, 1978). While still to be fully elucidated, synthesis of the basal lamina matrix would be seen as a function both of the stromal and epithelial tissues; though, the role of each tissue in dictating the involvement of the other in the biosynthetic process remains to be defined.

By virtue of its anatomic position it is not unexpected that the basal lamina should play a role of some significance in the tissue interactive inductive processes involved in regulation of biological behaviour within stromal and epithelial cell populations of tissues. Among various functions assigned to the basal lamina, of interest to this review is its projected role as a permissive substrate supporting and influencing the appropriate location and orientation of overlying epithelia. Adoption of particular cellular configurations have been postulated as determinants of cell sensitivity to given inductive regulatory signals, and as a consequence, of significance in the morphological and functional expression of epithelia (Folkman, 1977; Gospodarowicz *et al.*, 1978). A second projected role is that of a matrix through which the informational signal(s) between stromal and epithelial tissues must pass and where the various extracellular constituents of the matrix may act as mediators of particular inductive interactions. Grobstein (1975) has surmised that individual matrix components may not, alone, necessarily act as a complete lexicon of morphogenetic determinants operative for mediating or complementing activation of gene expression. Rather, an interplay of more than one matrix molecular species may be an essential requirement to influence the processus of differentiation, with these matrix constituents interacting both between themselves and with the cellular systems to generate new levels of order and information.

B. *In adult tissues*

A tendency has been to regard the stroma of adult tissues as an essentially inactive matrix involved primarily in maintenance of the histotypic organization of organs, and in provision of vascular and lymphatic supplies required for nutrition of the epithelium. Nevertheless, that the stroma should retain an active role in the adult organism may be anticipated. Many of the epithelia of adult organs are typical steady-state tissues maintaining throughout adult life a proliferative compartment and producing a flow of new cells where cell differentiation is obviously a continuous process, and where regulatory mechanisms, both intrinsic and extrinsic, should be operative.

Most stroma contain a basic complex of cellular elements (including fibroblasts, endothelial cells, nerve, and cells within the vascular system); matrix fibres (including collagen and elastic fibres); and an extracellular matrix composed of various macromolecules (including glycosaminoglycans and glycoproteins) (Mathews, 1975), these stromal constituents varying in amount and type from tissue to tissue and in different stages of development or in different pathological circumstance. Among the various subsets of factors present in adult stroma are several (viz. fibroblasts, matrix fibres and matrix proteins) established in embryonic mesenchymal tissues as being of significance in relation to epithelial proliferation, differentiation and morphogenesis during normal embryonic development.

Evidence has slowly, but progressively, accumulated over the past thirty years that, at least, certain of the epigenetic tissue regulatory mechanisms established prenatally may persist and operate in the maintenance of normal cellular growth and differentiation in adult tissues (Fleischmajer & Billingham, 1968; Tarin, 1972; Sengel, 1976; Cunha *et al.*, 1980) (Table 1). The data, though limited, suggests clearly that stroma, from neonate or adult tissues, as that from prenatal tissues, can display both permissive or instructive influences on epithelia; and, that the specificity of stromal action and the competence of epithelia to respond may vary from tissue to tissue. Furthermore, attention has been directed to the potential involvement of inductive interactive influences between stroma and epithelium in the cyclic morphologic and functional changes manifested by different organs of the reproductive system.

Permissive interactions have now been defined in a range of adult organs, including vagina, mammary glands, prostate, endometrium and skin (see Table 1), with stroma clearly effecting DNA synthesis, proliferative activity, and maintenance of organization and function in homotypic epithelia. As in embryonic tissue, the stroma of adult organs may also function in the hormonal response of epithelia with involvement of an interactive process dependent on the particular properties of the epithelium and stroma. The failure of isolated adult hormone-sensitive epithelia to respond to trophic hormones has become well established (Table 1), with the observation that such epithelia respond in characteristic manner only when in presence of an appropriate stroma.

That stroma of the adult organism may signal instructive information capable of reprogramming the phenotypic expression of heterotypic epithelium remains, however, to be fully elucidated. Never-

Table 1. *Role of stroma in expression of epithelial phenotype – a survey of some adult systems*

Stroma (Inducting tissue)	Epithelium (responding tissue)	Epithelial response	Reference
—	Mammary (mouse)	Unresponsive to trophic hormonal conditions: fails to organize into tubules or to attain normal secretory activity	Lasfargues (1957a, b)
Mammary (mouse)	Mammary (mouse)	Histiotypic organization	Herrman (1960)
Mammary + adipose tissue (mouse)	Mammary (mouse)	Histiotypic organization; milk fat secretory activity	Franks & Barton (1960), Franks *et al.* (1970)
Cornea	Cornea	Normal metabolic activity	Franks & Barton (1960)
—	Prostate (mouse; human)	Loss of differentiation; no ^{3}H-Tdr or amino acid incorporation; unresponsive to testosterone	Cunha *et al.* (1980)
Prostate (mouse)	Prostate (mouse)	Cytodifferentiation; testosterone stimulation	Briggaman & Wheeler (1968)
Prostate (mouse)	Urogenital sinus (embryonic mouse)	Glandular morphogenesis in presence of androgens	
—	Epidermis (human)	Degeneration; loss of ability to incorporate ^{3}H-Tdr	
Dermis (human); frozen-thawed dermis heterologous or heterospecific (guinea pig) dermis	Epidermis (human)	Epidermal organization retained; normal proliferative activity	

Table 1. (*cont.*)

Stroma (Inducting tissue)	Epithelium (responding tissue)	Epithelial response	Reference
Dermis – sole (guinea pig)	Skin – ear: trunk (guinea pig)	Morphological characteristics of sole epidermis expressed	Billingham & Silvers (1968)
Dermis – sole (guinea pig)	Oesophagus: tongue (guinea pig)	Characteristics of lingual or oesophageal epithelium retained	
Dermis – ear (guinea pig)	Skin – sole: trunk (guinea pig)	Morphological characteristics of ear epidermis expressed	
Dermis – ear (guinea pig)	Tongue (guinea pig): oesophagus (guinea pig/hamster) cheek-pouch (hamster)	Characteristics of lingual, oesophageal or cheek pouch epithelium retained	
Dermis – trunk (guinea pig)	Skin – sole: ear: oesophagus: tongue	Morphological characteristics of trunk epidermis expressed	
—	Vagina (mouse)	Unresponsive to oestradiol: decrease in proliferative activity: failure to keratinize	Flaxman *et al.* (1973)
—	Urothelium (rat)	Abnormal cytodifferentiation; no basal lamina formed	Hodges *et al.* (1976, 1977)
Bladder	Urothelium (rat)	Characteristic urothelial pattern of cytodifferentiation; basal lamina present at tissue interface	Hodges *et al.* (1976, 1977)

—	Endometrium (human)	Unresponsive to progesterone or oestradiol	Kirk et al. (1978)
Vagina (mouse) Cervix (mouse)	Uterine (neonate mouse)	Vaginal or cervical morphogenesis: possible functional differentiation	Cunha (1976), Cunha & Lung (1979)
Urogenital sinus mesenchyme (embryonic mouse)	Urothelium (mouse)	Expression of glandular phenotype of prostate: duct-aciner architecture: loss of alkaline phosphatase bladder marker: acquisition of non-specific estrase prostate-associated epithelial marker	Cunha et al. (1980)
Urogenital sinus mesenchyme (embryonic wild-type mouse)	Urothelium (androgen-insensitive from Tfm/Y mouse)	Formation of prostate-like acini: secretory activity: probable expression of androgen receptor activity	Cunha et al. (1980)
Gland-free mammary fat pads (mouse)	Mammary (embryonic mouse)	Morphogenesis	Sakakura et al. (1979a)
Gland-free mammary fat pads (mouse)	Lung Pancreas Salivary gland	No morphogenetic response No morphogenetic response	Sakakura et al. (1979a) Sakakura et al. (1979a)
—	Trachea (rat)	Loss of differentiated characters; fail to proliferate; no basal lamina formed	Heckman et al. (1978), Terzaghi & Klein-Szanto (1980)
Frozen-thawed trachea; bladder; oesophagus; small intestine (rat)		Characteristic mucocilliary differentiation; formation of basal lamina	

theless, there has emerged from those studies currently available, the view that in the adult, as in the prenatal tissue, stroma may control expression of basic epithelial architecture, histiotype, and functional activity, though expression of epithelial regional specificity may be a function of the epithelium itself (see Billingham & Silvers, 1968; Sengel, 1976; Cunha *et al.*, 1980). Such concepts have evolved from attempts to alter phenotypic expression of adult epithelia through association with heterotypic stromal inductors as assayed for epithelia of skin, tongue, oesophagus, cheek pouch (Billingham & Silvers, 1968, 1971); uterus and urothelium (Cunha & Lung, 1979; Cunha *et al.*, 1980) (Table 1). From this and other limited data (see section 2C) have been surmised the competence of adult epithelia to change, under the influences of given inductive signals, their potential for particular genomic expression achieving instead expression of new developmental programmes. The molecular mechanism(s) underlying such stromal-epithelial regulation of cellular proliferation, differentiation and function in neonate and adult tissue remains as for embryonic tissue, a matter of conjecture. However, some insight into the problem may come from investigation of the respective biosynthetic activities of epithelium and stroma as recently considered in accessory sex gland function in the adult (see Cunha & Lung, 1980); while, in another approach attention has been directed to the function and identity of mesenchymal cell-derived factors on epithelial differentiation as characterized in *in vitro* cell culture.

In summary, the interaction of external factors produced by stroma and epithelium both in embryonic and adult tissue would be seen as providing basic control systems regulating pathways of mitotic activity, structural and functional differentiation, and morphogenesis; whereas finer control mechanisms would be operative through internal factors involving series of cell–cell interactions dependent on the ability of these cells to generate such factors. These are areas of biological importance not only for an understanding of normal tissue behaviour but also of neoplasia.

The concept of permissive or instructive stromal regulatory functions on the induction, expression and maintenance of epithelial morphogenetic and functional activity imposes various crucial questions. In particular, and of interest to this review, is whether dysfunction of stromal-epithelial interactive processes, as a consequence of tissue ageing or of some pathological insult, could impose changes in specific control properties causally related to the formation and/or maintenance of the neoplastic cell state.

C. *In carcinogenesis*

The problem of tumour formation has been considered at various times in terms of anomalous differentiation (Conheim, 1889; Markert, 1968; Potter, 1969; Coggin & Anderson, 1974; Pierce, 1974; Braun, 1975; Pierce *et al.*, 1978). In this view, tumour formation would involve a breakdown of differentiative control due to loss or damage to basic genetic information; and, would represent defects in the characteristic expression of that information into normal form and function of given cell types. In such a situation, cells initiated into a neoplastic pathway would express a phenotype unable to manifest appropriate functional integration into normal systems. This resulting, either by failure of these cells to develop appropriate replication restraining mechanisms for a particular level of differentiation, or by their failure to recognize differentiation signals that would normally remove them from the proliferative cycle. However, the organized histological features of tumours and their tendency to express certain characteristics unique to the tissue of their origin would presume that tumours possess to some extent, at least, regulatory mechanisms of the tissue of origin, even though these may function inadequately. This would further postulate that if epigenetic mechanisms similar to those involved in normal cell differentiation are concerned in the promotion and maintenance of the neoplastic state, morphogenetic cell interactions could also play a significant role in the structural moulding of tumours (see also Dawe *et al.*, 1976). As a logical extension has been the proposal that if neoplasms represent a breakdown in tissue regulatory mechanisms then a reversibility of the neoplastic state might be anticipated, with characteristics of normal differentiation reappearing if the epigenetic pathways whereby they are expressed were to be reestablished (see Tarin, 1972; Braun, 1975; Pierce *et al.*, 1978). A number of studies have provided varying supportive evidence (see Hodges, 1981; Table 2) based on changes in the growth and development of tumours following coculture with inductively active embryonic tissues (Mathis & Seilern-Aspang, 1962; Seilern-Aspang *et al.*, 1963; Lustig & Lustig, 1968; Ellison *et al.*, 1969; Crocker & Vernier, 1972; De Cosse *et al.*, 1973, 1975; Rousseau-Merck *et al.*, 1977; Armstrong & Rosenau, 1978), or regenerating tissues (Seilern-Aspang & Kratochwil, 1962; Seilern-Aspang *et al.*, 1963). Further evidence of possible reversion to a normal phenotype has come from coculture studies of abnormal epithelium and normal stroma (Van Scott & Reinertson, 1961; Cooper & Pinkus, 1979) demonstrating induction of normal epidermal differentiation (kera-

Table 2. *Tumour response to inductively active embryonic tissues*

Tumour (effector tissue)	Inducing tissue	Tumour response	References
Sarcoma: chick, human	Chick embryo notochord	Development of cartilagenous tissue	Mathis & Seilern-Aspang (1962), Seilern-Aspang et al. (1963)
Renal tumour: rat	Rat or mouse neural tube	One instance of differentiation in 28 tumour cultures	Ellison et al. (1969)
Ascites tumour: mouse	Chick embryo notochord	Cellular organization	Mathis & Seilern-Aspang (1962)
Mammary tumour: mouse, Sarcoma 180 mouse tumour, human tumours	Chick embryo notochord, neural tube, mesonephros	Cellular organization	Lustig & Lustig (1968)
Undifferentiated renal tumour: human	Fetal mouse neural tissue	Generation of primitive nephron elements	Crocker & Vernier (1972)
Mammary tumour: mouse	Mouse embryo mammary mesenchyme	Differentiation of ducts; but tumour growth not different from that of controls when implanted back into syngeneic animals	De Cosse et al. (1973, 1975)
Nephroblastoma: human	Chick and mouse tissues: ureter, mesencephalon	Improved viability and architecture: some tubular organization	Rousseau-Merck et al. (1977)
Mammary carcinoma: human	Mouse and human fetal tissues: mammary or salivary gland mesenchyme	Growth and maintenance of neoplastic cells	Armstrong & Rosenau (1978)

tinization) and loss of neoplastic characteristics in basal cell carcinoma associated with normal dermis. Of particular significance, have been the studies of Mintz & Illmensee (1975) and Mintz (1978) demonstrating that teratoma cells will lose their neoplastic properties and differentiate into normal cells under appropriate environmental conditions. The evidence implicates, therefore, some alteration (or reversion) in gene expression within the neoplastic cell populations of tumours as a result of interaction between these cells and inductively active normal tissues. Furthermore, different populations of neoplastic cells may express varying degrees of competence to respond to the regulatory systems operative in normal development; and, the differentiation of tumours into essentially normal tissue types by correction of tissue control imbalances attests to neoplasia as basically a disease of differentiation.

As a logical extension have been the observations from indirect circumstantial sources and from direct experimental investigations providing evidence that an imbalance in the relationship between stroma and epithelium might be involved in the development of neoplasia (see Tarin, 1972). That disruption of the orderly sequence of stromal-epithelial inductive interactions, induced experimentally or through genetic defects, can lead to inappropriate states of differentiation and development resulting in impaired embryogenesis and dysmorphogenesis has been clearly established in the teratological literature (Saxén & Rapola, 1969). In considering the concept of failure in stromal-epithelial regulatory function as a significant pathogenetic factor in the origin of epithelial tumours two essentially opposing aspects are postulated. Namely, stromal-epithelial interactive dysfunction could be the consequence either (a) of a primary carcinogenic event on the epithelium with some secondary stromal function or change in stromal activity required to promote cellular transformation initiated in the epithelium into a neoplastic process and/or biologic characterization; or, (b) that epithelial carcinogenesis is an indirect effect resulting from carcinogen-induced changes of stroma which induce or permit abnormal epithelial behaviour. It has been further postulated that stroma may act as the target tissue mediating or complementing metabolic activation of, at least, certain carcinogens for specific action on the overlying epithelium.

The involvement of stromal elements in the development and differentiation of epithelial neoplasms has been an issue of debate for some time. That epithelial tumour promotion and progression may be a response to changes instituted in the connective tissue is by no means a new idea and much indirect circumstantial evidence has been established from histopathological studies of skin (Orr, 1938, 1963; Gillman *et al.*, 1955; Orr & Spencer, 1972; Tarin, 1972). This concept has received further attention in recent studies of benign prostatic hypertrophy (BPH) (Pradhan & Chandra, 1975; McNeal, 1978; Müntzing *et al.*, 1979; Wilson, 1979). Aberration in growth control resulting in proliferation of stromal and epithelial cells and formation of new glandular prostatic architecture characterizes the clinical condition of BPH. From indirect morphological evidence has been formulated the hypothesis that genesis and growth of nodular hyperplasia in BPH is a consequence of neo-inductive prostatic stromal properties to which the adult prostatic ductal epithelium is competent to respond (McNeal, 1978; Müntzing *et al.*, 1979). The inductive potential of BPH stroma remains, however, to be experimentally verified.

Further evidence for the suggestion that epigenetic mechanisms underlie the neoplastic state come from experimental studies of recombinants between stromal and epithelial tissues of different pathological constitution. Such use of heteropathic models permits more direct study of a dysfunction in the stromal-epithelial tissue regulatory mechanism, and has provided evidence that stromal influences are clearly implicated in the aetiology of non-neoplastic and neoplastic patterns of abnormal morphogenesis and cytodifferentiation. The effect of chemical carcinogens upon tissue interactions was first investigated along such experimental lines by Billingham *et al.* (1951) and Marchant & Orr (1953). Reciprocal grafting of methylcholanthrene (MCA)-treated (i.e. initiated) mouse epidermis to dermal beds in untreated areas of the skin, and of normal epidermis to dermal beds prepared in MCA-treated areas of skin resulted in a preferential development of carcinomas in the latter situation. The opinion was held that the primary carcinogen-induced lesion lay in the dermis and not in the epidermis. However, in a later study, Steinmuller (1971) using F_1 hybrid mice as the MCA-treated hosts and the inbred parent strain mice as the donors of the untreated epidermal grafts established that the carcinomas were derived from F_1 carcinogen-treated epithelium left behind after tissue separation. This was based on the observation that these carcinomas on transplantation were rejected by the parent strain recipients in contrast to their growth in F_1 recipients. Main (1972) subsequently suggested that rejection of the transplanted carcinomas by parent strain recipients could equally reflect rejection of the tumour-supporting stroma with which the graft recipients were not histocompatible as opposed to the neoplastic cells themselves. Though inconclusive, the studies did not invalidate the concept of an involvement in neoplasia of some alteration in the interrelationship between stroma and epithelium.

Various heterotypic recombinant studies have established pronounced differences in the growth-promoting activity of normal adult dermis as compared to that of either embryonic dermis or adult peritumoral dermis (Redler & Lustig, 1968; Sengel, 1976). The enhanced growth-promoting activity of adult peritumoral dermis has been offered as evidence that uncontrolled growth of epithelial tumours may not necessarily be an intrinsic property of epithelial cells but could be dependent on some pro-mitotic factor(s) originating from the stroma. Dysfunction of the tissue regulatory process would, as a consequence, reside in local patches of altered stromal properties. That changes in phenotypic expression in epithelia must be seen to be related to abnormal stromal function or to some derangement of the normal stromal function is further supported by two recent heteropathic recombinant studies. By reciprocal recombinations of stroma and epithelium prepared from vaginae of neonatally oestrogenized and untreated mice, Cunha *et al.* (1977) demonstrated that while vaginal epithelium from treated mice is irreversibly altered and continues to express ovary-independent hyperplasia even when associated with normal vaginal stroma, neonatal exposure to oestrogen also produces permanent alteration in the stromal properties which induces or elicits ovary-independent hyperplasia and parakeratosis in normal, untreated vaginal epithelium. The findings were seen to suggest that induction and maintenance of abnormal vaginal cytodifferentiation as expressed in ovary-independent hyperplasia, a non-neoplastic pathological disorder, may be determined in part by stromal effects of neonatal exposure to exogenous oestrogen.

Evidence for an interrelationship between stroma and epithelium in neoplasia has come from a study on methylnitrosourea-treated adult rat bladder (Hodges *et al.*, 1977). Reciprocal heteropathic combinations between tissues from normal and carcinogen-treated bladders established that expression of abnormal epithelial surface features characteristic of neoplasia in bladder could be induced in an untreated epithelium when associated with the stroma of the carcinogen-treated organ. Whereas, no reversal of the neoplastic state, as associated with morphological change, could be induced in carcinogen-treated epithelium associated with untreated stroma. Such results were seen to imply a carcinogen-induced stromal effect on the epithelium with induction of a new epithelial phenotype as a consequence of stromal dysfunction elicited by the carcinogen. However, whether such a tissue interactive interrelationship is involved in the initiation of the epithelial neoplastic process, other than a direct carcinogen-induced alteration in the epithelium, remains to be established.

By a transplantation process, Sakakura *et al.* (1979b) have demonstrated recently the localized influence of fetal mesenchyme implants on morphogenesis and growth of adult mammary epithelium. Such stimulation of focal proliferation in the mammary duct system was observed to accelerate local development of mammary carcinomas in mice carrying mammary tumour virus (MTV-S). It was opinioned that some factor of the fetal mesenchyme in inducing epithelial proliferation created an epithelial population particularly responsive to the transforming effect of MTV-S. While still inconclusive the study was seen to provide evidence for the view that stromal tissues may act as a promoter of tumour development. A similar, if less defined conclusion, resulted from the study of De Ome *et al.* (1959). This demonstrated that hyperplastic alveolar nodules, produced in female mice infected with mammary tumour virus, when transplanted into homologous recipients progressed to overt tumours only when placed in specific mammary fat.

The interrelationship between stroma and epithelium in carcinogenesis has been probably most extensively investigated by Dawe and his associates (Dawe & Lawe, 1959; Dawe *et al.*, 1962, 1966, 1976) using an embryonic tissue-viral tumour system. Submandibular salivary gland rudiments, separated into epithelial and mesenchymal components and treated with polyoma virus, were shown not to produce tumours when transplanted separately into syngeneic hosts. On recombination prior to transplantation, however, the capacity for neoplasia was restored and it was initially concluded that the integrity of the stromal-epithelial unit was critical for tumour formation. More recently (Dawe *et al.*, 1976), polyoma-induced neoplastic change in embryonic salivary gland epithelium maintained in an altered relationship to its mesenchyme (as established through transfilter recombination) was shown to result in a transformed epithelium which, on transplantation, produced undifferentiated carcinomas. This was in striking contrast to the differentiated adenomas produced by the morphologically intact salivary gland complex. Of relevance also, was the absence of morphogenetic activity but retention of proliferative activity by epithelium maintained transfilter to homotypic mesenchyme in non-infected cultures. This was seen to suggest that although neoplastic transformation of epithelium occurred in the absence of stromal-induced morphogenesis and was probably the consequence of a direct primary viral carcinogen-epithelial cell interaction, some stromal-epithelial interaction during tumour genesis would be essential for establishing the morphologic and biologic characteristic of the tumour as it develops.

One other *in vitro* recombinant system developed to demonstrate stromal-epithelial interaction in polyoma virus induction of ameloblastoma in mice (Main & Wahead, 1971) implicated an inductive action by infected mesenchyme in the proliferative response of uninfected tooth-germ epithelium. Attempts to transplant such cultures (Main *et al.*, 1974) produced only squamous cell carcinomas and not the ameloblastomas characteristic of *in vivo* induced neoplasia. Whether this could be viewed as a consequence of odontogenic mesenchyme destruction by cytolytic viral action and loss of mesenchyme-epithelial interactive function in tumour characterization remains to be established.

Other than its potential regulatory role in tumour-promoting manipulations, stroma encompassing the tumour may have further important relationships to tumour growth for support, proliferation and maintenance of neoplastic cells (Tarin, 1972). These in turn may contribute to the development of the collagenous stroma of tumours by synthesizing collagen proteins and by releasing growth-promoting factors that induce proliferation of host fibroblasts (Foulds, 1969), and angiogenic factors that stimulate growth of new blood vessels (Folkman & Cotran, 1976) invoked in the growth and histological development of tumours. Furthermore, neoplastic cells may express competence to exert epigenetic influences similar to the inductive interactions of normal development and establish a capacity to induce proliferative and differentiative activity in adjacent normal tissues and within neoplastic cell populations (Auersperg & Finnegan, 1974; Jimenez de Asua *et al.*, 1980).

3. CONCLUDING COMMENTS

In conclusion, involvement of epigenetic mechanisms must be considered as one of several possible key events in neoplasia. Of significance is the recognition of a multiphase mechanism in the process of neoplasia with the underlying principle that tumour initiation would be probably, or at least in general, a genetic event whereas tumour promotion and overt formation would derive from epigenetic action. More specifically, interrelationships between stroma and epithelium, and dysfunction of this particular tissue regulatory process, emerge as significant features both in non-neoplastic and neoplastic disease. Stromal inductive influences can be implicated in the carcinogenic process both in a permissive and instructive capacity, although their role at different stages of neoplasia from initiation to overt neoplasma remains to be defined in specific terms.

REFERENCES

AHUJA, M. R. & ANDERS, F. (1977). Cancer as a problem of gene regulation. *Recent Advances in Cancer Research: Cell Biology, Molecular Biology and Tumour Virology*, ed. R. C. Gallo, pp. 103–17. Cleveland: CRC Press.

ARMSTRONG, R. C. & ROSENAU, W. (1978). Cocultivation of human primary breast carcinomas and embryonic mesenchyme resulting in growth and maintenance of tumour cells. *Cancer Research*, **38**, 894–900.

AUERSPERG, N. & FINNEGAN, C. V. (1974). The differentiation and organization of tumours in vitro. In *Neoplasia and Cell Differentiation*, pp. 279–318. Basel: Karger.

BECKER, F. F. (1975). *Cancer: A Comprehensive Treatise*, vols. 1–4. New York and London: Plenum Press.

BILLINGHAM, R. E., ORR, J. W. & WOODHOUSE, D. L. (1951). Transplantation of skin components during chemical carcinogenesis with 20-methylcholanthrene. *British Journal of Cancer*, **5**, 417–32.

BILLINGHAM, R. E. & SILVERS, W. K. (1967). Studies on the conservation of epidermal specificities of skin and certain mucosas in adult mammals. *Journal of Experimental Medicine*, **125**, 429–46.

BILLINGHAM, R. E. & SILVERS, W. K. (1968). Dermoepidermal interactions and epithelial specificity. In *Epithelial-Mesenchymal Interactions*, ed. R. Fleischmajer & R. E. Billingham, pp. 252–66. Baltimore: Williams and Wilkins.

BILLINGHAM, R. E. & SILVERS, W. K. (1971). A biologist's reflections on dermatology. *Journal of Investigative Dermatology*, **57**, 227–40.

BRAUN, A. C. (1975). Differentiation and dedifferentiation. In *Cancer*, vol. 3, ed. F. F. Becker, pp. 3–20. New York and London: Plenum Press.

BRIGGAMAN, R. A. & WHEELER, C. E. (1968). Epidermal-dermal interactions in adult human skin: role of dermis epidermal maintenance. *Journal of Investigative Dermatology*, **51**, 454–65.

CAHAN, W. G. & SCHOTTENFELD, D. (1977). Proceedings of the International Workshop on multiple primary cancers. *Cancer*, **40**, 1785–985.

CLARK GRIFFEN, A. & SHAW, C. R. (eds.) (1979). *Carcinogens: Identification and Mechanisms of Action*. New York: Raven Press.

COGGIN, J. H. & ANDERSON, N. G. (1974). Cancer, differentiation and embryonic antigens: some central problems. *Advances in Cancer Research*, **19**, 105–65.

CONHEIM, J. F. (1889). New Sydenham Society, London. *Lectures on General Pathology*, **2**, 1.

COOMBS, M. M. (1980). Chemical carcinogenesis: a view at the end of the first half-century. *Journal of Pathology*, **130**, 117–45.

COOPER, M. & PINKUS, H. (1979). Intrauterine transplantation of rat basal cell carcinoma as a model for reconversion of malignant to benign growth. *Cancer Research*, **37**, 2544–52.

CROCKER, J. F. S. & VERNIER, R. L. (1972). Congenital nephroma of infancy: induction of renal structures by organ culture. *Journal of Pediatrics*, **80**, 69–73.

CUNHA, G. R. (1976). Epithelial-stromal interaction in the development of the urogenital tract. *International Review of Cytology*, **47**, 137–94.

CUNHA, G. R., CHUNG, L. W. K., SHANNON, J. M. & REESE, B. A. (1980). Stromal-epithelial interactions in sex differentiation. *Biology of Reproduction*, **22**, 19–42.

CUNHA, G. R., LUNG, B. & KATO, K. (1977). Role of the epithelial-stromal interaction during the development and expression of ovary independent vaginal hyperplasia. *Developmental Biology*, **56**, 52–67.

CUNHA, G. R. & LUNG, B. (1979). The importance of stroma in morphogenesis and functional activity of urogenital epithelium. *In Vitro*, **15**, 50–71.

CUNHA, G. R. & LUNG, B. (1980). Experimental analysis of male accessory sex gland development. In *Male Accessory Sex Glands*, ed. Spring-Mills & E. S. E. Hafex. pp. 39–59. Amsterdam: Elsevier/North-Holland Biomedical Press.

DAWE, C. J. & LAWE, L. W. (1959). Morphological changes in salivary-gland tissue of the newborn mouse exposed to parotid-tumour agent *in vitro*. *Journal of the National Cancer Institute*, **23**, 1157–77.

DAWE, C. J., LAWE, L. W., MORGAN, W. D. & SHAW, M. S. (1962). Morphologic responses to tumor viruses. *Federation Proceedings of the American Society for Experimental Biology*, **21**, 5–14.

DAWE, C. J., MORGAN, W. D. & SLATICK, M. S. (1966). Influence of epithelio-mesenchymal interactions on tumour induction by polyoma virus. *International Journal of Cancer*, **1**, 419–50.

DAWE, C. J., MORGAN, W. D., WILLIAMS, J. E. & SUMMEROW, J. P. (1976). Inductive epithelio-mesenchymal interaction: is it involved in the development of epithelial neoplasms? In *Progress in Differentiation Research*, ed. N. Müller-Bërat, pp. 305–18. Amsterdam: North-Holland.

DECOSSE, J. J., GOSSENS, C. L. & KUZMA, J. F. (1973). Breast cancer: induction of differentiation by embryonic tissue. *Science*, **181**, 1057–8.

DECOSSE, J. J., GOSSENS, C., KUZMA, J. F. & UNSWORTH, B. R. (1975). Embryonic inductive tissues that cause histologic differentiation of murine mammary carcinoma in vitro. *Journal of National Cancer Institute*, **54**, 913–22.

DE OME, K. B., FAULKIN, L. J., BERN, H. A. & BLAIR, P. B. (1959). Development of mammary tumors from hyperplastic alveolar nodules transplanted into gland-free mammary fat pads of female C3H mice. *Cancer Research*, **19**, 515–20.

DEUCHAR, E. M. (1975). *Cellular Interactions in Animal Development*. London: Chapman and Hall.

ELLISON, M. L., AMBROSE, E. J. & EASTY, G. C. (1969). Differentiation in a transplantable rat tumour maintained in organ culture. *Experimental Cell Research*, **55**, 198–204.

EPSTEIN, S. S. (1974). Environmental determinants of human cancer. *Cancer Research*, **34**, 2425–35.

FARBER, E. & CAMERON, R. (1980). The sequential analysis of cancer development. *Advances in Cancer Research*, **31**, 125–226.

FIDLER, I. J. (1978). Tumor heterogeneity and the biology of cancer invasion and metastasis. *Cancer Research*, **38**, 2651–60.

FLAXMAN, B. A., CHOPRA, D. P. & NEWMAN, D. (1973). Growth of mouse vaginal epithelial cells in vitro. *In Vitro*, **9**, 194–201.

FLEISCHMAJER, R. & BILLINGHAM, R. (eds.) (1968). *Epithelial-Mesenchymal Interactions*. Baltimore, Maryland: Williams & Wilkins.

FOLKMAN, J. (1977). Conformational control of cell and tumor growth. In *Recent Advances in Cancer Research: Cell Biology, Molecular Biology and Tumour Virology*, vol. 1, ed. R. C. Gallo, pp. 119–30. Cleveland: CRC Press.

FOLKMAN, J. & COTRAN, R. (1976). Relation of vascular proliferation to tumour growth. *International Review of Experimental Pathology*, **16**, 207–48.

FOULDS, L. M. (1969). *Neoplastic Development*, vol. 1. London and New York: Academic Press.

FRANKS, L. M. & BARTON, A. A. (1960). The effects of testosterone on the ultra-structure of the mouse prostate *in vivo* and in organ cultures. *Experimental Cell Research*, **19**, 35–40.

FRANKS, L. M., RIDDLE, P. W., CARBONELL, A. W. & GEY, G. O. (1970). A comparative study of the ultrastructure and lack of growth capacity of adult human prostate epithelium mechanically separated from its stroma. *Journal of Pathology*, **100**, 113–19.

GERMAN, J. (ed.) (1974). *Chromosomes and Cancer*. New York: John Wiley & Sons.

GILLMAN, T., PENN, J., BRONKS, D. & ROUX, M. (1955). Possible significance of abnormal dermal collagen and of epidermal regeneration in the pathogenesis of skin cancers. *British Journal of Cancer*, **9**, 272–83.

GOSPODAROWICZ, D., GREENBURG, G. & BIRDWELL, C. R. (1978). Determination of cellular shape by the extracellular matrix and its correlation with the control of cellular growth. *Cancer Research*, **38**, 4155–71.

GROBSTEIN, C. (1955). Tissue interaction in the morphogenesis of mouse embryonic rudiments *in vitro*. In *Aspects of Synthesis and Order in Growth*, ed. D. Rudnick, pp. 233–56. Princeton University Press.

GROBSTEIN, C. (1975). Developmental role of extracellular matrix: retrospective and prospective. In *Extracellular Matrix Influences on Gene Expression*, ed. H. C. Slavkin & R. C. Greulich, pp. 9–16. New York: Academic Press.

HAY, E. D. (1977). Cell-matrix interaction in embryonic induction. In *International Cell Biology*, pp. 50–7. New York: Rockefeller University Press.

HECKER, E. (1976). Definitions and terminology in cancer (tumor) etiology. An analysis aiming at proposals for a current internationally standardized termino-logy. *Gann*, **67**, 471–81.

HECKMAN, C. A., MARCHOK, A. C. & NETTESHEIM, P. (1978). Respiratory tract epithelium in primary culture: concurrent growth and differentiation during establishment. *Journal of Cell Science*, **32**, 269–91.

HERRMAN, H. (1960). Direct metabolic interactions between animal cells. *Science*, **132**, 529–32.

HODGES, G. M. (1981). Growth, differentiation and function of tumours in organ culture. In *Phenomenon of Control of Growth in Neoplastic and Differentiative Systems*, ed. G. V. Sherbert. Basel: S. Karger

HODGES, G. M., HICKS, R. M. & SPACEY, G. D. (1976). Scanning electron microscopy of cell surface changes in methylnitrosurea (MNU)-treated rat bladder *in vivo* and *in vitro*. *Differentiation*, **6**, 143–50.

HODGES, G. M., HICKS, R. M. & SPACEY, G. D. (1977). Epithelial-stromal interactions in normal and chemical carcinogen-treated adult bladder. *Cancer Research*, 3720–30.

HUXLEY, J. S. (1958). *Biological Aspects of Cancer*. New York: Harcourt.

JIMENEZ, DE ASUA, L., LEVI-MONTALCINI, R., SHIELDS, R. & IACOBELLI, S. (eds.) (1980). *Control Mechanisms in Animal Cells: Specific Growth Factors*. New York: Raven Press.

KARKINEN-JÄÄSKELÄINEN, M., SAXÉN, L. & WEISS, L. (eds.) (1977). *Cell Interactions in Differentiation*. London: Academic Press.

KEFALIDES, N. A. (ed.) (1978). *First International Symposium on the Biology and Chemistry of Basement Membranes*. New York: Academic Press.

KEFALIDES, N. A. (1979). Biochemistry and metabolism of basement membranes. *International Review of Cytology*, **61**, 167–228.

KIRK, D., KING, R. J., HEYES, J., PEACHY, L., HIRSCH, P. J. & TAYLOR, R. W. T. (1978). Normal human endometrium in cell culture. I. Separation and characterization of epithelial and stromal components in vitro. *In Vitro*, **14**, 651–62.

KLEIN, G. (1979). Immune and non-immune control of neoplastic development: contrasting effects of host and tumor evolution. In *Accomplishments in Cancer Research*, ed. J. G. Fortner & J. E. Rhoads, pp. 123–46. Philadelphia, Toronto: J. B. Lippincott.

KNUDSON, A. C. JR (1977). Genetics and etiology of human cancer. *Advances in Human Genetics*, **8**, 1–66.

KOPROWSKI, H. (ed.) (1977). *Neoplastic Transformation: Mechanisms and Consequences*. Berlin: Dahlem Konferenzen.

LASFARGUES, E. Y. (1957a). Cultivation and behaviour in vitro of the normal mammary epithelium of the adult mouse. *Anatomical Record*, **127**, 117–25.

LASFARGUES, E. Y. (1957b). Cultivation and behaviour in vitro of the normal mammary epithelium of the adult mouse. II. Observations on the secretory activity. *Experimental Cell Research*, **13**, 553–62.

LASH, J. W. & BURGER, M. M. (eds.) (1977). *Cell and Tissue Interactions*. New York: Raven Press.

LUSTIG, E. S. & LUSTIG, L. (1968). Action in vitro of the embryonic inducers on experimental and human tumours. In *Cancer Cells in Culture*, ed. H. Katsuta, pp. 135–42. Baltimore: University Park Press.

McNEAL, J. E. (1978). Origin and evolution of benign prostatic enlargement. *Investigative Urology*, **15**, 340–5.

MAIN, J. H. P. (1972). Developmental considerations; carcinogenesis and oncology. In *Developmental Aspects of Oral Biology*, ed. H. C. Slavkin & A. Bavetta, pp. 385–405.

MAIN, J. H. P., McCOMB, R. J. & MOCK, D. (1974). Transplantation studies on polyoma virus-infected cultures of mouse tooth germs. *Journal of the National Cancer Institute*, **52**, 951–61.

MAIN, J. H. P. & WAHEAD, M. A. (1971). Epitheliomesenchymal interactions in the proliferative response evoked by polyoma virus in odontogenic epithelium in vitro. *Journal of the National Cancer Institute*, **47**, 711–26.

MARCHANT, J. & ORR, J. W. (1953). Further attempts to analyse the role of epidermis and deeper tissues in experimental chemical carcinogenesis by transplantation and other methods. *British Journal of Cancer*, **7**, 329–41.

MARKERT, C. L. (1968). Neoplasia: a disease of cell differentiation. *Cancer Research*, **28**, 1908–14.

MATHEWS, M. B. (1975). *Connective Tissue: Macromolecular Structure and Evolution.* Berlin and New York: Springer-Verlag.

MATHIS, G. & SEILERN-ASPANG, F. (1962). Das Verhalten entarte Zellen (Ehrlich-Ascites-Tumor der Maus) im Embryonafeld. *Naturwissenschaften*, **5**, 111.

MILLER, E. C. (1978). Some current perspectives on chemical carcinogenesis in humans and experimental animals. *Cancer Research*, **38**, 1479–96.

MINTZ, B. (1978). Genetic mosaicism and in vivo analysis of neoplasia and differentiation. In *Cell Differentiation and Neoplasia*, ed. G. F. Saunder, pp. 27–56. New York: Raven Press.

MINTZ, B. & ILLMENSEE, K. (1975). Normal genetically mosaic mice produced from malignant teratocarcinoma cells. *Proceedings of the National Academy of Sciences, USA*, **72**, 3585–9.

MÜNTZING, J., LILJEKVIST, J. & MURPHY, G. P. (1979). Chalones and stroma as possible growth-limiting factors in the rat ventral prostate. *Investigative Urology*, **16**, 399–402.

ORR, J. W. (1938). The changes antecedent to tumour formation during the treatment of mouse skin with carcinogenic hydrocarbons. *Journal of Pathology and Bacteriology*, **46**, 495–515.

ORR, J. W. (1963). The role of the stroma in epidermal carcinogenesis. *National Cancer Institute Monographs*, **10**, 531–7.

ORR, J. W. & SPENCER, A. T. (1972). Transplantation studies of the role of the stroma in epidermal carcinogenesis. In *Tissue Interactions in Carcinogenesis*, ed. D. Tarin, pp. 291–303. London: Academic Press.

PIERCE, G. B. (1974). Neoplasms, differentiation and mutations. *American Journal of Pathology*, **77**, 103–18.

PIERCE, G. B., SHIKES, R. & FINK, L. M. (1978). *Cancer: A Problem of Developmental Biology.* New Jersey: Prentice Hall.

PITOT, H. C. (1977a). The natural history of neoplasia. *American Journal of Pathology*, **89**, 401–12.

PITOT, H. C. (1977b). The stability of events in the natural history of neoplasia. *American Journal of Pathology*, **89**, 703–16.

POSTE, G. & NICOLSON, G. L. (eds.) (1977). *Cell Surface Interactions in Embryogenesis.* Amsterdam: North-Holland Division of ASP Biological and Medical Press.

POTTER, V. R. (1969). Recent trends in cancer biochemistry: the importance of studies on fetal tissue. *Proceedings of the Canadian Cancer Research Conference*, **8**, 9–30.

PRADHAN, B. K. & CHANDRA, K. (1975). Morphogenesis of nodular hyperplasia-prostate. *Journal of Urology*, **113**, 210–13.

PREHN, R. T. (1972). *Neoplasia*. In *Principles of Pathobiology*, ed. M. E. LaVia & R. B. Hill, Jr, pp. 191–241. Oxford University Press.

PREHN, R. T. (1976). Tumor progression and homeostasis. *Advances in Cancer Research*, **23**, 203–36.

REDLER, P. & LUSTIG, E. S. (1968). Differences in the growth promoting effect of normal and peritumoral dermis in epidermis in vitro. *Developmental Biology*, **17**, 679–91.

ROUSSEAU-MERCK, M. F., LOMBARD, M. N., NEZELOF, C. & MOULY, H. (1977). Limitation of the potentialities of nephroblastoma differentiation *in vitro*. *European Journal of Cancer*, **13**, 163–70.

SAKAKURA, T., NISHIZUKA, Y. & DAWE, C. J. (1979a). Capacity of mammary fat pads of adult C3H/HeMs mice to interact morphogenetically with fetal mammary epithelium. *Journal of the National Cancer Institute*, **63**, 733–6.

SAKAKURA, T., SAKAGAMI, Y. & NISHIZUKA, Y. (1979b). Acceleration of mammary cancer development by grafting of fetal mammary mesenchymes in C3H mice. *Gann*, **70**, 459–66.

SAXÉN, L. (1977). Morphogenetic tissue interactions: An introduction. In *Cell Interactions in Development*, ed. M. Karkinen-Jaaskelainen, L. Saxén & L. Weiss, pp. 145–51. New York: Academic Press.

SAXÉN, L. (1979). Mechanisms of morphogenetic tissue interactions: the message of transfilter experiments. In *Differentiation and Neoplasia*, vol. II, ed. R. G. Mckinnell, M. A. DiBerardino, M. Blumfeld, & R. D. Bergad, pp. 148–54. Amsterdam. North-Holland.

SAXÉN, L. & RAPOLA, J. (1969). *Congenital Defects*. New York: Holt, Rinehart and Winston.

SEARLE, C. E. (ed.) (1976). *Chemical Carcinogens*. ACS Monograph Series. No. 173.

SEILERN-ASPANG, F., HONUS, E. & KRATOCHWIL, K. (1963). Die Verknorpelung Mesenchlicher Sarkome durch Chrondrogene Induktoren der Huhnerchorda. *Acta Biologica et Medica Germanica*, **11**, 281–5.

SEILERN-ASPANG, F. & KRATOCHWIL, K. (1962). Induction and differentiation of an epithelial tumour in the newt (Triturus cristatus). *Journal of Embryology and Experimental Morphology*, **10**, 337–41.

SENGEL, P. (1976). *Morphogenesis of Skin*. Cambridge University Press.

SLAVKIN, H. C. & GREULICH, R. C. (eds.) (1975). *Extracellular Matrix Influences on Gene Expression*. New York: Academic Press.

SMITHERS, D. W. (1962). Cancer: an attack on cytologism. *Lancet*, **1**, 493–9.

STEINMULLER, D. (1971). A reinvestigation of epidermal transplantation during chemical carcinogenesis. *Cancer Research*, **31**, 2080–4.

TARIN, D. (ed.) (1972). *Tissue Interactions in Carcinogenesis*. London: Academic Press.

TEMIN, H. M. (1977). RNA viruses and cancer. *Cancer*, **39**, 422–8.

TERZAGHI, M. & KLEIN-SZANTO, A. J. P. (1980). Differentiation of normal and cultured preneoplastic tracheal epithelial cells in rats: importance of epithelial-mesenchymal interactions. *Journal of the National Cancer Institute*, **65**, 1039–48.

THOMAS, J. A. (ed.) (1970). *Organ Culture*. New York and London: Academic Press.

VAN, SCOTT, E. J. & REINERTSON, R. P. (1961). The modulating influence of stromal environment on epithelial cells studied in human autotransplants. *Journal of Investigative Dermatology*, **36**, 109–31.

WESSELLS, N. K. (1977). *Tissue Interactions and Development*. Merlo Park, California: W. A. Benjamin Inc.

WILLIS, R. A. (1973). *Pathology of Tumours*, 5th edn. London: Butterworths.

WILSON, J. D. (1979). The pathogenesis of benign prostatic hypertrophy. *American Journal of Medicine*, **68**, 745–56.

INDEX

Small intestine, structure of 286
Sodium azide 59
Sodium pump 198
Solute flux 73
Somatostatin 45, 49, 114
Space full response 251
Spatial relationship in surface
      epithelium 285–299
Sphincters, in plasmadesmata 65
Squamous epithelium 232
Stem cells, number of 287
Steroids 42, 44, 133
   as inductors 143
Stomatal guard cells, plasmodesmata in 70
Stress fibre 233
Stroma
   and mitotic activity 342
   and tumour formation 333
Suberin 59
Substance P 45
Sulphation 43
Symplastic connections 58, 61
Synapse, electrical 1
Synaptic cleft 49
Systemic circulation 47

Tachyphyllaxis 52
Tannic acid 63
Target cell 41
T-cell 40
Teratocarcinoma 195
Testosterone 133

Tight junctions 23, 85, 114, 149, 265, 270
TPA (tumour promoting agent) 301–314
Transducers 81, 82
Transmembrane channels 273
Transmission, of inductive signals 336
Trophectoderm 149, 153
Trophoblast 168
Trysin 6
Tumour angiogenesis 253, 274
   induction of blood vessels by 253
Tumour formation
   in regulatory dysfunction 333
Tumour promotion 301
   and latent malignancy 312
Turgor pressure 77

Uncoupling effect, of calcium ions 311
Uncoupling effect, of CO_2 106
Uridine nucleotides as markers 138

Vascular epithelium, function of 263
Vasodilation 46
Vasopressin 51
Villus 286, 290
Vinblastine, as a mitogen 277
Viral particles, movement of 59
VIP 46

Wound response 251

Xylem 251